WORKSHOPS IN COMPUTING

Series edited by C. J. van Rijsbergen

Also in this series

Z User Workshop, London 1992
Proceedings of the Seventh Annual Z User Meeting, London, 14–15 December 1992
J.P. Bowen and J.E. Nicholls (Eds.)

Interfaces to Database Systems (IDS92)
Proceedings of the First International Workshop on Interfaces to Database Systems, Glasgow, 1–3 July 1992
Richard Cooper (Ed.)

AI and Cognitive Science '92
University of Limerick, 10–11 September 1992
Kevin Ryan and Richard F.E. Sutcliffe (Eds.)

Theory and Formal Methods 1993
Proceedings of the First Imperial College Department of Computing Workshop on Theory and Formal Methods, Isle of Thorns Conference Centre, Chelwood Gate, Sussex, UK, 29–31 March 1993
Geoffrey Burn, Simon Gay and Mark Ryan (Eds.)

Algebraic Methodology and Software Technology (AMAST'93)
Proceedings of the Third International Conference on Algebraic Methodology and Software Technology, University of Twente, Enschede, The Netherlands, 21–25 June 1993
M. Nivat, C. Rattray, T. Rus and G. Scollo (Eds.)

Logic Program Synthesis and Transformation
Proceedings of LOPSTR 93, International Workshop on Logic Program Synthesis and Transformation, Louvain-la-Neuve, Belgium, 7–9 July 1993
Yves Deville (Ed.)

Database Programming Languages (DBPL-4)
Proceedings of the Fourth International Workshop on Database Programming Languages – Object Models and Languages, Manhattan, New York City, USA, 30 August–1 September 1993
Catriel Beeri, Atsushi Ohori and Dennis E. Shasha (Eds.)

Music Education: An Artificial Intelligence Approach, Proceedings of a Workshop held as part of AI-ED 93, World Conference on Artificial Intelligence in Education, Edinburgh, Scotland, 25 August 1993
Matt Smith, Alan Smaill and Geraint A. Wiggins (Eds.)

Rules in Database Systems
Proceedings of the 1st International Workshop on Rules in Database Systems, Edinburgh, Scotland, 30 August–1 September 1993
Norman W. Paton and M. Howard Williams (Eds.)

Semantics of Specification Languages (SoSL)
Proceedings of the International Workshop on Semantics of Specification Languages, Utrecht, The Netherlands, 25–27 October 1993
D.J. Andrews, J.F. Groote and C.A. Middelburg (Eds.)

Security for Object-Oriented Systems
Proceedings of the OOPSLA-93 Conference Workshop on Security for Object-Oriented Systems, Washington DC, USA, 26 September 1993
B. Thuraisingham, R. Sandhu and T.C. Ting (Eds.)

Functional Programming, Glasgow 1993
Proceedings of the 1993 Glasgow Workshop on Functional Programming, Ayr, Scotland, 5–7 July 1993
John T. O'Donnell and Kevin Hammond (Eds.)

Z User Workshop, Cambridge 1994
Proceedings of the Eighth Z User Meeting, Cambridge, 29–30 June 1994
J.P. Bowen and J.A. Hall (Eds.)

6th Refinement Workshop
Proceedings of the 6th Refinement Workshop, organised by BCS-FACS, London, 5–7 January 1994
David Till (Ed.)

Incompleteness and Uncertainty in Information Systems
Proceedings of the SOFTEKS Workshop on Incompleteness and Uncertainty in Information Systems, Concordia University, Montreal, Canada, 8–9 October 1993
V.S. Alagar, S. Bergler and F.Q. Dong (Eds.)

Rough Sets, Fuzzy Sets and Knowledge Discovery
Proceedings of the International Workshop on Rough Sets and Knowledge Discovery (RSKD'93), Banff, Alberta, Canada, 12–15 October 1993
Wojciech P. Ziarko (Ed.)

continued on back page...

A. Ponse, C. Verhoef and S.F.M. van Vlijmen (Eds)

Algebra of Communicating Processes

Proceedings of ACP94, the First Workshop on the Algebra of Communicating Processes, Utrecht, The Netherlands, 16–17 May 1994

Published in collaboration with the
British Computer Society

BCS

Springer-Verlag
London Berlin Heidelberg New York
Paris Tokyo Hong Kong
Barcelona Budapest

A. Ponse, PhD
Programming Research Group,
University of Amsterdam, Kruislaan 403,
1098 SJ Amsterdam, The Netherlands

C. Verhoef, PhD
Department of Mathematics and Computing Science,
Eindhoven University of Technology,
PO Box 513, 5600 MB Eindhoven, The Netherlands

S.F.M. van Vlijmen, Drs.
Department of Philosophy,
Utrecht University, Heidelberglaan 8,
3584 CS Utrecht, The Netherlands

ISBN-13: 978-3-540-19909-0 e-ISBN-13: 978-1-4471-2120-6
DOI: 10.1007/978-1-4471-2120-6

British Library Cataloguing in Publication Data
A catalogue record for this book is available from the British Library

Typesetting: Camera ready by contributors

34/3830-543210 Printed on acid-free paper

Preface

ACP, the Algebra of Communicating Processes, is an algebraic approach to the study of concurrent processes, initiated by Jan Bergstra and Jan Willem Klop in the early eighties. These proceedings comprise the contributions to ACP94, the first workshop devoted to ACP. The workshop was held at Utrecht University, 16–17 May 1994.

These proceedings are meant to provide an overview of current research in the area of ACP. They contain fifteen contributions. The first one is a classical paper on ACP by J.A. Bergstra and J.W. Klop: The Algebra of Recursively Defined Processes and the Algebra of Regular Processes, Report IW 235/83, Mathematical Centre, Amsterdam, 1983. It serves as an introduction to the remainder of the proceedings and, indeed, as a general introduction to ACP. An extended abstract of this paper is published under the same title in the ICALP'84 proceedings. Of the remaining contributions, three were submitted by the invited speakers and the others were selected by the programme committee.

As for the presentations, Jos Baeten, Rob van Glabbeek, Jan Friso Groote, and Frits Vaandrager were each invited to deliver a lecture. A paper relating to Frits Vaandrager's lecture has already been submitted for publication elsewhere and is not, therefore, included in these proceedings. Gabriel Ciobanu, one of our guests, gave an impression of his work in an extra lecture. Furthermore, ten presentations were given on the basis of selected papers.

The first day of ACP94 was devoted to the language μCRL, a formalism extending ACP with algebraic data specification, and to the incorporation of real time in ACP. Expressiveness issues and case studies characterized the second and last day of the workshop.

We wish to express our gratitude to the University of Amsterdam, Utrecht University, Eindhoven University of Technology, and the "Nationale Faciliteit Informatica" (NFI) for their financial support of ACP94. We gratefully acknowledge the programme committee and subreferees for their careful reviewing. Finally, we appreciated the friendly and smooth cooperation of all those who contributed to the preparation of this volume.

Alban Ponse
Chris Verhoef
Bas van Vlijmen
Utrecht, July 1994

Invited Speakers

J.C.M. Baeten	Eindhoven University of Technology
R.J. van Glabbeek	Stanford University
J.F. Groote	Utrecht University
F.W. Vaandrager	CWI and University of Amsterdam

Programme Committee

I. Bethke	CWI and Utrecht University
J.W. Klop	CWI and Free University Amsterdam
S. Mauw	Eindhoven University of Technology
J.C. Mulder	Eindhoven University of Technology
A. Ponse	University of Amsterdam
C. Verhoef	Eindhoven University of Technology
S.F.M. van Vlijmen	Utrecht University

Organizing Committee

A. Ponse	University of Amsterdam
C. Verhoef	Eindhoven University of Technology
S.F.M. van Vlijmen	Utrecht University

Contents

The Algebra of Recursively Defined Processes and the Algebra of Regular Processes*

J.A. Bergstra
Programming Research Group, University of Amsterdam
Kruislaan 403, 1098 SJ Amsterdam, The Netherlands
Department of Philosophy, Utrecht University
Heidelberglaan 8, 3584 CS Utrecht, The Netherlands

J.W. Klop
Department of Software Technology, CWI
P.O. Box 94079, 1090 GB Amsterdam, The Netherlands
Dept. of Mathematics and Computer Science, Free University Amsterdam
de Boelelaan 1081, 1081 HV Amsterdam, The Netherlands

Abstract

We introduce recursively defined processes and regular processes, both in presence and absence of communication. It is shown that both classes are process algebras. An interpretation of CSP in the regular processes is presented. As an example of recursively defined processes, bag and stack are discussed in detail. It is shown that the bag cannot be recursively defined without merge.
We introduce fixed point algebras which have interesting applications in several proofs. An example is presented of a fixed point algebra which has an undecidable word problem.

Key Words & Phrases: concurrency, nondeterministic process, merge, process algebra, regular processes, recursively defined processes, fixed point algebra.

Introduction

ACP, algebra of communicating processes, was introduced in BERGSTRA & KLOP [4]. It combines a purely algebraic formulation of a part of MILNER's CCS [13] with an algebraic presentation of the denotational semantics of processes as given by DE BAKKER & ZUCKER in [1, 2]; moreover it includes two laws on communication of atomic actions which are also present in HENNESSY [8].

*This is a reproduction of Report IW 235/83, Mathematisch Centrum, Amsterdam, 1983. The authors are grateful to Wan Fokkink for organizing its conversion to a form acceptable for the publisher. An extended abstract of this paper appeared in: J. Paredaens, editor, *Proceedings 11th ICALP*, Antwerpen, volume 172 of *Lecture Notes in Computer Science*, pages 82–95. Springer-Verlag, 1984.

The ingredients of ACP are the following:

- A finite set A of so-called *atomic actions*, including a constant δ for *deadlock* (or *failure*). With $\underline{A}$ we denote A - $\{\delta\}$, the *proper* actions.
- A mapping $. \mid . : A \times A \to A$, called the *communication function.* If $a \mid b = c$ then c is the action that results from *simultaneously* executing a and b. Processes will cooperate by sharing actions rather than sharing data.
- A subset H of A (usually H contains the actions which must communicate with other actions in order to be executable). The elements of H are called *subatomic actions.*
- A *signature* of operations $\cdot, +, \|, ⫽ , \mid, \delta, \partial_H$. (For $x \cdot y$ we will often write xy.)

The *axioms* of ACP are these:

$x + y = y + x$	A1
$x + (y + z) = (x + y) + z$	A2
$x + x = x$	A3
$(x + y) \cdot z = x \cdot z + y \cdot z$	A4
$(x \cdot y) \cdot z = x \cdot (y \cdot z)$	A5
$x + \delta = x$	A6
$\delta \cdot x = \delta$	A7
$a \mid b = b \mid a$	C1
$(a \mid b) \mid c = a \mid (b \mid c)$	C2
$\delta \mid a = \delta$	C3
$x \| y = x ⫽ y + y ⫽ x + x \mid y$	CM1
$a ⫽ x = a \cdot x$	CM2
$(ax) ⫽ y = a(x \| y)$	CM3
$(x + y) ⫽ z = x ⫽ z + y ⫽ z$	CM4
$(ax) \mid b = (a \mid b) \cdot x$	CM5
$a \mid (bx) = (a \mid b) \cdot x$	CM6
$(ax) \mid (by) = (a \mid b) \cdot (x \| y)$	CM7
$(x + y) \mid z = x \mid z + y \mid z$	CM8
$x \mid (y + z) = x \mid y + x \mid z$	CM9
$\partial_H(a) = a \quad \text{if } a \notin H$	D1
$\partial_H(a) = \delta \quad \text{if } a \in H$	D2
$\partial_H(x + y) = \partial_H(x) + \partial_H(y)$	D3
$\partial_H(x \cdot y) = \partial_H(x) \cdot \partial_H(y)$	D4

These axioms reflect in an algebraic way that $+$ represents *choice*, $\cdot$ represents *sequential composition* and $\|$ the *merge* operator.

The operations $⌊\!⌊$ (*left merge*) and $|$ (*communication merge*) are auxiliary ones. Our primary interest remains for $+, \cdot, \|$. The process $x ⌊\!⌊ y$ is like $x \| y$, but takes its first step from x, and $x \mid y$ is like $x \| y$ but requires the first action to be a communication (between a first step of x and a first step of y).

1 Preliminaries

1.1 Models of ACP

The axioms of ACP allow for an enormous variety of models. In [3, 4, 5] we investigated the model A^∞. In the present paper we take into consideration graph theoretic models as well. Especially we consider *finitely branching* graphs.

Four types of models thus emerge:

(i) A_ω, the initial model of ACP seen as an equational specification over the signature with a constant for each atom.

(ii) A_ω mod n (also written as A_n) for $n \geq 1$: a homomorphic image of A_ω obtained by identifying two processes in A_ω if their trees coincide up to depth n.

(iii) A^∞; this is the projective limit of the structures A_n.

(iv) graph theoretic models.

More information on these matters can be found in [3–6].

1.2 Restricted signatures

It is useful to consider a smaller set of operations on processes, for instance: only $+$ and $\cdot$. Then one may forget δ and consider structures

$$\underline{A}_\omega(+,\cdot),\ \underline{A}_n(+,\cdot),\ \underline{A}^\infty(+,\cdot)$$

where $\underline{A} = A - \{\delta\}$.

Under the assumption that $a \mid b = \delta$ for all $a, b \in A$, we may add $\|$ and $⌊\!⌊$ to the signature of these algebras, thus obtaining

$$\underline{A}_\omega(+,\cdot,\|,⌊\!⌊),\ \underline{A}_n(+.\cdot,\|,⌊\!⌊) \text{ and } \underline{A}^\infty(+,\cdot,\|,⌊\!⌊).$$

Of course these structures can be constructed immediately without any reference to communication. Let PA be the following axiom system:

$$x + y = y + x \quad \text{A1}$$
$$x + (y + z) = (x + y) + z \quad \text{A2}$$
$$x + x = x \quad \text{A3}$$
$$(x + y) \cdot z = x \cdot z + y \cdot z \quad \text{A4}$$
$$(x \cdot y) \cdot z = x \cdot (y \cdot z) \quad \text{A5}$$

$$x \parallel y = x \mathbin{\|\!\_} y + y \mathbin{\|\!\_} x \quad \text{M1}$$
$$a \mathbin{\|\!\_} x = a \cdot x \quad \text{M2}$$
$$(ax) \mathbin{\|\!\_} y = a(x \parallel y) \quad \text{M3}$$
$$(x + y) \mathbin{\|\!\_} z = x \mathbin{\|\!\_} z + y \mathbin{\|\!\_} z \quad \text{M4}$$

Then $\underline{A}_\omega(+, \cdot, \|, \mathbin{\|\!\_})$ is just the initial algebra of PA.

1.3 Linear terms and guarded terms

Let $X_1, \ldots, X_n$ be variables ranging over processes. Given a (restricted) signature of operators from $+, \cdot, \|, \mathbin{\|\!\_}, |, \partial_H, \delta$ two kinds of terms containing variables $X_1, \ldots, X_n$ are of particular importance:

(i) *Linear terms.* Linear terms are inductively defined as follows:

- atoms a, δ and variables X_i are linear terms,
- if T_1 and T_2 are linear terms then so are $T_1 + T_2$ and aT_1 (for $a \in A$).

An equation $T_1 = T_2$ is called linear if T_1, T_2 are linear.

(ii) *Guarded terms.* The *unguarded* terms are inductively defined as follows:

- X_i is unguarded,
- if T is unguarded then so are $T + T'$, $T \cdot T'$, $\partial_H(T)$, $T \parallel T'$, $T \mathbin{\|\!\_} T'$, $T \mid T'$ (for every T').

A term T is guarded if it is not unguarded.

1.4 Process graphs

Process graphs (or, as we will sometimes call them: transition diagrams) constitute a very useful tool for the description of processes. In this section we will consider finite process graphs (possibly containing cycles). Finite process graphs over A will find a semantics in A^∞ via a system of recursion equations.

A *process graph* g for an action alphabet A is a rooted directed graph with edges labeled by elements of A. (Here g may be infinite and may contain cycles.)

Let g be a finite process graph over A. We show how to find a semantics of g in A^∞. To each node s of g with a positive outdegree, attach a process name X_s. Then the following system of guarded linear equations arises:

$$X_s = \sum_{(a,t)\in U} a \cdot X_t + \sum_{a\in V} a \qquad (E_X)$$

where $U = \{(a,y) \mid g : s \xrightarrow{a} t \;\&\; t \text{ has positive outdegree}\}$,
$V = \{a \mid \exists t \; g : s \longrightarrow t \;\&\; t \text{ has outdegree } 0\}$.

This system E_X has a unique solution in A^∞ and with s_0 the root of g, we define

$$[\![g]\!] = p_{s_0}$$

where $< p_s >$ solves E_X.

1.5 Operations on process graphs

We assume that $. \mid .$ is defined as a communication function: $A \times A \to A$. Now let g_1, g_2 be two process graphs for A. We define new process graphs as follows:

$g_1 + g_2$ results by glueing together the roots of g_1 and g_2,

$g_1 \cdot g_2$ results by glueing together the root of g_2 and all endpoints of g_1,

$\partial_H(g_1)$ results by replacing all labels $a \in H$ by δ in g_1,

$g_1 \parallel g_2$ is the cartesian product of the node sets of g_1 and g_2 provided with labeled edges as follows:

(i) $a : (s_1, s_2) \longrightarrow (s_1', s_2)$ if in g_1 we have $a : s_1 \longrightarrow s_1'$

(ii) $a : (s_1, s_2) \longrightarrow (s_1, s_2')$ if in g_2 we have $a : s_2 \longrightarrow s_2'$

(iii) $a : (s_1, s_2) \longrightarrow (s_1', s_2')$ if for some $b, c \in A$ we have $b \mid c = a$ and $b : s_1 \longrightarrow s_1'$ in g_1, $c : s_2 \longrightarrow s_2'$ in g_2.

$g_1 \,⌊\!⌊\, g_2$ is defined like $g_1 \parallel g_2$, but leaving out all transitions of types (ii) and (iii) if s_1 is the root of g_1.

$g_1 \mid g_2$ is defined like $g_1 \parallel g_2$ but leaving out all transitions of types (i) and (ii) if s_1 resp. s_2 is the root of g_1, g_2.

Of course we have $[\![g_1 + g_2]\!] = [\![g_1]\!] + [\![g_2]\!]$ etc.

2 Regular Processes

2.1 The algebra of regular processes

For $p \in A^\infty$ we define the collection Sub(p) of *subprocesses* of p as follows.

$$p \in \mathrm{Sub}(p)$$

$$ax \in \mathrm{Sub}(p) \Longrightarrow x \in \mathrm{Sub}(p), \text{ provided } a \neq \delta$$

$$ax + y \in \mathrm{Sub}(p) \Longrightarrow x \in \mathrm{Sub}(p), \text{ provided } a \neq \delta$$

Definition. $p \in A^\infty$ is *regular* if Sub(p) is finite.

Notation. $r(A^\infty)$ denotes the collection of regular processes in A^∞.

Theorem 2.1.1. (i) *If p is regular then there is a finite process graph g with $[\![g]\!] = p$, and conversely.*
(ii) *The class of regular processes is closed under the operations $+, \cdot, \|, \lfloor\!\lfloor, |, \partial_H$. Hence $r(A^\infty)$ is a subalgebra of A^∞.*
(iii) *$r(A^\infty)$ contains exactly the solutions of finite systems of guarded linear equations.*

Proof. (i) and (iii) are standard; (ii) is an immediate consequence of the fact that the operations $+, \cdot, \|, \lfloor\!\lfloor, |, \partial_H$ acting on graphs preserve finiteness. (Cf. [6], Section 2.2) □

2.2 CSP program algebras

In this subsection we illustrate the use of the algebras $r(A^\infty)$ by giving an interpretation of simplified CSP programs in such algebras.

Let Σ be an algebraic signature and let X be a set of variables. A *CSP component program* S is defined by:

$$S ::= b \mid b\&x := t \mid b\&C!t \mid b\&C?x \mid S_1; S_2 \mid S_1 \Box S_2 \mid \textbf{while } b \textbf{ do } S \textbf{ od}.$$

Here b is a boolean (quantifier free) expression. The action b is a guard, which can only be passed when it evaluates to **true**; $b\&p$ can only be performed if b is true. It is usual to abbreviate **true** & p to p. All variables x must occur in X. Further, C is an element of a set of channel names,

A *CSP program* P is a construct of the form $[S_1 \parallel \ldots \parallel S_K]$ with the S_i CSP-component programs.

Remark. Originally the CPS syntax indicates restrictions: the S_i must work with different variables, the channels are used to interconnect specific pairs of components. (See HOARE [9].)

However, from our point of view these restrictions are just guide-lines on how to obtain a properly modularised system (semantically their meaning is not so clear).

Let a CSP program $P = [S_1 \parallel \ldots \parallel S_n]$ be given. We will evaluate an *intermediate semantics* for it by embedding it in a process algebra.

First we fix a set of atomic actions; these are:

(i)	$b_1, \neg b_1, b_1 \wedge b_2$	if b_1, b_2 occur in P
(ii)	$b\&x := t$	if x and t occur in P, for all b from (i)
(iii)	$b\&C!t$	if $C!t$ occurs in P, for all b from (i)
(iv)	$b\&C?x$	if $C?x$ occurs in P, for all b from (i)

Let us call this alphabet of actions $A_{\mathrm{CSP-P}}$. If we delete all actions of the form $b\&C!t$ or $b\&C?x$ we obtain A_p. So A_p contains the proper actions that evaluation of P can involve, while $A_{\mathrm{CSP-P}}$ contains the subatomic actions as well. H contains the actions of the form $b\&C!t$ and $b\&C?x$.

Next we fix a communication function. All communications lead to δ, except the following ones:

$$b_1\&C!t \mid b_2\&C?x = (b_1 \wedge b_2)\&x := t.$$

We will first find an image $[\![P]\!]$ of P in $A^{\infty}_{\mathrm{CSP-P}}$. This is done using the notation of μ-calculus. We use an inductive definition for subprograms of the component programs first:

$$[\![b]\!] = b$$

$$[\![b\&x := t]\!] = b\&x := t$$

$$[\![b\&C!t]\!] = b\&C!t$$

$$[\![b\&C?x]\!] = b\&C?x$$

$$[\![S_1; S_2]\!] = [\![S_1]\!] \cdot [\![S_2]\!]$$

$$[\![S_1 \square S_2]\!] = [\![S_1]\!] + [\![S_2]\!]$$

$$[\![\textbf{while } b \textbf{ do } S \textbf{ od}]\!] = \mu x(b \cdot [\![S]\!] \cdot x + \neg b).$$

Here $\mu x(b{\cdot}[\![S]\!]{\cdot}x{+}\neg b)$ is the unique solution of the equation $X = b{\cdot}[\![S]\!]{\cdot}X{+}\neg b$. It is easily seen that the solution $\underline{X}$ is regular whenever $[\![S]\!]$ is regular. Inductively one finds that $[\![S]\!]$ is regular for each component program S.

Finally for the program P we obtain:

$$[\![P]\!] = [\![[S_1 \parallel \ldots \parallel S_n]]\!] = \partial_H([\![S_1]\!] \parallel \ldots \parallel [\![S_n]\!]).$$

We can now draw two interesting conclusions:

(i) $[\![P]\!]$ is regular;

(ii) $[\![P]\!]$ can just as well be (recursively) defined in $\underline{A}^{\infty}_{P}(+,\cdot)$ (so without any mention of communication).

Proof. (i) $[\![S_i]\!]$ is regular because it is defined using linear recursion equations only. Consequently the $[\![S_i]\!]$ are in $r(A^{\infty}_{\mathrm{CSP-P}})$ and so is $[\![P]\!]$ because $r(A^{\infty}_{\mathrm{CSP-P}})$ is a subalgebra of $A^{\infty}_{\mathrm{CSP-P}}$.
(ii) follows from (i) and Theorem 2.1.1. (iii). □

Remark. In general one must expect that a recursive definition of $[\![P]\!]$ not involving merge will be substantially more complex than the given one with merge.

3 Recursively Defined Processes

3.1 The algebra of recursively defined processes

Let $X = \{X_1, \ldots, X_n\}$ be a set of process names (variables). We will consider terms over X composed from atoms $a \in A$ and the operators $+, \cdot, \|, \lfloor\!\lfloor, |, \partial_H$.

A system E_X of *guarded fixed point equations* for X is a set of n equations

$$X_i = T_i(X_1, \ldots, X_n), \ i = 1, \ldots, n,$$

with $T_i(X_1, \ldots, X_n)$ a guarded term.

Theorem 3.1.1. *Each system E_X of guarded fixed point equations has a unique solution in $(A^\infty)^n$.*

Proof. See DE BAKKER & ZUCKER [1, 2]; essentially E_X is seen as an operator $(A^\infty)^n \longrightarrow (A^\infty)^n$ which under suitable metrics is a contraction and has exactly one fixed point, by Banach's fixed point theorem. □

Definition. $p \in A^\infty$ is called *recursively definable* if there exists a system E_X of guarded fixed point equations over X with solution $(p, q_1, \ldots, q_{n-1})$.

Proposition 3.1.2. *The recursively defined processes constitute a subalgebra of A^∞.*

Proof. Let $E_X = \{X_i = T_i(X) \mid i = 1, \ldots, n\}$ and $E_Y = \{Y_j = S_j(Y) \mid j = 1, \ldots, m\}$. Let $E_Z = E_X \cup E_Y \cup \{Z = T_1(X) \| S_1(Y)\}$. Now if E_X defines p and E_Y defines q, then E_Z defines $p \| q$. Likewise for the other operations. □

Notation. With $R(A^\infty)$ we denote the subalgebra of recursively defined processes.

Remark. For algebras with restricted signatures the above construction of a subalgebra of recursively defined processes is equally valid. Of course, the equations will then use the restricted signatures only. This leads to algebras like

$$R(\underline{A}^\infty(+, \cdot)) \ \text{ and } \ R(\underline{A}^\infty(+, \cdot, \|, \lfloor\!\lfloor)).$$

3.2 Recursive definitions and finitely generated process algebras

Let $p_1, \ldots, p_n$ be processes in A^∞. Then $A_\omega(p_1, \ldots, p_n)$ will denote the subalgebra of A^∞ generated by $p_1, \ldots, p_n$.

Theorem 3.2.1. *Let $\underline{X}_1, \ldots, \underline{X}_n$ be solutions of the system of guarded fixed point equations E_X. Then $A_\omega(\underline{X}_1, \ldots, \underline{X}_n)$ is closed under taking subprocesses.*

Proof. Let $p \in A_\omega(\underline{X}_1, \ldots, \underline{X}_n)$. Then for some term T we have $p = T(\underline{X}_1, \ldots, \underline{X}_n)$; after substitutions corresponding to $X_i = T_i(X_1, \ldots, X_n)$ we may assume that T is guarded.

On the basis of ACP one can rewrite $T(X_1, \ldots, X_n)$ into the form

$$\Sigma a_i \cdot R_i(X_1, \ldots, X_n) + \Sigma b_i.$$

Consequently all immediate subprocesses of p, i.e. the $R_i(\underline{X}_1, \ldots, \underline{X}_n)$, are in $A_\omega(\underline{X}_1, \ldots, \underline{X}_n)$ as well. □

This theorem gives a useful criterion for recursive definability (to be used in Section 5):

Corollary 3.2.2. (i) *Let* $p \in R(\underline{A}^\infty(+, \cdot, \|, \lfloor\!\lfloor))$. *Then* Sub$(p)$ *is finitely generated using* $+, \cdot, \|, \lfloor\!\lfloor, a \in A$.
(ii) *Likewise for the restricted signature of* $+, \cdot, a \in A$. □

3.3 Finitely branching processes

Definition. Let $p \in A^\infty$. (i) Then $\mathcal{G}_p$ is the *canonical process graph* of p, defined as follows.

The set of *nodes* of $\mathcal{G}_p$ is Sub$(p) \cup \{o\}$. Here o is a termination node. The *root* of $\mathcal{G}_p$ is p. The (labeled and directed) *edges* of $\mathcal{G}_p$ are given by:

(1) if $a \in$ Sub(p) then $a \xrightarrow{a} o$ is an edge,

(2) if $ax \in$ Sub(p) then $ax \xrightarrow{a} x$ is an edge,

(3) if $ax + y \in$ Sub(p) then $ax + y \xrightarrow{a} x$ is an edge.

(ii) Let $p \xrightarrow{a_0} p_1 \xrightarrow{a_1} \ldots$ be a maximal path in $\mathcal{G}_p$ (i.e. infinite or terminating in o). Then $a_0 a_1 \ldots$ is a *trace* of p.

(iii) p is *perpetual* if all its traces are infinite.

(iv) $\|p\|$, the *breadth* of p, is the outdegree of the root of $\mathcal{G}_p$. Here $\|p\| \in \mathbb{N}$, or $\|p\|$ is infinite.

(v) p is *finitely branching* if for all $q \in$ Sub(p), $\|q\|$ is finite.

(vi) p is *uniformly finitely branching* if $\exists n \in \mathbb{N}\ \forall q \in$ Sub(p) $\|q\| < n$.

The proof of the following proposition is routine and omitted.

Proposition. *The uniformly finitely branching processes constitute a subalgebra of* A^∞. □

The next theorem gives further criteria for recursive definability of processes.

Theorem 3.3.1. (i) *Recursively defined processes are finitely branching.*
(ii) *Moreover, processes recursively defined using only* $+, \cdot$ *are uniformly finitely branching.*
(iii) *There exists a process* $p \in R(\underline{A}^\infty(+, \cdot, \|, \lfloor\!\lfloor))$ *which is not uniformly finitely branching.*

Proof. (i), (ii): straightforward.
(iii): Consider the solution $\underline{X}$ of

$$X = a + b(Xc \parallel Xd).$$

Define with induction on n the following processes:

$$\begin{cases} p_0 = a \\ p_{n+1} = p_n \cdot c \parallel p_n \cdot d. \end{cases}$$

Claims: (1) $\forall n \in \mathbb{N}\ \exists q_n \in \mathrm{Sub}(\underline{X})\ \partial_{\{b\}}(q_n) = p_n$

(2) $\|p_n\| = 2^n$

(3) $\|\partial_H(x)\| \leq \|x\|$

Here (1) states that the p_n are 'almost' subprocesses of $\underline{X}$. Claim (3) is the general and obvious fact that the projection operator ∂_H certainly cannot increase the breadth of its argument. Combined with the observation of Claim (2) that the breadths of the p_n are unbounded, the claim yields the result.

We will now prove Claim (1) and (2). (The proof of Claim (3) is straightforward.)

Proof of Claim (1).

Let

$$\begin{cases} q_0 = \underline{X} \\ q_{n+1} = q_n \cdot c \parallel q_n \cdot d. \end{cases}$$

We will prove that $\partial_{\{b\}}(q_n) = p_n$, for all $n \geq 0$.
$n = 0$: $\partial_{\{b\}}(q_0) = a + \delta(\ldots) = a = p_0$.
Induction hypothesis: $\partial_{\{b\}}(q_n) = p_n$ and $q_n \in \mathrm{Sub}(\underline{X})$.
To prove: $\partial_{\{b\}}(q_{n+1}) = p_{n+1}$ and $q_{n+1} \in \mathrm{Sub}(\underline{X})$.
Since $q_n \in \mathrm{Sub}(\underline{X})$, we have $q_n c \in \mathrm{Sub}(\underline{X}c)$ and $q_n d \in \mathrm{Sub}(\underline{X}d)$.
So $q_{n+1} = q_n c \parallel q_n d \in \mathrm{Sub}(\underline{X}c \parallel \underline{X}d) \subseteq \mathrm{Sub}(a + b(\underline{X}c \parallel \underline{X}d)) = \mathrm{Sub}(\underline{X})$.
Furthermore, $\partial_{\{b\}}(q_{n+1}) = \partial_{\{b\}}(q_n c \parallel q_n d) =$ (since there is no nontrivial communication) $\partial_{\{b\}}(q_n c) \parallel \partial_{\{b\}}(q_n d) = \partial_{\{b\}}(q_n)c \parallel \partial_{\{b\}}(q_n)d = p_n c \parallel p_n d = p_{n+1}$.

Proof of Claim (2). We will give a sketch of the proof.
Define the set D of "inside-out traces" as follows:

$$\begin{cases} a \in D \\ \sigma, \tau \in D \Longrightarrow \sigma c \tau d, \sigma d \tau c \in D. \end{cases}$$

Now consider e.g. $p_2 = [(ac \parallel ad)c] \parallel [(ac \parallel ad)d]$.
D contains some traces of p_2, such as *acaddadacc*. Traces in D arise from an "inside-out evaluation" of the merges in the unevaluated expression for p_2, as suggested by the following figure:

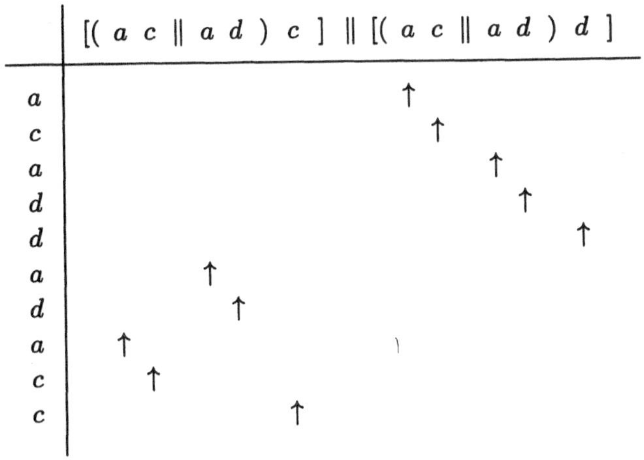
[(a c ‖ a d) c] ‖ [(a c ‖ a d) d]
a
c
a
d
d
a
d
a
c
c

Moreover, traces in D can be evaluated in precisely one way. (This does not hold for traces not in D; e.g. *aacdaadcdc* may be obtained starting from each of the four occurrences of a in the expression for p_2.) Hence the four summands of p_2, corresponding to the four occurrences of a in the expression for p_2, are different since they contain a trace which is characteristic for them. So $\|p_2\| = 4$. Likewise one proves the general statement in Claim (2). □

Theorem 3.3.2. *Let E_X be a system of guarded fixed point equations over $+, \cdot, A, X$. Suppose the solutions $\underline{X}$ are perpetual. Then they are regular.*

Proof. Since the $\underline{X}_i$ in $\underline{X} = \{\underline{X}_1, \ldots, \underline{X}_m\}$ are perpetual, we have $\underline{X}_i \cdot p = \underline{X}_i$ for every $p \in A^\infty$. Therefore every product $X_i \cdot t$ in E_X may be replaced by X_i without altering the solution vector $\underline{X}$. This leads to a system E'_X where only prefix multiplication is used, or in other words, containing only linear equations (see 1.3). Hence the solutions $\underline{X}$ of E'_X are regular, by Theorem 2.1.1.(i). □

Corollary 3.3.3. *Let p be a finitely branching and perpetual process. Let* Sub(p) *be generated using $+, \cdot$ by a finite subset $X \subseteq$* Sub(p).

Then p is regular.

Proof. Say $\mathcal{X} = \{q_1, \ldots, q_m\}$. Since p is finitely branching, and hence also the q_i are finitely branching, we can find guarded expressions (using $+, \cdot$ only) $T(X_1, \ldots, X_n)$ and $T_i(X_1, \ldots, X_{m_i})$ such that

$$\begin{cases} p = T(p_1, \ldots, p_n) \\ q_i = T_i(q_{i1}, \ldots, q_{im_i}), \ i = 1, \ldots, m. \end{cases}$$

Here the p_k $(k = 1, \ldots, n)$ and q_{ij} $(i = 1, \ldots, m;\ j = 1, \ldots, m_i)$ are by definition in Sub(p); therefore the p_k and q_{ij} can be expressed in $q_1, \ldots, q_m$. So there are *guarded* $+, \cdot$-terms T' and T'_i such that

$$\begin{cases} p = T'(q_1, \ldots, q_m) \\ q_i = T'_i(q_1, \ldots, q_m), \ i = 1, \ldots, m. \end{cases}$$

Since p is perpetual, every subprocess of p is perpetual; in particular the q_i $(i = 1, \ldots, m)$. By the preceding theorem p and the q_i are now regular. □

Remark. The condition 'finitely branching' is necessary in this Corollary, as the following example shows. Consider

$$p = \sum_{i=1}^{\infty} a^i b^\omega;$$

more precisely, p is the projective sequence $(p_1, p_2, \ldots, p_n, \ldots)$ with

$$p_n = \sum_{i=1}^{n} a^i b^{n-i}.$$

Then the canonical transition diagram of p is

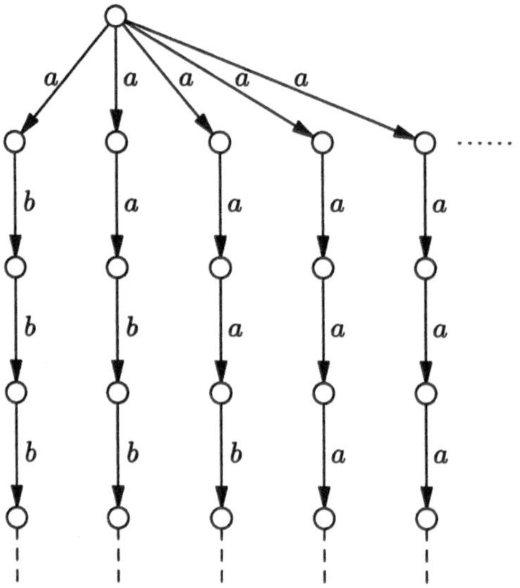

Now p is perpetual and $\mathrm{Sub}(p) = \{p\} \cup \{a^n b^\omega \mid n \geq 0\}$, so $\mathrm{Sub}(p)$ is generated by its finite subset $\{p, b^\omega\}$; yet p is not regular.

3.4 Interesting examples of recursive definitions

We will consider BAG, COUNTER and STACK. Let D be a finite set of data values. Let $A = D \cup \underline{D}$, where $\underline{D} = \{\underline{d} \mid d \in D\}$. Let us first consider a bag B over D; its actions are

d : add d to the bag
$\underline{d}$: take d from the bag.

The initial state of B is empty. Thus the behaviour of B is some process in A^∞.

Similarly the stack S is represented by a process in A^∞.

A counter C is a process in $\{0, p, s\}^\infty$ where the actions $0, p, s$ have the following meaning:

$0:$ assert that C has value 0
$p:$ add one to the counter
$s:$ subtract one from the counter (if possible).

Of course these descriptions are rather informal; a much more precise definition could be given along the lines of BERGSTRA & KLOP [7].

Here we are interested in recursive definitions for B, S and C:

$B = \sum_{d\in D} d \cdot (\underline{d} \parallel B)$
$S = \sum_{d\in D} d \cdot T_d \cdot S$ $T_d = \underline{d} + \sum_{b\in D} b \cdot T_b \cdot T_d$ for all $d \in D$
$C = (0 + s \cdot H) \cdot C$ $H = p + s \cdot H \cdot H$

Remarks. The equation for B has been discussed in detail in [7].

The recursive definition of S is equivalent to one of HOARE [10].

The equations for C are similar to those for S when $D = \{s\}$ and p stands for $\underline{s}$. It only has the extra option for testing on value zero.

4 Undecidability of the Word Problem in Fixed Point Algebras

As in 3.2, for $p_1, \ldots, p_n \in A^\infty$ we denote with $A_\omega(p_1, \ldots, p_n)$ the subalgebra of A^∞ generated by $p_1, \ldots, p_n$.

Let $X_1, \ldots, X_n$ be the set of new names for processes, and let $\underline{X}_1, \ldots, \underline{X}_n$ be processes in A^∞. Then with $A_\omega(\underline{X}_1, \ldots, \underline{X}_n)$ we denote an algebra as above but with $X_1, \ldots, X_n$ added to the signature.

Remark. Let us denote with $A_\omega[X_1, \ldots, X_n]$ the free ACP algebra generated over new names $X_1, \ldots, X_n$. For each set of interpretations $\underline{X}_1, \ldots, \underline{X}_n$ there is a homomorphism

$$\phi : A_\omega[X_1, \ldots, X_n] \to A_\omega(\underline{X}_1, \ldots, \underline{X}_n).$$

Now suppose that E_X is a system of guarded fixed point equations for $X = \{X_1, \ldots, X_n\}$. Then

$$A_\omega[X_1, \ldots, X_n]/E_X$$

is the algebra obtained by dividing out the congruence generated by E_X. On the other hand, let $\underline{X}_1, \ldots, \underline{X}_n$ be the unique solutions of E_X in A^∞. There is again a homomorphism

$$\phi : A_\omega[X_1, \ldots, X_n]/E_X \to A_\omega(\underline{X}_1, \ldots, \underline{X}_n).$$

Both algebras $A_\omega[X_1, \ldots, X_n]/E_X$ and $A_\omega(\underline{X}_1, \ldots, \underline{X}_n)$ may be vastly different however. Being an initial algebra of a finite specification, $A_\omega[X_1, \ldots, X_n]/E_X$ is semicomputable. As we shall see, $A_\omega(\underline{X}_1, \ldots, \underline{X}_n)$ is in general cosemicomputable and we will present an example where it is not computable indeed.

Definition. A *fixed point algebra* is an algebra

$$A_\omega(\underline{X}_1, \ldots, \underline{X}_n)$$

where the $\underline{X}_i$ are solutions in A^∞ of some system of guarded fixed point equations E_X.

Definition. Consider an algebra $A_\omega(\underline{X}_1, \ldots, \underline{X}_n)$. The *word problem* for this algebra consists in deciding for given terms T and R over $+, \cdot, \|, ⌊\!⌊, |$, $\partial_H, A, X_1, \ldots, X_n$ whether or not

$$A_\omega(\underline{X}_1, \ldots, \underline{X}_n) \models T = R.$$

Theorem 4.1. *For each fixed point algebra $A_\omega(\underline{X}_1, \ldots, \underline{X}_n)$ the word problem is co-r.e.*

Proof. Consider $A_\omega(\underline{X}_1, \ldots, \underline{X}_n)$ and let $\underline{X}_i$ solve E_X. Let T, R be two terms. Then

$$A_\omega(\underline{X}_1, \ldots, \underline{X}_n) \models T \neq R \Longleftrightarrow$$
$$\exists k \; A_\omega(\underline{X}_1, \ldots, \underline{X}_n)/\equiv_k \; \models T \neq R.$$

Here $\equiv_k$ is the congruence on A^∞ defined as follows: if $p = (p_1, p_2, \ldots)$ and $q = (q_1, q_2, \ldots)$ then $p \equiv_k q \Longleftrightarrow p_i = q_i$ for $i = 1, .., k$.

Now $A_\omega(\underline{X}_1, \ldots, \underline{X}_n)/\equiv_k$ is a finite algebra which can be uniformly computed from E_X and k; in this algebra $T \neq R$ can be effectively decided.

Consequently inequality of T and R is semicomputable and equality is cosemicomputable (co-r.e). □

Theorem 4.2. *There is a fixed point algebra $A_\omega(\underline{X}_1, \ldots, \underline{X}_n)$ with undecidable word problem.*

Proof. Let K be a recursively enumerable but not recursive subset of ℕ. The elements of K can be recognized by a register machine on three counters in which the argument n is initially stored in the first counter (see HOPCROFT & ULLMAN [11]).

The counters will be denoted by x, y, z; we use α as a metavariable ranging over x, y, z. The register machine has the following instructions:

$\alpha := \alpha + 1$; **goto** j
$\alpha := \alpha \dot{-} 1$; **goto** j
if $\alpha = 0$ **then goto** j_1 **else goto** j_2
stop.

A program for the register machine is a numbered sequence $I_1, \ldots, I_k$ of such instructions, where obviously $j, j_1, j_2 \in \{1, \ldots, k\}$ for all instructions.

Let P be a program which recognizes K in the sense that $P(n,0,0) \longrightarrow$ **stop** iff $n \in K$. Let l be the number of instructions of P. We will define a fixed point algebra in which P can be represented. The alphabet A, the set of subatomic action H and the set X of variables for the system of recursion equations for this fixed point algebra are as follows:

$$A = \{s_\alpha, p_\alpha, 0_\alpha, b, \mathbf{stop}, \delta\}$$
$$H = A \ - \ \{b, \mathbf{stop}, \delta\}$$
$$X = \{C_\alpha, H_\alpha, B, X_i \mid \alpha = x, y, z;\ i = 1, \ldots, l\}.$$

The communication function $. \mid .$ is given by

$$s_\alpha \mid s_\alpha \ = \ p_\alpha \mid p_\alpha \ = \ 0_\alpha \mid 0_\alpha \ = \ b \ (\alpha = x, y, z)$$

All other communications equal δ.

Before giving the system of equations E_X we define a map $\nless \ \ngtr$ from register machine instructions to process algebra expressions over A, X:

$$\nless \alpha := \alpha + 1;\ \mathbf{goto}\ j \ngtr \ = \ s_\alpha \cdot X_j$$
$$\nless \alpha := \alpha \dot{-} 1;\ \mathbf{goto}\ j \ngtr \ = \ (0_\alpha + p_\alpha) \cdot X_j$$
$$\nless \mathbf{if}\ \alpha = 0\ \mathbf{then\ goto}\ j\ \mathbf{else\ goto}\ j' \ngtr \ = \ 0_\alpha \cdot X_j \ + \ p_\alpha \cdot s_\alpha \cdot X_{j'}$$
$$\nless \mathbf{stop} \ngtr = \mathbf{stop}$$

Let E_X be the system of guarded recursion equations:

$$\begin{cases} C_\alpha \ = \ (0_\alpha \ + \ s_\alpha \cdot H) \cdot C & (\alpha = x, y, z) \\ H_\alpha \ = \ (p_\alpha \ + \ s_\alpha \cdot H) \cdot H & (\alpha = x, y, z) \\ B \ = \ b \cdot B & \\ X_j \ = \ \nless I_j \ngtr & (j \ = \ 1, \ldots, l) \end{cases}$$

(Note that the $\nless I_j \ngtr$ contain variables from $X_1, \ldots, X_l$.)

Further, note that the expression $H_\alpha^n \cdot C_\alpha$ denote the state of counter α containing n, and consider the following term representing P:

$$\partial_H(X_1 \parallel H_x^n \cdot C_x \| C_y \| C_z).$$

We claim that

$$A_\omega(\underline{C}_x, \underline{H}_x, \underline{C}_y, \underline{H}_y, \underline{C}_z, H_z, \underline{X}_1, ..., \underline{X}_l, \underline{B}) \models \partial_H(X_1 \parallel H_x^n\, C_x \| C_y \| C_z) = B$$

if and only if the computation $P(n,0,0)$ diverges.

The straightforward proof of the claim is omitted.

Thus we find that the predicate $n \notin K$ is one-one reducible to the word problem of the fixed point algebra $A_\omega(\underline{C}_x, \ldots, \underline{B})$, which shows that this word problem is undecidable. □

5 Technical Aspects of Different Recursive Definition Mechanisms

In this section we will provide information about particular recursive definition mechanisms. We summarize the results in a sequence of theorems:

Theorem 5.1. *C (counter) and S (stack) cannot be defined by means of a single equation over* $A^\infty(+,\cdot)$.

Theorem 5.2. *B (bag) cannot be recursively defined over* $A^\infty(+,\cdot)$ *(provided its domain of values contains at least two elements).*

Theorem 5.3. *If* $\underline{X}$ *is recursively defined over* $A^\infty(+,\cdot,\|,\mathbin{\|\!\_})$ *and* $\underline{X} \notin A_\omega$ *then* $\underline{X}$ *has an infinite regular (i.e. eventually periodic) trace.*

Theorem 5.4. *There is a process* $p \in \{a,b\}^\infty$ *which cannot be recursively defined in* $\{a,b\}^\infty(+,\cdot,\|,\mathbin{\|\!\_})$ *but which can be recursively defined in* $\{a,b,c,d,\delta\}^\infty(+,\cdot,\|,\mathbin{\|\!\_},|,\partial_H)$ *where H and the communication function are appropriately chosen.*

We will give the proofs of these theorems in the order of increasing length of the proof.

Proof of Theorem 5.1. Immediately, by Theorem 3.3.2 and the fact that C and S are clearly not regular. □

Proof of Theorem 5.4. Consider the alphabet $A = \{a,b,c,d,\delta\}$, with $H = \{c,d\}$ as set of subatomic actions and with communication function given by:

$$c \mid c = a;\ d \mid d = b;\ \text{other communications equal } \delta.$$

Let

$$p = ba(ba^2)^2(ba^3)^2(ba^4)^2\ldots$$

and consider the system of equations

$$\begin{cases} X &= cXc + d \\ Y &= dXY \\ Z &= dXcZ. \end{cases}$$

It turns out that $p = \partial_H(dcY \| Z)$. To prove this, consider the processes

$$p_n \quad = \quad \partial_H(dc^nY \| Z)$$

for $n \geq 1$. Now we claim that for all $n \geq 1$:

$$p_n \quad = \quad ba^nba^{n+1}p_{n+1}$$

which immediately yields the result. Proof of the claim:

$$\begin{aligned} p_n \quad = \quad & \partial_H(dc^nY \| Z) \ = \ \partial_H(dc^nY \| dXcZ) \ = \ ba^n\partial_H(Y \| Xc^ncZ) \ = \\ & ba^n\partial_H(dXY \| (cXc+d)c^{n+1}Z) \ = \ ba^nb\partial_H(XY \| c^{n+1}Z) \ = \\ & ba^nba^{n+1}\partial_H(Xc^{n+1}Y \| Z) \ = \ ba^nba^{n+1}\partial_H(dc^{n+1}Y \| Z) \ = \\ & ba^nba^{n+1}p_{n+1}. \end{aligned}$$

The fact that p cannot be recursively defined without communication follows immediately from Theorem 5.3 whose proof will follow now. □

Proof of Theorem 5.3.

To obtain information about traces of recursively defined processes, we need the concept of a *trace generator* of a term $T(X_1, \ldots, X_n)$. If T is closed, i.e. contains no variables X_i, the trace generators of T as defined below are just the usual traces; if T is open then its trace generators may also contain variables. First we need a 'normal form' of terms:

Definition. (i) On the set of terms built from $+, \cdot, \|, \lfloor\!\lfloor, a \in A$ we define the following reduction rules (which may be applied in a context):

$$\begin{array}{ll} x + x \to x & \\ (x+y)z \to xz + yz & \\ z(x+y) \to zx + zy & (*) \\ x \| y \to x \lfloor\!\lfloor y + y \lfloor\!\lfloor x & \\ a \lfloor\!\lfloor x \to ax & \\ (ax) \lfloor\!\lfloor y \to a(x \| y) & \\ (x+y) \lfloor\!\lfloor z \to x \lfloor\!\lfloor z + y \lfloor\!\lfloor z. & \end{array}$$

A term in which none of these reduction rules can be applied, is in *trace normal form*.

(ii) Let $T \to \ldots \to T'$ be a reduction according to the rules above such that no further step is possible, i.e. T' is in trace normal form. Let

$$T' = \sum_{i=1}^{k} \tau_i$$

where the τ_i are indecomposable w.r.t. $+$.

Then the τ_i $(i = 1, \ldots, k)$ are the *trace generators* of T. (So a trace normal form is a sum of trace generators.)

The reduction rules above correspond to the axioms A3,4 and M1-4, except for the rule $(*)$. Note that we work modulo A1,2,5 (associativity of $+, \cdot$ and commutativity of $+$).

To show that the trace generators are well-defined by (ii) of the definition, one needs the following fact whose proof is standard and will be omitted (cf. [4] for a similar proof):

Proposition. (i) *Every reduction using the rules in the preceding definition must terminate.*
(ii) *All reductions with the same start terminate eventually in the same result.* □

Example. (i) $a(b+b+ca)a$ reduces to the trace normal form $aba + acaa$, hence has trace generators (here also traces) aba, $acaa$.
(ii) $X(a+b) \| c$ reduces to

$$Xa \lfloor\!\lfloor c + Xb \lfloor\!\lfloor c + cXa + cXb,$$

hence has trace generators $Xa \lfloor\!\lfloor c$, $Xb \lfloor\!\lfloor c$, cXa and cXb.

In fact we are only interested in the prefix of a trace generator up to the first variable:

Definition. Let τ be a trace generator of the form $w(\ldots(X\ldots$ where $w \in A^*$ (i.e. w is a term built from $a \in A$ by $\cdot$ only) and where w is followed by some brackets (possibly none) followed by the variable X.

Then the trace generator wX is called a *prefix* of τ.

Example. $aaabbX$ is a prefix of the trace generator $aaabb((XX) \mathbin{\|\!\lfloor} b)$.

Proposition. *Let T, S be terms such that T contains a trace generator with prefix wX, and S contains a trace generator vY.*

Then $T[X := S]$, the term resulting from substituting S for the occurrences of X in T, contains a trace generator with prefix wvY.

Proof. Elementary. Note that the left-linearity of $\|\!\lfloor$ is used as well as the auxiliary rule (*) needed for computing trace generators. □

Example. $S(X) \equiv b^2(X^2 \mathbin{\|\!\lfloor} b) + c$ when substituted in $T(X) \equiv a^3[(X(X \mathbin{\|\!\lfloor} a)) \mathbin{\|\!\lfloor} a]$ yields as one of its trace generators

$$a^3b^2[((X^2 \mathbin{\|\!\lfloor} b)b^2((X^2 \mathbin{\|\!\lfloor} b) \mathbin{\|\!\lfloor} a)) \mathbin{\|\!\lfloor} a]$$

which has indeed a^3b^2X as a prefix.

We can now finish the proof of Theorem 5.3. Let $E_X = \{X_i = T_i(X) \mid i = 1, \ldots, n\}$ be a system of guarded equations defining $\underline{X} = \{\underline{X}_1, \ldots, \underline{X}_n\}$ where $\underline{X}_1$ has an infinite trace. Define a directed graph $\mathcal{G}$ on $X = \{X_1, \ldots, X_n\}$, with edges labeled by $w \in A^*$ as follows:

$X_i \xrightarrow{w_{ij}} X_j$ is a labeled edge of $\mathcal{G}$ if:

$w_{ij}X_j$ is a prefix of a trace generator of $T_i(X)$.

We may suppose that every $T_i(X)$ $(i = 1, \ldots, n)$ contains some variable (otherwise the trivial equation $X_i = T_i$ could be eliminated first). Hence $\mathcal{G}$ contains no endnodes. Therefore $\mathcal{G}$, being finite, must contain a path starting with X_1 and eventually cyclic, e.g.:

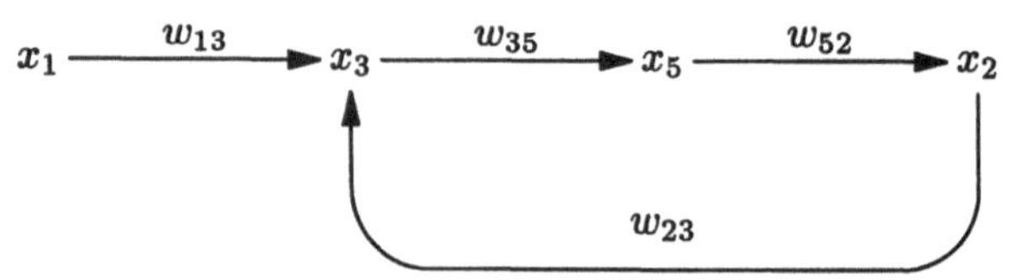

But then, by the previous Proposition, repeated substitution leads to a process $\underline{X}_1$ with an eventually periodic trace; in our example: $w_{13}(w_{35}w_{52}w_{23})^\omega$. □

Example. If E_X is

$$\begin{cases} X_1 &= a(X_2 \mathbin{\|\!\lfloor} X_3) + a \\ X_2 &= bc(X_3 \parallel X_3) \\ X_3 &= aaX_1X_3 \end{cases}$$

then

$$X_1 \xrightarrow{a} X_2 \xrightarrow{bc} X_3 \xrightarrow{aa} X_1$$

hence $\underline{X}_1$ contains a trace $(abcaa)^\omega$.

Proof of Theorem 5.2.

The behaviour of a bag B was defined above (Section 3.4):

$$B \; = \; \sum_{d \in D} d(\underline{d} \parallel B).$$

In this subsection we will consider the case that $D = \{a\}$, and the case $D = \{a, b\}$. (The results for the last case generalize at once to the case $D = \{a_1, \ldots, a_n\}$.)
In the first case

$$B \; = \; a(\underline{a} \parallel B) \qquad (^*)$$

is equivalent to the following definition without $\parallel$:

$$\begin{cases} B \; = \; aCB & (^{**}) \\ C \; = \; \underline{a} + aCC \end{cases}$$

as can be seen by realizing that the behaviour of a bag with singleton value domain is identical to that of a stack over the same domain. Indeed, both recursive definitions yield the transition diagram (or process graph):

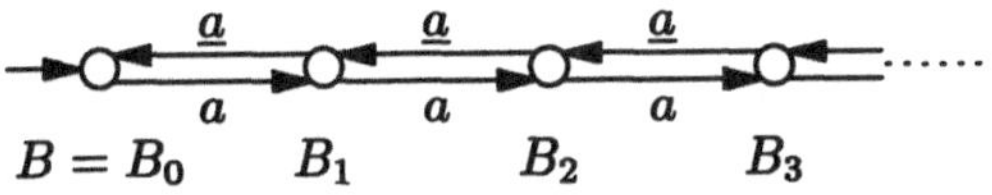

According to Theorem 3.2.1, the subprocesses B_n $(n \geq 0)$ of B are finitely generated (using $(^*)$) by $+, \cdot, \parallel, \mathbin{|\!\!\lfloor}, B, a, \underline{a}$. Indeed one easily verifies that

$$B_n \; = \; \underline{a}^n \parallel B.$$

Using $(^{**})$, the same theorem says for the restricted signature without $\parallel, \mathbin{|\!\!\lfloor}$, that the B_n are finitely generated by $+, \cdot, B, C, a, \underline{a}$. Indeed:

$$B_n \; = \; C^n \cdot B.$$

Before considering the case that D is not a singleton, say $D = \{a, b\}$ (the general case follows by a simple argument) and showing that the bag then needs $\parallel, \mathbin{|\!\!\lfloor}$ for its recursive definition, note that $(^*)$ can be rewritten as

$$B \; = \; (a\underline{a}) \mathbin{|\!\!\lfloor} B.$$

The intuition here is that B is the 'ω-merge' of $a\underline{a}$, i.e.

$$B \; = \; a\underline{a} \parallel a\underline{a} \parallel a\underline{a} \parallel a\underline{a} \parallel \ldots.$$

(For trace theory, the ω-merge occurs e.g. in [12].) To be precise:

Let p be a process. Then the ω-merge of p, written as $p^{\underline{\omega}}$, is the limit of the iteration sequence

$$p,\ p \parallel p,\ p \parallel p \parallel p,\ \ldots$$

as defined in [3], where the existence of the limit is shown. It is easy to prove that this limit is also obtained by the guarded fixed point equation

$$X \;=\; p \mathbin{\|\!\!\_} X.$$

(Note that $X = p \parallel X$ would not do as it is not guarded and has no unique solution.)

Next consider the bag B over $\{a,b\}$, that is:

$$B \;=\; a(\underline{a} \parallel B) + b(\underline{b} \parallel B).$$

Some alternative definitions are

$$B \;=\; a(\underline{a} \parallel B) \parallel b(\underline{b} \parallel B),$$

or

$$B \;=\; (a\underline{a} + b\underline{b}) \mathbin{\|\!\!\_} B,$$

or

$$B \;=\; (a\underline{a} \parallel b\underline{b}) \mathbin{\|\!\!\_} B,$$

or

$$\begin{cases} B &= X \parallel Y \\ X &= a(\underline{a} \parallel X) \\ Y &= b(\underline{b} \parallel Y) \end{cases}$$

or

$$\begin{cases} B &= X_1 \parallel Y_1 \\ X_1 &= aX_2X_1 \\ X_2 &= \underline{a} + aX_2X_2 \\ Y_1 &= bY_2Y_1 \\ Y_2 &= \underline{b} + bY_2Y_2 \end{cases}$$

(The last two systems are guarded after an appropriate substitution.)
The last system of equations is of interest since it shows that $R(A^\infty(+,\cdot))$ is not closed under $\parallel$ (after the result below is proved).

We will show that B cannot recursively be defined over $+,\cdot$, i.e. $B \notin R(A^\infty(+,\cdot))$. We start with some observations about B. Its canonical process graph is:

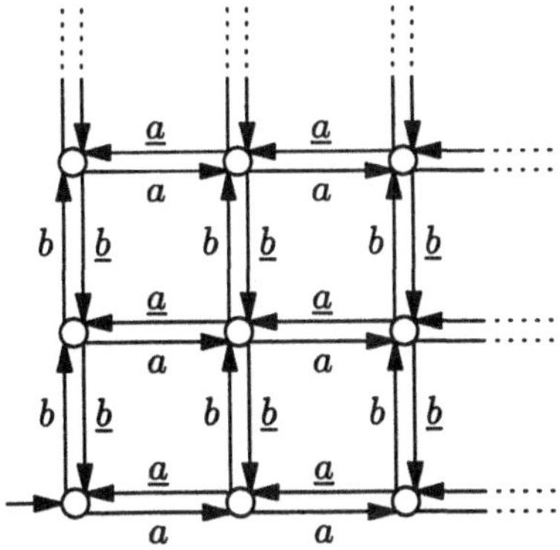

and its subprocesses are the $B_{m,n}$ $(m,n \geq 0)$ where $B = B_{0,0}$; the $B_{m,n}$ satisfy for all $m,n \geq 0$:

$$B_{m,n} \quad = \quad aB_{m+1,n} + \underline{a}B_{m-1,n} + bB_{m,n+1} + \underline{b}B_{m,n-1}$$

with the understanding that summands in which a negative subscript appears, must vanish.

The subprocesses $B_{m,n}$ are by Theorem 3.2.1 generated by $B, a, b, \underline{a}, \underline{b}$ via $+, \cdot, \|, ⌊⌊$; indeed it is easy to compute that

$$B_{m,n} \quad = \quad a^m \parallel b^n \parallel B.$$

Graphically we display the $B_{m,n}$ in the "a,b-plane":

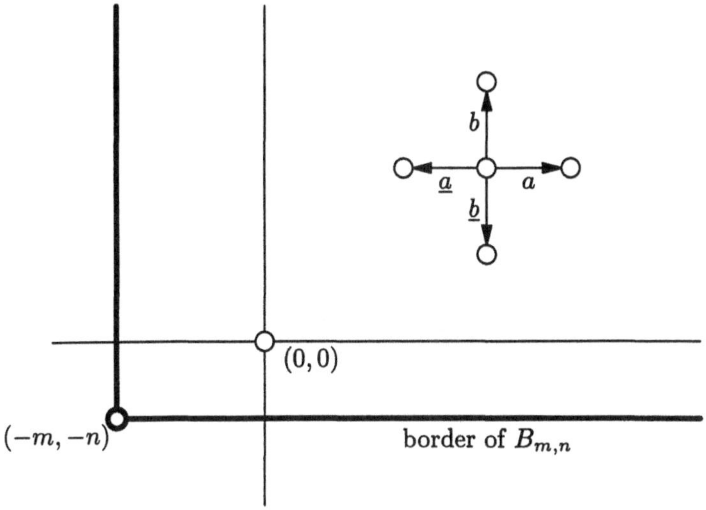

in which the starting node of $B_{m,n}$ is at $(0,0)$ and all traces of $B_{m,n}$ stay confined in the indicated quadrant.

Suppose for a proof by contradiction that $B \in R(A^\infty(+,\cdot))$. Then, by Corollary 3.2.2, the collection of subprocesses $B_{m,n}$ $(m,n \geq 0)$ is finitely generated using $+,\cdot$ only by say $\underline{X}_1,\ldots,\underline{X}_k$. Let the $B_{m,n}$ therefore be given by

$$B_{m,n} \;=\; T_{m,n}(\underline{X})$$

where $T_{m,n}(X)$ are terms involving only $+,\cdot,a,\underline{a},b,\underline{b},X$.
(Here $X = (X_1,\ldots,X_k)$ contains the variables of the system of recursive definitions yielding solutions $\underline{X}$ and used to define B.)

We may assume that every occurrence of X_i in $T_{m,n}$ is immediately preceded by some $u \in A = \{a,\underline{a},b,\underline{b}\}$. If not, we expand the corresponding $\underline{X}_i$ as

$$\underline{X}_i \;=\; a\underline{X}_{i1} + \underline{a}\underline{X}_{i2} + b\underline{X}_{i3} + \underline{b}\underline{X}_{i4}$$

(some summands possibly vanishing) and replace $\underline{X}_i$ by its subprocesses $\underline{X}_{i1},\ldots,\underline{X}_{i4}$ in the set of generators $\underline{X}$.

Further, we may take $T_{m,n}$ to be in normal form w.r.t. rewritings

$$(x+y)z \rightarrow xz + yz.$$

Now consider an occurrence of X_i in $T_{m,n}$. Then X_i is contained in a subterm of the form uX_iP, $u \in A$, P maybe vanishing. Take P maximal so, i.e. uX_iP is not a proper subterm of some uX_iPQ.

Then it is easy to see that $\underline{X}_i\underline{P}$ (where $\underline{P}$ is P after substituting $\underline{X}_j$ for $X_j, j = 1,\ldots,k$) is a subprocess of $B_{m,n}$, i.e.

$$\underline{X}_i\underline{P} \;=\; B_{k,l}$$

for some k,l.

Thus we find that all generators are left-factors of some subprocess of B. If such a left-factor $\underline{X}_i$ is perpetual then clearly in the factorization $\underline{X}_i\underline{P} = B_{k,l}$ we have already $\underline{X}_i = B_{k,l}$. For proper factorizations (i.e. where $\underline{X}_i$ is not perpetual) we have the following remarkable properties:

Proposition. *Let $PQ = B_{m,n}$ be a factorization of a subprocess of B. Suppose P is not perpetual. Then:*
(i) all finite traces of P end in the same point of the a,b-plane;
(ii) P determines n,m and Q uniquely (i.e. if moreover $PQ' = B_{m',n'}$, then $Q = Q'$ and $n,m = n',m'$).

Proof. (i) Consider the following figure:

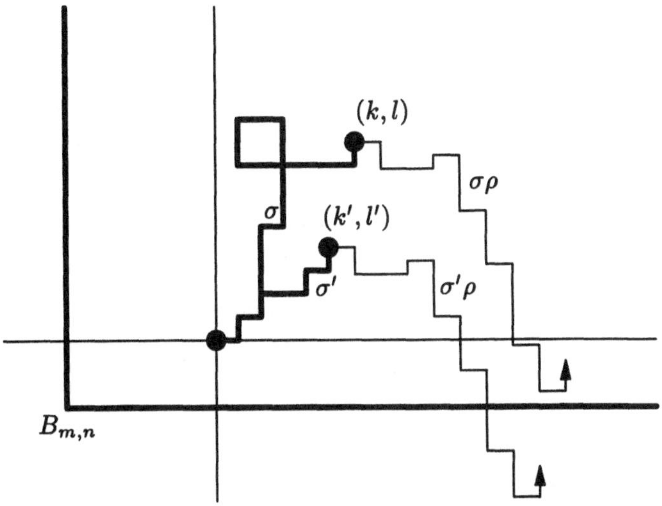

Suppose P has traces σ, σ' ending in different points (k,l) and (k',l'). Then Q has a trace ρ such that $\sigma\rho$ leads to the border of $B_{m,n}$. However, then $\sigma'\rho$ exceeds this border, contradicting the assumption $PQ = B_{m,n}$.
(ii) To see that $B_{m,n}$ is uniquely determined, let $PQ' = B_{m',n'}$. Say P's finite traces terminate in (k,l). Now consider a trace ρ in P which avoids this 'exit point'. (Here the argument breaks down for the case of a singleton value domain $D = \{a\}$.)

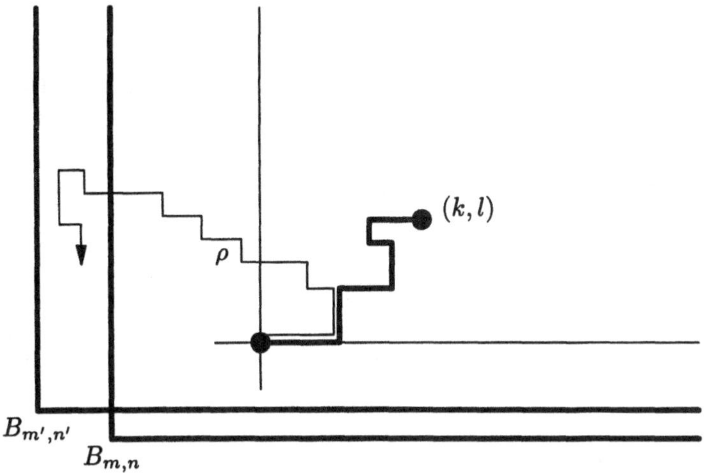

Since (k,l) is P's only exit point, ρ is confined to stay in P as long as it avoids (k,l). But then a trace ρ as in the figure which enters the symmetrical difference of the areas occupied in the a,b-plane by $B_{m,n}$ and $B_{m',n'}$ leads to an immediate contradiction.

The unicity of Q is proved by similar arguments. (Note that Q is itself a subprocess of B.) □

A corollary of the preceding Proposition is that in the equations $B_{m,n} = T_{m,n}(\underline{X})$ every $\underline{X_i P}$ (as defined above) can be replaced by B_{k_i,l_i} *depending on* i *alone*. Therefore the set of generators can be taken to consist of a finite subset of the collection of $B_{m,n}$, say $\{B_{k_i,l_i} \mid i = 1, \ldots, p\}$.

However, by Corollary 3.3.3 B must then be regular, an evident contradiction. Hence B cannot be recursively defined with $+$ and $\cdot$ alone. □

Remark. For the case of the 'general' bag defined by

$$B = \sum_{d \in D} d(\underline{d} \parallel B)$$

where D contains at least two elements, the non-eliminability of $\parallel$, $\lfloor\!\lfloor$ follows from the above by the following argument. Let $\phi : D \to \{a, b\}$ be a surjection. Then ϕ extends in the obvious way to a map from $(D \cup \underline{D})^\infty$ to $\{a, \underline{a}, b, \underline{b}\}^\infty$, by replacing each atom $u \in D \cup \underline{D}$ by $\phi(u)$. This extended mapping is easily shown to be a homomorphism w.r.t. all operations. Hence a recursive definition without $\parallel$, $\lfloor\!\lfloor$ for the general bag would by this 'collapsing' mapping yield a similar recursive definition for the bag over $\{a, \underline{a}, b, \underline{b}\}$.

References

[1] J.W. de Bakker and J.I. Zucker. Denotational semantics of concurrency. In *Proceedings of the* 14th *Annual ACM Symposium on Theory of Computing, San Francisco, California*, pages 153–158. ACM, 1982.

[2] J.W. de Bakker and J.I. Zucker. Processes and the denotational semantics of concurrency. *Information and Control*, 54(1/2):70–120, 1982.

[3] J.A. Bergstra and J.W. Klop. Fixed point semantics in process algebras. Report IW 206/82, Mathematisch Centrum, Amsterdam, 1982.

[4] J.A. Bergstra and J.W. Klop. Process algebra and mutual exclusion. Report IW 218/83, Mathematisch Centrum, Amsterdam, 1983.

[5] J.A. Bergstra and J.W. Klop. A process algebra for the operational semantics of static data flow networks. Report IW 222/83, Mathematisch Centrum, Amsterdam, 1983.

[6] J.A. Bergstra and J.W. Klop. An abstraction mechanism for process algebras. Report IW 231/83, Mathematisch Centrum, Amsterdam, 1983.

[7] J.A. Bergstra and J.W. Klop. An algebraic specification method for processes over a finite action set. Report IW 232/83, Mathematisch Centrum, Amsterdam, 1983.

[8] M. Hennessy. A term model for synchronous processes. *Information and Control*, 51(1):58–75, 1981.

[9] C.A.R. Hoare. Communicating sequential processes. *Communications of the ACM*, 21(8):666–677, 1978.

[10] C.A.R. Hoare. A model for communicating sequential processes. In R.M. McKeag and A.M. Macnaghton, editors, *On the construction of programs – an advanced course*, pages 229–243. Cambridge University Press, 1980.

[11] J.E. Hopcroft and J.D. Ullman. *Introduction to Automata Theory, Languages and Computation.* Addison-Wesley, 1979.

[12] T. Ito and Y. Nishitani. On universality of concurrent expressions with synchronization primitives. *Theoretical Computer Science*, 19:105–115, 1982.

[13] R. Milner. *A Calculus of Communicating Systems*, volume 92 of *Lecture Notes in Computer Science.* Springer-Verlag, 1980.

The Syntax and Semantics of μCRL

Jan Friso Groote
Department of Philosophy, Utrecht University
Heidelberglaan 8, 3584 CS Utrecht, The Netherlands

Alban Ponse
Programming Research group, University of Amsterdam
Kruislaan 403, 1098 SJ Amsterdam, The Netherlands

Abstract

A simple specification language based on CRL (*Common Representation Language*) and therefore called μCRL (*micro* CRL) is proposed. It has been developed to study processes with data. So the language contains only basic constructs with an easy semantics. To obtain executability, *effective* μCRL has been defined. In effective μCRL equivalence between closed *data-terms* is decidable and the operational behaviour is finitely branching and computable. This makes effective μCRL a good platform for tooling activities.

Key Words & Phrases: Specification Language, Abstract Data Types, Process Algebra, Operational Semantics.
1985 Mathematics Subject Classification: 68N99.
1987 CR Categories: D.2.1, D.3.1, D.3.3.
Note: The authors are supported by the European Communities under RACE project no. 1046, Specification and Programming Environment for Communication Software (SPECS). The first author is also supported by ESPRIT Basic Research Action 3006 (CONCUR). This document does not necessarily reflect the view of the SPECS project.

1 Introduction

In telecommunication applications the necessity of the use of formal methods has been observed several times. For that purpose several specification languages have been developed (SDL [9], LOTOS [25], PSF [30] and CRL [36]). These languages are designed to optimise usability. However, they turn out to be rather complicated, especially as far as their semantic basis is concerned. An enormous amount of manpower has already been invested into tooling these languages. But, although some major achievements have been made, this turns out to be hard and results often lag behind expectations.

This document presents the original defining text of μCRL such as it appeared in [15] and [12], with some small textual changes. Since its development, μCRL has been provided with a proof system [16, 17] and it has been shown how proofs in this proofsystem can be computer checked [35]. These techniques have been applied to a wealth of examples such as [8, 13, 14, 28, 29]. Also some more fundamental investigations have been made, and extensions to μCRL have been proposed [7, 34, 18]. For an overview and some conclusions about the current and future developments, see [17].

In this paper we define a language called μCRL (*micro* CRL, where CRL stands for Common Representation Language [36]) as it consists of the essence of CRL. It has been developed under the assumption that an extensive study of the basic constructs of specification languages will yield fundamental insights that are hard to obtain via the languages mentioned above. These insights may assist further development of these languages. So our language is indeed very small although its definition still requires quite some pages. As μCRL only contains core constructs, it may not be so well suited as an actual specification language.

An advantage of our 'simple' approach is that when in the future several constructs that are not included in the language will be well understood and will have a concise and natural semantics, we can add them to the language without a time and manpower consuming redesign of existing but not optimally devised features.

The language μCRL consists of data and processes. The data part contains equational specifications: one can declare sorts and functions working upon these sorts, and describe the meaning of these functions by equational axioms. The process part contains processes described in the style of CCS [31], CSP [22] or ACP [3, 4], where the particular process syntax has been taken from ACP. It basically consists of a set of uninterpreted actions that may be parameterised by data. These actions can represent all kinds of real world activities, depending on the usage of the language. There are sequential, alternative and parallel composition operators. Furthermore, recursive processes are specified in a simple way.

An important feature is executability. To obtain this, we define *effective* μCRL. In effective μCRL it is required that the equations specifying data constitute a semi-complete term rewriting system. This implies that data equivalence is decidable. Moreover, the specification of recursive processes must be guarded and sums over data sorts must be finite. This guarantees that the operational behaviour of every effective μCRL specification is finitely branching and computable. We believe that effective μCRL is an excellent base for building tools.

Acknowledgements. The idea for μCRL comes from Jan Bergstra, who had also a pervasive influence on its current form, especially in keeping the language small. We further thank Jos Baeten, Michel Dauphin, Arie van Deursen, Willem Jan Fokkink, Bertrand Gruson, Jan Gustafsson, Georg Karner, Martin Kooij, Henri Korver, Sjouke Mauw, Emma van der Meulen, Jan Rekers and Gert Veltink for their valuable comments.

2 The syntax of μCRL

In this section we present the syntax of μCRL. It contains two major components, namely data specified by a many sorted term rewriting system and processes which are based on process algebra [4]. The syntax is defined in the BNF formalism. Syntactical categories are written in italics and we use a '.' to end each BNF clause. In reasoning about the syntax of μCRL we use the symbol $\equiv$ to denote syntactic equivalence.

2.1 Names

We assume the existence of a set $\mathcal{N}$ of *names* that are used to denote sorts, variables, functions, processes and labels of actions. The *names* in $\mathcal{N}$ are sequences over an alphabet not containing

> $\bot, +, \|, \mathbin{\|\!\lfloor}, |, \lhd, \rhd, \cdot, \delta, \tau, \partial, \rho, \Sigma, \surd, \times, \rightarrow, :, =, (,), \{, \}$, ‘,’,
> a space and a newline.

The space and the newline serve as separators between names and are used to lay out specifications. The symbol $\bot$ is used in the description of the semantics and the other symbols have special functions. Moreover, $\mathcal{N}$ does not contain the reserved keywords **sort**, **proc**, **var**, **act**, **func**, **comm**, **rew** and **from**.

2.2 Lists

In the sequel *X-list*, $\times$-*X-list*, and *space-X-list* for any syntactical category X are defined by the following BNF syntax:

$$\begin{aligned} X\textit{-list} &::= X \mid X\textit{-list}, X. \\ \times\textit{-}X\textit{-list} &::= X \mid \times\textit{-}X\textit{-list} \times X. \\ \textit{space-}X\textit{-list} &::= X \mid \textit{space-}X\textit{-list}\ X. \end{aligned}$$

Lists are often described by the (informal) use of dots, e.g. $b_1 \times ... \times b_m$ with $m \geq 1$ is a $\times$-*X-list* where $b_1, ..., b_m$ are expressions in the syntactical category X. Note that lists cannot be empty.

2.3 Sort specifications

A *sort-specification* consists of a list of *names* representing sorts, preceded by the keyword **sort**.

$$\textit{sort-specification} \quad ::= \quad \textbf{sort}\ \textit{space-name-list}.$$

2.4 Function specifications

A *function-specification* consists of a list of function declarations. A *function-declaration* consists of a *name-list* (the names play the role of constant and function names), the sorts of their parameters and their target sort:

$$\begin{aligned} \textit{function-specification} &::= \textbf{func}\ \textit{space-function-decl-list}. \\ \textit{function-decl} &::= \textit{name-list} : \rightarrow \textit{name} \\ &\mid\ \ \textit{name-list} : \times\textit{-name-list} \rightarrow \textit{name}. \end{aligned}$$

2.5 Rewrite specifications

A *rewrite-specification* is given by a many sorted term rewriting system. Its syntax is given by the following BNF grammar:

rewrite-specification ::= *variable-decl-section*
  rewrite-rules-section.

In a *variable-decl-section* all variables that are used in a *rewrite-rules-section* must be declared. In such a declaration, it is also stated what the sort of a variable is. A variable declaration section may be empty.

variable-decl-section ::= **var** *space-variable-decl-list*
  | .

In a *variable-decl*aration, the *name-list* contains the declared variables and the *name* denotes their sort:

variable-decl ::= *name-list* : *name*.

Data-terms are defined in the standard way. The *name* without brackets in the syntax represents a variable or a constant.

data-term ::= *name*
  | *name*(*data-term-list*).

The equations in a *rewrite-rules-section* define the meaning of functions operating on data. The syntax of a *rewrite-rules-section* is given by:

rewrite-rules-section ::= **rew** *space-rewrite-rule-list*.
rewrite-rule ::= *name* = *data-term*
  | *name*(*data-term-list*) = *data-term*.

2.6 Process expressions and process specifications

In this section we first define what *process-exp*ressions look like. Then we define how these expressions can be used to construct *process-specifications*.

*Process-exp*ressions are defined via the following syntax explicitly taking care of the precedence among operators:

process-exp ::= *parallel-exp*
  | *parallel-exp* + *process-exp*.

parallel-exp ::= *merge-parallel-exp*
  | *comm-parallel-exp*
  | *cond-exp*
  | *cond-exp* $\mathbin{\|\!\lfloor}$ *cond-exp*.

merge-parallel-exp ::= *cond-exp* $\|$ *merge-parallel-exp*

	$\mid$	*cond-exp* $\|$ *cond-exp*.
comm-parallel-exp	::=	*cond-exp* $\mid$ *comm-parallel-exp*
	$\mid$	*cond-exp* $\mid$ *cond-exp*.
cond-exp	::=	*dot-exp*
	$\mid$	*dot-exp* $\triangleleft$ *data-term* $\triangleright$ *dot-exp*.
dot-exp	::=	*basic-exp*
	$\mid$	*basic-exp* $\cdot$ *dot-exp*.
basic-exp	::=	δ
	$\mid$	τ
	$\mid$	∂({*name-list*}, *process-exp*)
	$\mid$	τ({*name-list*}, *process-exp*)
	$\mid$	ρ({*renaming-decl-list*}, *process-exp*)
	$\mid$	Σ(*single-variable-decl*, *process-exp*)
	$\mid$	*name*
	$\mid$	*name*(*data-term-list*)
	$\mid$	(*process-exp*).

The + is the alternative composition. A *process-expression* $p+q$ behaves exactly as the argument that performs the first step.

The merge or parallel composition operator ($\|$) interleaves the behaviour of both arguments except that some actions in the arguments may communicate, which means that they happen at exactly the same moment and result in a communication action. In a *comm-specification* it can be declared which actions communicate. The left merge ($\lfloor\!\lfloor$) behaves exactly as the parallel operator, except that its first step must originate from its left argument only. The communication merge ($\mid$) also behaves as the parallel operator, but now the first action must be a communication between both components. The left merge and the communication merge are added to allow proof theoretic reasoning. It is not expected that they will be used in specifications. In the sequel the syntactical category *parallel-expression* also refers to *merge-parallel-expression* and *comm-parallel-expression*.

The *conditional* construct *dot-expression*$\triangleleft$*data-term*$\triangleright$ *dot-expression* is an alternative way to write an **if - then - else**-expression and is introduced by HOARE cs. [23] (see also [2]). The *data-term* is supposed to be of the standard sort of the Booleans (**Bool**). The $\triangleleft$-part is executed if the *data-term* evaluates to true (T) and the $\triangleright$-part is executed if the *data-term* evaluates to false (F).

The sequential composition operator '$\cdot$' says that first its left hand side can perform actions, and if it terminates then the second argument continues.

The constant δ describes the process that cannot do anything, especially, it cannot terminate. For instance, the process $\delta \cdot p$ can never perform an action of p. We also expect that δ is not used in specifications, but in reasoning δ is very handy to indicate that at a certain place a deadlock occurs.

The constant τ represents some internal activity that cannot be observed by the environment. It is therefore called the internal action.

The encapsulation operator ∂ is used to prevent actions of which the *name* is mentioned in its first argument from happening. This enables one to force actions into a communication.

The hiding operator, also denoted by a τ, is used to rename actions of which the *name* is mentioned into an internal action.

The renaming operator ρ is more general. It renames the *names* of actions according to the scheme in its first argument. A *renaming-decl*aration is given by the following syntax:

renaming-decl ::= *name* $\rightarrow$ *name*.

The first mentioned *name* is renamed to the second one.

The sum operator is used to declare a variable of a specific sort for use in a *process-exp*ression. A *single-variable-decl*aration is defined by:

single-variable-decl ::= *name* : *name*.

The scope of the variable is exactly the *process-exp*ression mentioned in the sum operator. The behaviour of this construct is a choice between the behaviours of *process-exp*ression in which each value of the sort of the variable has been substituted for the variable.

The constructs *name* and *name*(*data-term-list*) are either process instantiations or actions: *name* refers to a declared process (or to an action) and *data-term-list* contains the arguments of the process identifier (or the action).

The syntax of *process-exp*ressions says that $\cdot$ binds strongest, the conditional construct binds stronger than the parallel operators which in turn bind stronger than +.

A *process-specification* is a list of (parameterised) names, which are used as process identifiers, that are declared together with their bodies.

process-specification	::=	**proc** *space-process-decl-list*.
process-decl	::=	*name* = *process-exp*
	|	*name*(*single-variable-decl-list*) = *process-exp*.

2.7 Action specification

In an *action-specification* all actions that are used are declared. Actions may be parameterised by data, and in that case we must declare on which sorts an action depends. An *action-specification* has the following form:

action-specification	::=	**act** *space-action-decl-list*.
action-decl	::=	*name*
	|	*name-list* : $\times$-*name-list*.

2.8 Communication specification

A *comm-specification* prescribes how actions may communicate. It only describes communication on the level of *names* of actions, e.g. if it is specified that *in* | *out* = *com* then each action $in(t_1, ..., t_k)$ can communicate with

$out(t'_1, ..., t'_m)$ to $com(t_1, ..., t_k)$ provided $k = m$ and t_i and t'_i denote the same data-element for $i = 1, ..., k$.

$$\begin{array}{rcl} \textit{comm-specification} & ::= & \textbf{comm}\ \textit{space-comm-decl-list}. \\ \textit{comm-decl} & ::= & \textit{name} \,|\, \textit{name} = \textit{name}. \end{array}$$

In the last rule the | is a language symbol and should not be confused with the | used in sets and the BNF-syntax.

2.9 Specifications

Specifications are entities in which data, processes, actions etc. can be declared. The syntax of a *specification* is:

$$\begin{array}{rcl} \textit{specification} & ::= & \textit{sort-specification} \\ & | & \textit{function-specification} \\ & | & \textit{rewrite-specification} \\ & | & \textit{action-specification} \\ & | & \textit{comm-specification} \\ & | & \textit{process-specification} \\ & | & \textit{specification specification}. \end{array}$$

2.10 The standard sort Bool

In every *specification* the following function and sort declarations must be included. The reason for this special treatment of the sort **Bool** is that we want to guarantee that true and false as booleans are different. This can only be achieved if the names for true, false and the sort of booleans are predetermined.

$$\begin{array}{ll} \textbf{sort} & \textbf{Bool} \\ \textbf{func} & T :\to \textbf{Bool} \\ & F :\to \textbf{Bool} \end{array}$$

2.11 An example

As an example we give a *specification* of a data transfer process. *Data-elements* of sort D are transferred from *in* to *out*.

$$\begin{array}{ll} \textbf{sort} & \textbf{Bool} \\ \textbf{func} & T, F :\to \textbf{Bool} \\ \textbf{sort} & D \\ \textbf{func} & d1, d2, d3 :\to D \\ \textbf{act} & in, out : D \\ \textbf{proc} & TR = \sum(x : D, in(x) \cdot out(x) \cdot TR) \end{array}$$

2.12 The from construct

For a *process-expression* or a *data-term* t, we write t **from** E for a *specification* E where we mean the *process-expression* or *data-term* t as defined in E. Often, it is clear from the context to which *specification* E the item t belongs. In this case we generally write t without explicit reference to E.

3 Static semantics

Not every *specification* is necessarily correctly defined. It may be that objects are not declared, that they are declared at several places or are not used in a proper way. In this section we define under which circumstances a *specification* does not have these problems and hence has a correct *static semantics*. Furthermore, we define some functions that will be used in the definition of the semantics of μCRL.

3.1 The signature of a specification

The signature of a specification is an important ingredient in defining the static semantics. It consists of a five-tuple of which each component is a set containing all elements of a main syntactical category declared in a *specification* E.

Definition 3.1. Let E be a *specification*. The *signature* $Sig(E) = (Sort, Fun, Act, Comm, Proc)$ of E is defined as follows:

- If $E \equiv \textbf{sort}\ n_1 \ldots n_m$ with $m \geq 1$, then $Sig(E) \stackrel{\text{def}}{=} (\{n_1, \ldots, n_m\}, \emptyset, \emptyset, \emptyset, \emptyset)$.

- If $E \equiv \textbf{func}\ fd_1 \ldots fd_m$ with $m \geq 1$, then $Sig(E) \stackrel{\text{def}}{=} (\emptyset, Fun, \emptyset, \emptyset, \emptyset)$, where

$$\begin{array}{lll} Fun & \stackrel{\text{def}}{=} & \{n_{ij} :\rightarrow S_i \mid fd_i \equiv n_{i1}, \ldots, n_{il_i} :\rightarrow S_i, 1 \leq i \leq m, 1 \leq j \leq l_i\} \\ & \cup & \{n_{ij} : S_{i1} \times \ldots \times S_{ik_i} \rightarrow S_i \mid \\ & & \quad fd_i \equiv n_{i1}, \ldots, n_{il_i} : S_{i1} \times \ldots \times S_{ik_i} \rightarrow S_i, \\ & & \quad 1 \leq i \leq m, 1 \leq j \leq l_i\}. \end{array}$$

- If E is a *rewrite-specification*, then $Sig(E) \stackrel{\text{def}}{=} (\emptyset, \emptyset, \emptyset, \emptyset, \emptyset)$.

- If $E \equiv \textbf{act}\ ad_1 \ldots ad_m$ with $m \geq 1$, then $Sig(E) \stackrel{\text{def}}{=} (\emptyset, \emptyset, Act, \emptyset, \emptyset)$, where

$$\begin{array}{lll} Act & \stackrel{\text{def}}{=} & \{n_i \mid ad_i \equiv n_i, 1 \leq i \leq m\} \\ & \cup & \{n_{ij} : S_{i1} \times \ldots \times S_{ik_i} \mid \\ & & \quad ad_i \equiv n_{i1}, \ldots, n_{il_i} : S_{i1} \times \ldots \times S_{ik_i}, 1 \leq i \leq m, 1 \leq j \leq l_i\}. \end{array}$$

- If $E \equiv \textbf{comm}\ cd_1 \ldots cd_m$ with $m \geq 1$, then $Sig(E) \stackrel{\text{def}}{=} (\emptyset, \emptyset, \emptyset, \{cd_i \mid 1 \leq i \leq m\}, \emptyset)$.

- If $E \equiv \textbf{proc}\ pd_1 \ldots pd_m$ with $m \geq 1$, then $Sig(E) \stackrel{\text{def}}{=} (\emptyset, \emptyset, \emptyset, \emptyset, \{pd_i \mid 1 \leq i \leq m\})$.

- If $E \equiv E_1\ E_2$ with $Sig(E_i) = (Sort_i, Fun_i, Act_i, Comm_i, Proc_i)$ for $i = 1, 2$, then
$Sig(E) \stackrel{\text{def}}{=} (Sort_1 \cup Sort_2, Fun_1 \cup Fun_2, Act_1 \cup Act_2, Comm_1 \cup Comm_2,$
$Proc_1 \cup Proc_2)$.

Definition 3.2. Let $Sig = (Sort, Fun, Act, Comm, Proc)$ be a signature. We write

$Sig.Sort$ for $Sort$,
$Sig.Fun$ for Fun,
$Sig.Act$ for Act,
$Sig.Comm$ for $Comm$,
$Sig.Proc$ for $Proc$.

3.2 Variables

Variables play an important role in specifications. The next definition says which *names* can play the role of a variable without confusion with defined constants. Moreover, variables must have an unambiguous and declared sort.

Definition 3.3. Let Sig be a signature. A set $\mathcal{V}$ containing elements $\langle x : S\rangle$ with x and S *names*, is called a *set of variables* over Sig iff for each $\langle x : S\rangle \in \mathcal{V}$:

- for each *name* S' and *process-expression* p it holds that $x :\rightarrow S' \notin Sig.Fun$, $x \notin Sig.Act$ and $x = p \notin Sig.Proc$,
- $S \in Sig.Sort$,
- for each *name* S' such that $S' \not\equiv S$ it holds that $\langle x : S'\rangle \notin \mathcal{V}$.

Definition 3.4. Let *var-dec* be a *variable-decl-section*. The function *Vars* is defined by:

$$Vars(\textit{var-dec}) \stackrel{\text{def}}{=} \begin{cases} \emptyset & \text{if } \textit{var-dec} \text{ is empty,} \\ \{\langle x_{ij} : S_i\rangle \mid 1 \leq i \leq m,\ 1 \leq j \leq l_i\} & \\ & \text{if for some } m \geq 1\ \textit{var-dec} \equiv \\ & \textbf{var}\ x_{11}, ..., x_{1l_1} : S_1\ ...\ x_{m1}, ..., x_{ml_m} : S_m. \end{cases}$$

In the following definitions we give functions yielding the sort and the variables in a *data-term* t. If for some reason no answer can be obtained, for instance because an undeclared *name* appears in t, a $\perp$ results. Of course this only works properly if $\perp$ does not occur in *names*.

Definition 3.5. Let t be a *data-term* and Sig a signature. Let $\mathcal{V}$ be a set of variables over Sig. We define:

$$sort_{Sig,\mathcal{V}}(t) \stackrel{\text{def}}{=} \begin{cases} S & \text{if } t \equiv x \text{ and } \langle x : S \rangle \in \mathcal{V}, \\ S & \text{if } t \equiv n,\ n :\rightarrow S \in Sig.Fun \\ & \quad \text{and for no } S' \not\equiv S\ n :\rightarrow S' \in Sig.Fun, \\ S & \text{if } t \equiv n(t_1, ..., t_m), \\ & \quad n : sort_{Sig,\mathcal{V}}(t_1) \times ... \times sort_{Sig,\mathcal{V}}(t_m) \rightarrow S \in Sig.Fun, \\ & \quad \text{and for no } S' \not\equiv S \\ & \quad n : sort_{Sig,\mathcal{V}}(t_1) \times ... \times sort_{Sig,\mathcal{V}}(t_m) \rightarrow S' \in Sig.Fun, \\ \perp & \text{otherwise.} \end{cases}$$

Definition 3.6. Let Sig be a signature, $\mathcal{V}$ a set of variables over Sig and let t be a *data-term*.

$$Var_{Sig,\mathcal{V}}(t) \stackrel{\text{def}}{=} \begin{cases} \{\langle x : S \rangle\} & \text{if } t \equiv x \text{ and } \langle x : S \rangle \in \mathcal{V}, \\ \emptyset & \text{if } t \equiv n \text{ and } n :\rightarrow S \in Sig.Fun, \\ \bigcup_{1 \le i \le m} Var_{Sig,\mathcal{V}}(t_i) & \text{if } t \equiv n(t_1, ..., t_m), \\ \{\perp\} & \text{otherwise.} \end{cases}$$

We call a *data-term* t *closed* w.r.t. a signature Sig and a set of variables $\mathcal{V}$ iff $Var_{Sig,\mathcal{V}}(t) = \emptyset$. Note that $Var_{Sig,\mathcal{V}}(t) \subseteq \mathcal{V} \cup \{\perp\}$ for any *data-term* t.

3.3 Static semantics

A *specification* must be internally consistent. This means that all objects that are used must be declared exactly once and are used such that the sorts are correct. It also means that action, process, constant and variable *names* cannot be confused. Furthermore, it means that communications are specified in a functional way and that it is guaranteed that the rewrite rules satisfy a usual condition that the variables that are used at the right hand side of a equality sign must also occur at the left hand side. Because all these properties can be statically decided, a *specification* that is internally consistent is called SSC (*Statically Semantically Correct*). For a better understanding of the next definition, it may be helpful to read definition 3.8 first.

Definition 3.7. Let Sig be a signature and $\mathcal{V}$ be a set of variables over Sig. We define the predicate 'is SSC w.r.t. Sig' inductively over the syntax of a *specification*.

- A *specification* **sort** $n_1 ... n_m$ with $m \ge 1$ is SSC w.r.t. Sig iff all *names* $n_1, ..., n_m$ are pairwise different.
- A *specification* **func** $n_{11}, ..., n_{1l_1} : S_{11} \times ... \times S_{1k_1} \rightarrow S_1$

  $\vdots$

  $n_{m1}, ..., n_{ml_m} : S_{m1} \times ... \times S_{mk_m} \rightarrow S_m$

  with $m \ge 1$, $l_i \ge 1$, $k_i \ge 0$ for $1 \le i \le m$ is SSC w.r.t. Sig iff

  - for all $1 \le i \le m$ the *names* $n_{i1}, ..., n_{il_i}$ are pairwise different,

- for all $1 \leq i < j \leq m$ it holds that if $n_{ik} \equiv n_{jk'}$ for some $1 \leq k \leq l_i$ and $1 \leq k' \leq l_j$, then either $k_i \neq k_j$, or $S_{il} \not\equiv S_{jl}$ for some $1 \leq l \leq k_i$,
- for all $1 \leq i \leq m$ and $1 \leq j \leq k_i$ it holds that $S_{ij} \in Sig.Sort$ and $S_i \in Sig.Sort$.

• A *specification* of the form: *var-dec*
rew-rul

where *var-dec* is a *variable-decl-section* and *rew-rul* is a *rewrite-rules-section* is SSC w.r.t. *Sig* iff

- *var-dec* is SSC w.r.t. *Sig*,
- *rew-rul* is SSC w.r.t. *Sig* and *Vars*(*var-dec*).

⋆ The empty *variable-decl-section* is SSC w.r.t. *Sig*.

A *variable-decl-section* **var** $n_{11}, ..., n_{1k_1} : S_1$

$$\vdots$$

$n_{m1}, ..., n_{mk_m} : S_m$

with $m \geq 1$, $k_i \geq 1$ for $1 \leq i \leq m$ is SSC w.r.t. *Sig* iff

- $n_{ij} \not\equiv n_{i'j'}$ whenever $i \neq i'$ or $j \neq j'$ for $1 \leq i \leq m$, $1 \leq i' \leq m$, $1 \leq j \leq k_i$ and $1 \leq j' \leq k_{i'}$,
- the set $Vars(\textbf{var}\ n_{11}, ..., n_{1k_1} : S_1 \ ...\ n_{m1}, ..., n_{mk_m} : S_m)$ is a set of variables over *Sig*.

⋆ A *rewrite-rules-section* **rew** $rw_1 \ ...\ rw_m$ with $m \geq 1$ is SSC w.r.t. *Sig* and $\mathcal{V}$ iff

- if $rw_i \equiv n = t$ for some $1 \leq i \leq m$, then
  - ∗ $n :\rightarrow sort_{Sig,\emptyset}(t) \in Sig.Fun$,
  - ∗ t is SSC w.r.t. *Sig* and $\emptyset$,
- if $rw_i \equiv n(t_1, ..., t_{k_i}) = t$ for some $1 \leq i \leq m$ and $k_i \geq 1$, then
  - ∗ $n : sort_{Sig,\mathcal{V}}(t_1) \times ... \times sort_{Sig,\mathcal{V}}(t_{k_i}) \rightarrow sort_{Sig,\mathcal{V}}(t) \in Sig.Fun$,
  - ∗ t, t_j $(1 \leq j \leq k_i)$ are SSC w.r.t. *Sig* and $\mathcal{V}$,
  - ∗ $Var_{Sig,\mathcal{V}}(t) \subseteq \bigcup_{1 \leq j \leq k_i} Var_{Sig,\mathcal{V}}(t_j)$.

⋆ A *data-term* n with n a *name* is SSC w.r.t. *Sig* and $\mathcal{V}$ iff $\langle n : S \rangle \in \mathcal{V}$ for some S, or $n :\rightarrow sort_{Sig,\mathcal{V}}(n) \in Sig.Fun$.

A *data-term* $n(t_1, ..., t_m)$ $(m \geq 1)$ is SSC w.r.t. *Sig* and $\mathcal{V}$ iff $n : sort_{Sig,\mathcal{V}}(t_1) \times ... \times sort_{Sig,\mathcal{V}}(t_m) \rightarrow sort_{Sig,\mathcal{V}}(n(t_1, ..., t_m)) \in Sig.Fun$ and all t_i $(1 \leq i \leq m)$ are SSC w.r.t. *Sig* and $\mathcal{V}$.

• A *specification* **act** $ad_1 \ ...\ ad_m$ with $m \geq 1$ is SSC w.r.t. *Sig* iff

- for all $1 \leq i \leq m$ the *action-declaration* ad_i is SSC w.r.t. *Sig*,
- for all $1 \leq i < j \leq m$ it holds that $Sig(\textbf{act}\ ad_i).Act \cap Sig(\textbf{act}\ ad_j).Act = \emptyset$.

⋆ An *action-declaration* n is SSC w.r.t. *Sig* iff for each *name* S' it holds that $n :\to S' \notin Sig.Fun$.

An *action-declaration* $n_1, ..., n_m : S_1 \times ... \times S_k$ with $k, m \geq 1$ is SSC w.r.t. *Sig* iff

- for all $1 \leq i < j \leq m$ it holds that $n_i \not\equiv n_j$,
- for all $1 \leq i \leq k$ it holds that $S_i \in Sig.Sort$,
- for all $1 \leq i \leq m$ and for each *name* S' it holds that $n_i : S_1 \times ... \times S_k \to S' \notin Sig.Fun$.

• A *specification* **comm** $n_{11} \,|\, n_{12} = n_{13}\ ...\ n_{m1} \,|\, n_{m2} = n_{m3}$ with $m \geq 1$ is SSC w.r.t. *Sig* iff

- for each $1 \leq i < j \leq m$ it is not the case that $n_{i1} \equiv n_{j1}$ and $n_{i2} \equiv n_{j2}$, or $n_{i1} \equiv n_{j2}$ and $n_{i2} \equiv n_{j1}$,
- for each $1 \leq i \leq m$ either $n_{i1} \in Sig.Act$ or there is a $k \geq 1$ such that $n_{i1} : S_1 \times ... \times S_k \in Sig.Act$,
- for each $1 \leq i \leq m$, $k \geq 1$ and *names* $S_1, ..., S_k$ it holds that if $n_{i1} : S_1 \times ... \times S_k \in Sig.Act$ then $n_{i2} : S_1 \times ... \times S_k \in Sig.Act$ and $n_{i3} : S_1 \times ... \times S_k \in Sig.Act$,
- for each $1 \leq i \leq m$, $k \geq 1$ and *names* $S_1, ..., S_k$ it holds that if $n_{i2} : S_1 \times ... \times S_k \in Sig.Act$ then $n_{i1} : S_1 \times ... \times S_k \in Sig.Act$ and $n_{i3} : S_1 \times ... \times S_k \in Sig.Act$,
- for each $1 \leq i \leq m$ it holds that if $n_{i1} \in Sig.Act$ then $n_{i2} \in Sig.Act$ and $n_{i3} \in Sig.Act$,
- for each $1 \leq i \leq m$ it holds that if $n_{i2} \in Sig.Act$ then $n_{i1} \in Sig.Act$ and $n_{i3} \in Sig.Act$.

• A *specification* **proc** $pd_1\ ...\ pd_m$ with $m \geq 1$ is SSC w.r.t. *Sig* iff

- for each $1 \leq i < j \leq m$:
  * if $pd_i \equiv n_i = p_i$ and $pd_j \equiv n_j = p_j$ then $n_i \not\equiv n_j$,
  * if for some $k \geq 1$ it holds that $pd_i \equiv n_i(x_1 : S_1, ..., x_k : S_k) = p_i$ and $pd_j \equiv n_j(x'_1 : S_1, ..., x'_k : S_k) = p_j$ then $n_i \not\equiv n_j$,
  * for all *names* S' it holds that $n_i :\to S_i \notin Sig.Fun$,
- if $pd_i \equiv n_i = p_i$ $(1 \leq i \leq m)$, then $n_i \notin Sig.Act$ and p_i is SSC w.r.t. *Sig* and $\emptyset$,
- if $pd_i \equiv n_i(x_{i1} : S_{i1}, ..., x_{ik_i} : S_{ik_i}) = p_i$ $(1 \leq i \leq m)$, then
  * $n_i : S_{i1} \times ... \times S_{ik_i} \notin Sig.Act$,
  * for all *names* S' it holds that $n_i : S_{i1} \times ... \times S_{ik_i} \to S' \notin Sig.Fun$,
  * the *names* $x_{i1}, ..., x_{ik_i}$ are pairwise different and $\{\langle x_{ij} : S_{ij} \rangle \mid 1 \leq j \leq k_i\}$ is a set of variables over *Sig*,
  * p_i is SSC w.r.t. *Sig* and $\{\langle x_{ij} : S_{ij} \rangle \mid 1 \leq j \leq k_i\}$.

⋆ A *process-expression* $p_1 + p_2$, *parallel-expressions* $p_1 \parallel p_2$, $p_1 \mathbin{⌊\!⌊} p_2$, $p_1 \mid p_2$, a *dot-expression* $p_1 \cdot p_2$ are SSC w.r.t. *Sig* and $\mathcal{V}$ iff

- p_1 is SSC w.r.t. *Sig* and $\mathcal{V}$,
- p_2 is SSC w.r.t. *Sig* and $\mathcal{V}$.

A *cond-expression* $p_1 \triangleleft t \triangleright p_2$ is SSC w.r.t. *Sig* and $\mathcal{V}$ iff

- p_1 is SSC w.r.t. *Sig* and $\mathcal{V}$,
- p_2 is SSC w.r.t. *Sig* and $\mathcal{V}$,
- t is SSC w.r.t. *Sig* and $\mathcal{V}$ and $sort_{Sig,\mathcal{V}}(t) = \mathbf{Bool}$.

The *basic-expressions* δ and τ are SSC w.r.t. *Sig* and $\mathcal{V}$.

The *basic-expressions* $\partial(\{n_1, ..., n_m\}, p)$ and $\tau(\{n_1, ..., n_m\}, p)$ with $m \geq 1$ are SSC w.r.t. *Sig* and $\mathcal{V}$ iff

- for all $1 \leq i < j \leq m$ $n_i \not\equiv n_j$,
- for $1 \leq i \leq m$ either $n_i \in Sig.Act$ or $n_i : S_1 \times ... \times S_k \in Sig.Act$ for some $k \geq 1$ and *names* $S_1, ..., S_k$,
- p is SSC w.r.t. *Sig* and $\mathcal{V}$.

The *basic-expression* $\rho(\{n_1 \to n'_1, ..., n_m \to n'_m\}, p)$ is SSC w.r.t. *Sig* and $\mathcal{V}$ iff

- for $1 \leq i \leq m$ either $n_i \in Sig.Act$ or $n_i : S_1 \times ... \times S_k \in Sig.Act$ for some $k \geq 1$ and *names* $S_1, ..., S_k$,
- for each $1 \leq i < j \leq m$ it holds that $n_i \not\equiv n_j$,
- for $1 \leq i \leq m$, $k \geq 1$ and *names* $S_1, .., S_k$ it holds that if $n_i : S_1 \times ... \times S_k \in Sig.Act$, then also $n'_i : S_1 \times ... \times S_k \in Sig.Act$,
- for $1 \leq i \leq m$ it holds that if $n_i \in Sig.Act$, then also $n'_i \in Sig.Act$,
- p is SSC w.r.t. *Sig* and $\mathcal{V}$.

A *basic-expression* $\Sigma(x : S, p)$ is SSC w.r.t. *Sig* and $\mathcal{V}$ iff

- $\mathcal{V} \backslash \{\langle x : S' \rangle \mid S' \text{ a } name\} \cup \{\langle x : S \rangle\}$ is a set of variables over *Sig*,
- p is SSC w.r.t. *Sig* and $\mathcal{V} \backslash \{\langle x : S' \rangle \mid S' \text{ a } name\} \cup \{\langle x : S \rangle\}$.

A *basic-expression* n is SSC w.r.t. *Sig* and $\mathcal{V}$ iff $n = p \in Sig.Proc$ for some *process-expression* p or $n \in Sig.Act$.

A *basic-expression* $n(t_1, ..., t_m)$ with $m \geq 1$ is SSC w.r.t. *Sig* and $\mathcal{V}$ iff

- $n(x_1 : sort_{Sig,\mathcal{V}}(t_1), ..., x_m : sort_{Sig,\mathcal{V}}(t_m)) = p \in Sig.Proc$ for some *names* $x_1, ..., x_m$ and *process-expression* p, or $n : sort_{Sig,\mathcal{V}}(t_1) \times ... \times sort_{Sig,\mathcal{V}}(t_m) \in Sig.Act$,
- for $1 \leq i \leq m$ the *data-term* t_i is SSC w.r.t. *Sig* and $\mathcal{V}$.

A *basic-expression* (p) is SSC w.r.t. *Sig* and $\mathcal{V}$ iff p is SSC w.r.t. *Sig* and $\mathcal{V}$.

- A *specification* E_1 E_2 is SSC w.r.t. *Sig* iff
  - E_1 and E_2 are SSC w.r.t. *Sig*,

- $Sig(E_1).Sort \cap Sig(E_2).Sort = \emptyset$,
- if $n : S_1 \times ... \times S_m \rightarrow S \in Sig(E_1).Fun$ for some $m \geq 0$ then $n : S_1 \times ... \times S_m \rightarrow S' \notin Sig(E_2).Fun$ for any *name* S',
- $Sig(E_1).Act \cap Sig(E_2).Act = \emptyset$,
- if $n_1 \,|\, n_2 = n_3 \in Sig(E_1).Comm$ then for any *names* n_3' and n_3'' $n_1 \,|\, n_2 = n_3' \notin Sig(E_2).Comm$ and $n_2 \,|\, n_1 = n_3'' \notin Sig(E_2).Comm$,
- if $pd_1 \in Sig(E_1).Proc$ and $pd_2 \in Sig(E_2).Proc$, then
  * if $pd_1 \equiv n_1 = p_1$ and $pd_2 \equiv n_2 = p_2$, then $n_1 \not\equiv n_2$,
  * if for some $m \geq 1$ $pd_1 \equiv n_1(x_1 : S_1, ..., x_m : S_m) = p_1$ and $pd_2 \equiv n_2(x_1' : S_1, ..., x_m' : S_m) = p_2$, then $n_1 \not\equiv n_2$.

Definition 3.8. Let E be a *specification*. We say that E is SSC iff E is SSC w.r.t. $Sig(E)$.

The following lemma is helpful in checking that the predicate 'is SSC' is correctly defined.

Lemma 3.9. *Let Sig be a signature and $\mathcal{V}$ be a set of variables over Sig. Let t be a data-term that is SSC* w.r.t. *Sig and $\mathcal{V}$. Then $sort_{Sig,\mathcal{V}}(t) \neq \perp$ and $\perp \notin Var_{Sig,\mathcal{V}}(t)$.*

3.4 The communication function

The following definition helps us in guaranteeing that the communication function is commutative and associative. This implies that the merge is also commutative and associative which allows us to write parallel processes without brackets as is done in the syntax (cf. LOTOS [25] where this is not the case).

Definition 3.10. Let Sig be a signature. The set $Sig.Comm^*$ is defined by:

$$Sig.Comm^* \stackrel{\text{def}}{=} \{n_1 \,|\, n_2 = n_3,\ n_2 \,|\, n_1 = n_3 \mid n_1 \,|\, n_2 = n_3 \in Sig.Comm\}.$$

So, in $Sig.Comm^*$ communication is always commutative. We say that a *specification* E is *communication-associative* iff

$$\begin{aligned} &n_1 \,|\, n_2 = n,\ n \,|\, n_3 = n' \in Sig(E).Comm^* \ \Rightarrow \\ &\qquad \exists n'' :\ n_2 \,|\, n_3 = n'', n_1 \,|\, n'' = n' \in Sig(E).Comm^*. \end{aligned}$$

With the condition that E is SSC this exactly implies that communication is associative.

4 Well-formed μCRL specifications

We define what well-formed specifications are. We only provide well-formed *specifications* with a semantics. Well-formedness is a decidable property.

Definition 4.1. Let E be a *specification* that is SSC. We say that E has *no empty sorts* iff for all $S \in Sig(E).Sort$ there is a *data-term* t that is SSC w.r.t. $Sig(E)$ and $\emptyset$ such that $sort_{Sig(E),\emptyset}(t) \equiv S$.

Definition 4.2. Let E be a *specification*. E is called *well-formed* iff

- E is SSC,
- E is communication-associative,
- E has no empty sorts,
- **Bool** $\in Sig(E).Sort$,
- $T :\rightarrow$ **Bool** $\in Sig(E).Fun$ and
- $F :\rightarrow$ **Bool** $\in Sig(E).Fun$.

5 Algebraic semantics

In this section we present the semantics of well-formed μCRL specifications. Given a signature Sig we introduce the class of Sig-algebras. Then for a well-formed *specification* E with $Sig(E) = Sig$, we define the subclass of Sig-algebras that form a model for the data part of E and in which the terms T and F of sort **Bool** are interpreted different. Then given such a model, we give an operational semantics for *process-expressions* in E.

5.1 Algebras

First we adapt the standard definitions of algebras etc. to μCRL (see e.g. [11] for these definitions).

Definition 5.1. Let E be a well-formed *specification*. A *$Sig(E)$-algebra* $\mathbb{A}$ is a structure containing

- for each $S \in Sig(E).Sort$ a non-empty domain $D(\mathbb{A}, S)$,
- for each $n :\rightarrow S \in Sig(E).Fun$ a constant $C(\mathbb{A}, n) \in D(\mathbb{A}, S)$,
- for each $n : S_1 \times ... \times S_m \rightarrow S \in Sig(E).Fun$ a function $F(\mathbb{A}, n : S_1 \times ... \times S_m)$ from $D(\mathbb{A}, S_1) \times ... \times D(\mathbb{A}, S_m)$ to $D(\mathbb{A}, S)$.

For two elements $a_1 \in D(\mathbb{A}, S_1)$ and $a_2 \in D(\mathbb{A}, S_2)$, we write $a_1 = a_2$ iff $S_1 \equiv S_2$ and a_1 and a_2 represent exactly the same element.

Definition 5.2. Let E be a well-formed *specification* and let $\mathbb{A}$ be a $Sig(E)$-algebra. We define the interpretation $[\![\cdot]\!]_{\mathbb{A}}$ from *data-terms* that are SSC w.r.t. $Sig(E)$ and $\emptyset$ into the domains of $\mathbb{A}$ as follows:

- if $t \equiv n$, then $[\![t]\!]_{\mathbb{A}} \stackrel{\text{def}}{=} C(\mathbb{A}, n)$,

- if $t \equiv n(t_1, ..., t_m)$ for some $m \geq 1$, then
$[\![t]\!]_{\mathbb{A}} \stackrel{\text{def}}{=} F(\mathbb{A}, n : sort_{Sig(E),\emptyset}(t_1) \times ... \times sort_{Sig(E),\emptyset}(t_m))([\![t_1]\!]_{\mathbb{A}}, ..., [\![t_m]\!]_{\mathbb{A}})$.

We say that a $Sig(E)$-algebra $\mathbb{A}$ is *minimal* iff for each $a \in D(\mathbb{A}, S)$ and $S \in Sig(E).Sort$, there is some *data-term* t that is SSC w.r.t. $Sig(E)$ and $\emptyset$ such that $[\![t]\!]_{\mathbb{A}} = a$. For *data-terms* t_1, t_2 that are SSC w.r.t. $Sig(E)$ and $\emptyset$ we write $\mathbb{A} \models t_1 = t_2$ iff $[\![t_1]\!]_{\mathbb{A}} = [\![t_2]\!]_{\mathbb{A}}$.

Definition 5.3. Let E be a well-formed *specification* and let $\mathbb{A}$ be a minimal $Sig(E)$-algebra. A function r mapping pairs of a sort S and an element from $D(\mathbb{A}, S)$ to *data-terms* that are SSC w.r.t. to $Sig(E)$ and $\emptyset$ is called a *representation function* of E and $\mathbb{A}$ iff $\mathbb{A} \models t = r(sort_{Sig(E),\emptyset}(t), [\![t]\!]_{\mathbb{A}})$ for each *data-term* t that is SSC w.r.t. $Sig(E)$ and $\emptyset$.

5.2 Substitutions

We define substitutions on *data-terms*. These substitutions are immediately extended to *process-expressions* because this is required for the definition of the operational semantics.

Definition 5.4. Let E be a well-formed *specification* and $\mathcal{V}$ a set of variables over $Sig(E)$. Let *Term* be the set of *data-terms* that are SSC w.r.t. $Sig(E)$ and $\mathcal{V}$. A *substitution* σ over $Sig(E)$ and $\mathcal{V}$ is a mapping

$$\sigma : \mathcal{V} \to Term$$

such that for each $\langle x : S\rangle \in \mathcal{V}$ it holds that $sort_{Sig(E),\mathcal{V}}(\sigma(\langle x : S\rangle)) = S$. Substitutions are extended to *data-terms* by:

$$\sigma(x) \stackrel{\text{def}}{=} \sigma(\langle x : S\rangle) \quad \text{if } \langle x : S\rangle \in \mathcal{V} \text{ for some } name\ S,$$
$$\sigma(n) \stackrel{\text{def}}{=} n \quad \text{if } n :\to S \in Sig(E).Fun,$$
$$\sigma(n(t_1, ..., t_m)) \stackrel{\text{def}}{=} n(\sigma(t_1), ..., \sigma(t_m)).$$

Definition 5.5. Let E be a well-formed *specification* and $\mathcal{V}$ a set of variables over $Sig(E)$. Let σ be a substitution over $Sig(E)$ and $\mathcal{V}$. We extend σ to *process-expressions* that are SSC w.r.t. $Sig(E)$ and $\mathcal{V}$ as follows:

- If $p_1 \Box p_2$ is a *process-expression*, a *parallel-expression* or a *dot-expression* ($\Box \in \{+, \|, \mathbin{\|\!\_}, |, \cdot\}$), then $\sigma(p_1 \Box p_2) \stackrel{\text{def}}{=} \sigma(p_1) \Box \sigma(p_2)$,

- $\sigma(p_1 \lhd t \rhd p_2) \stackrel{\text{def}}{=} \sigma(p_1) \lhd \sigma(t) \rhd \sigma(p_2)$ for a *cond-expression* $p_1 \lhd t \rhd p_2$,

- $\sigma(\delta) \stackrel{\text{def}}{=} \delta$ and $\sigma(\tau) \stackrel{\text{def}}{=} \tau$ for *basic-expressions* δ and τ,

- if $\Box(gl, p)$ is a *basic-expression* ($\Box \in \{\partial, \tau, \rho\}$), then
$\sigma(\Box(gl, p)) \stackrel{\text{def}}{=} \Box(gl, \sigma(p))$,

- $\sigma(\Sigma(x:S,p)) \stackrel{\text{def}}{=} \Sigma(x:S,\sigma'(p))$ where σ' is defined by

$$\sigma'(\langle x':S'\rangle) \stackrel{\text{def}}{=} \begin{cases} \langle x:S\rangle & \text{if } x' \equiv x \\ \sigma(\langle x':S'\rangle) & \text{otherwise,} \end{cases}$$

  for a *basic-expression* $\Sigma(x:S,p)$,

- $\sigma(n(t_1,...,t_m)) \stackrel{\text{def}}{=} n(\sigma(t_1),...,\sigma(t_m))$ for a *basic-expression* $n(t_1,...,t_m)$,

- $\sigma(n) \stackrel{\text{def}}{=} n$ for a *basic-expression* n,

- $\sigma((p)) \stackrel{\text{def}}{=} (\sigma(p))$ for a *basic-expression* (p).

The validity of the following lemma gives us confidence that substitutions are indeed correctly defined.

Lemma 5.6. *Let E be a well-formed specification and $\mathcal{V}$ a set of variables over $Sig(E)$. Let σ be a substitution over $Sig(E)$ and $\mathcal{V}$.*

- *For any data-term t that is SSC* w.r.t. *$Sig(E)$ and $\mathcal{V}$, $\sigma(t)$ is also a data-term that is SSC* w.r.t. *$Sig(E)$ and $\mathcal{V}$. Moreover, $sort_{Sig(E),\mathcal{V}}(t) \equiv sort_{Sig(E),\mathcal{V}}(\sigma(t))$.*

- *For any process-expression p that is SSC* w.r.t. *$Sig(E)$ and $\mathcal{V}$, $\sigma(p)$ is a process-expression that is SSC* w.r.t. *$Sig(E)$ and $\mathcal{V}$.*

5.3 Boolean preserving models

A $Sig(E)$-algebra $\mathbb{A}$ is a model of a well-formed *specification* E iff the equations defining the data in E hold in $\mathbb{A}$. Moreover, we say that $\mathbb{A}$ is *boolean preserving* iff T and F of sort **Bool** represent exactly the two different elements of $D(\mathbb{A},\textbf{Bool})$. Note that there are specifications which have no boolean preserving models of E, for instance a specification containing the equation $T = F$. For μCRL we are only interested in the minimal $Sig(E)$-algebras that are boolean preserving.

First we define the function *rewrites* that extracts the rewrite clauses together with declared variables from a *specification*.

Definition 5.7. We define the function *rewrites* on a *specification* E inductively as follows:

- If $E \equiv$ *sort-spec* with *sort-spec* a *sort-specification*,
  then $rewrites(E) \stackrel{\text{def}}{=} \emptyset$.

- If $E \equiv$ *func-spec* with *func-spec* a *function-specification*,
  then $rewrites(E) \stackrel{\text{def}}{=} \emptyset$.

- If $E \equiv V\ R$ with V a *variable-decl-section* and R a *rewrite-rules-section* with $R \equiv \mathbf{rew}\ rd_1\ ...\ rd_m$ for some $m \geq 1$, then

$$rewrites(E) \stackrel{\text{def}}{=} \{\langle \{rd_i \mid 1 \leq i \leq m\}, Vars(V) \rangle\}.$$

- If $E \equiv$ *act-spec* with *act-spec* an *action-specification*, then $rewrites(E) \stackrel{\text{def}}{=} \emptyset$.

- If $E \equiv$ *comm-spec* with *comm-spec* a *comm-specification*, then $rewrites(E) \stackrel{\text{def}}{=} \emptyset$.

- If $E \equiv$ *proc-spec* with *proc-spec* a *process-specification*, then $rewrites(E) \stackrel{\text{def}}{=} \emptyset$.

- If $E \equiv E_1\ E_2$ where E_1 and E_2 are *specifications*, then $rewrites(E) \stackrel{\text{def}}{=} rewrites(E_1) \cup rewrites(E_2)$.

Definition 5.8. Let E be a well-formed *specification*. A $Sig(E)$-algebra $\mathbb{A}$ is a *model* of E, notation $\mathbb{A} \models_D E$, iff whenever $t = t' \in R$ with $\langle R, \mathcal{V} \rangle \in rewrites(E)$, then for any substitution σ over $Sig(E)$ and $\mathcal{V}$ such that $Var_{Sig(E),\mathcal{V}}(\sigma(t)) = Var_{Sig(E),\mathcal{V}}(\sigma(t')) = \emptyset$ it holds that $\mathbb{A} \models \sigma(t) = \sigma(t')$.

We write $\mathbb{A} \models_D E$ with a subscript D because the model only concerns the data in E.

Definition 5.9. Let E be a well-formed *specification*. A $Sig(E)$-algebra $\mathbb{A}$ is called *boolean preserving* w.r.t. E iff

- it is not the case that $\mathbb{A} \models T = F$,

- $|D(\mathbb{A}, \mathbf{Bool})| = 2$, i.e. T and F are exactly the two elements of sort **Bool**.

5.4 The process part

In this section we define for each *process-expression* p that is SSC w.r.t. $Sig(E)$ and $\emptyset$, and each minimal model $\mathbb{A}$ of E that preserves the booleans and where E is some well-formed *specification*, a meaning in terms of a referential transition system (cf. the operational semantics in [3, 33, 36]).

Definition 5.10. A transition system $\mathcal{A}$ is a quadruple $(S, L, \longrightarrow, s)$ where

- S is a set of *states*,
- L is a set of *labels*,
- $\longrightarrow \subseteq S \times L \times S$ is a *transition relation*,
- $s \in S$ is the *initial state*.

Elements $(s', l, s'') \in \longrightarrow$ are generally written as $s' \xrightarrow{l} s''$.

Definition 5.11. Let E be a well-formed *specification*, $\mathbb{A}$ be a minimal model of E that is boolean preserving and r be a representation function of E and $\mathbb{A}$. Let p be a *process-expression* that is SSC w.r.t. $Sig(E)$ and $\emptyset$. The *meaning* of p **from** E in $\mathbb{A}$ with representation function r is the *referential* transition system $\mathcal{A}(\mathbb{A}, r, p \textbf{ from } E)$ defined by

$$(S, L, \longrightarrow, s)$$

where

- $S \stackrel{\text{def}}{=} \{q \mid q \text{ a } \textit{process-expression} \text{ that is SSC w.r.t. } Sig(E) \text{ and } \emptyset\} \cup \{\surd\}$,
- $L \stackrel{\text{def}}{=} \{n(t_1, ..., t_m) \mid m \geq 0, n \in Sig(E).Act \text{ and for } 1 \leq i \leq m \text{ it holds that } t_i \equiv r(S_i, a) \text{ for some } a \in D(\mathbb{A}, S_i) \text{ where } S_i \equiv sort_{Sig(E),\emptyset}(t_i)\} \cup \{\tau, \surd\}$,
- $s \stackrel{\text{def}}{=} p$,
- $\longrightarrow$ is the transition relation that contains exactly all transitions provable using the rules below (see for provability e.g. [19]). Let p, p', q, q' range over the set $S \setminus \{\surd\}$, P is a *process-expression* that is SSC w.r.t. $Sig(E)$ and some set of variables over $Sig(E)$, l ranges over the set L of labels, n, n_1, n_2 are *names*, $m \geq 0$ and $t_1, ..., t_m, u_1, ..., u_m$ are *data-terms* (note that there is no rule for δ):

  - $\surd \xrightarrow{\surd} \delta$.
  - $\tau \xrightarrow{\tau} \surd$.
  - $n \xrightarrow{n()} \surd$ if $n \in Sig(E).Act$,
    - $n(u_1, ..., u_m) \xrightarrow{n(t_1, ..., t_m)} \surd$ with $m \geq 1$ if
      - $n : sort_{Sig(E),\emptyset}(u_1) \times ... \times sort_{Sig(E),\emptyset}(u_m) \in Sig(E).Act$,
      - $t_i \equiv r(sort_{Sig(E),\emptyset}(u_i), [\![u_i]\!]_{\mathbb{A}})$.
  - $\dfrac{p \xrightarrow{l} p'}{n \xrightarrow{l} p'}$ if $n = p \in Sig(E).Proc$,
    - $\dfrac{p \xrightarrow{l} \surd}{n \xrightarrow{l} \surd}$ if $n = p \in Sig(E).Proc$,
    - $\dfrac{\sigma(P) \xrightarrow{l} p'}{n(u_1, ..., u_m) \xrightarrow{l} p'}$ with $m \geq 1$ if

  - $n(x_1 : sort_{Sig(E),\emptyset}(u_1), ..., x_m : sort_{Sig(E),\emptyset}(u_m)) = P \in Sig(E).Proc$,
  - there is a substitution σ over $Sig(E)$ and $\{\langle x_1 : sort_{Sig(E),\emptyset}(u_1)\rangle, ..., \langle x_m : sort_{Sig(E),\emptyset}(u_m)\rangle\}$ such that $\sigma(\langle x_i : sort_{Sig(E),\emptyset}(u_i)\rangle) \equiv u_i$ for $1 \leq i \leq m$,

- $\dfrac{\sigma(P) \xrightarrow{l} \surd}{n(u_1, ..., u_m) \xrightarrow{l} \surd}$ with $m \geq 1$ if
  - $n(x_1 : sort_{Sig(E),\emptyset}(u_1), ..., x_m : sort_{Sig(E),\emptyset}(u_m)) = P \in Sig(E).Proc$,
  - there is a substitution σ over $Sig(E)$ and $\{\langle x_1 : sort_{Sig(E),\emptyset}(u_1)\rangle, ..., \langle x_m : sort_{Sig(E),\emptyset}(u_m)\rangle\}$ such that $\sigma(\langle x_i : sort_{Sig(E),\emptyset}(u_i)\rangle) \equiv u_i$ for $1 \leq i \leq m$.

- $\dfrac{p \xrightarrow{l} p'}{p + q \xrightarrow{l} p'}$,
  - $\dfrac{p \xrightarrow{l} \surd}{p + q \xrightarrow{l} \surd}$,
  - $\dfrac{q \xrightarrow{l} q'}{p + q \xrightarrow{l} q'}$,
  - $\dfrac{q \xrightarrow{l} \surd}{p + q \xrightarrow{l} \surd}$.
- $\dfrac{p \xrightarrow{l} p'}{p \cdot q \xrightarrow{l} p' \cdot q}$,
  - $\dfrac{p \xrightarrow{l} \surd}{p \cdot q \xrightarrow{l} q}$.
- $\dfrac{p \xrightarrow{l} p'}{p \triangleleft t \triangleright q \xrightarrow{l} p'}$ if $\mathbb{A} \models t = T$,
  - $\dfrac{p \xrightarrow{l} \surd}{p \triangleleft t \triangleright q \xrightarrow{l} \surd}$ if $\mathbb{A} \models t = T$,
  - $\dfrac{q \xrightarrow{l} q'}{p \triangleleft t \triangleright q \xrightarrow{l} q'}$ if $\mathbb{A} \models t = F$,
  - $\dfrac{q \xrightarrow{l} \surd}{p \triangleleft t \triangleright q \xrightarrow{l} \surd}$ if $\mathbb{A} \models t = F$.

- $\dfrac{p \xrightarrow{l} p'}{p \parallel q \xrightarrow{l} p' \parallel q}$,
  - $\dfrac{q \xrightarrow{l} q'}{p \parallel q \xrightarrow{l} p \parallel q'}$,
  - $\dfrac{p \xrightarrow{l} \surd}{p \parallel q \xrightarrow{l} q}$,
  - $\dfrac{q \xrightarrow{l} \surd}{p \parallel q \xrightarrow{l} p}$,
  - $\dfrac{p \overset{n_1(t_1,\ldots,t_m)}{\longrightarrow} p' \quad q \overset{n_2(t_1,\ldots,t_m)}{\longrightarrow} q'}{p \parallel q \overset{n(t_1,\ldots,t_m)}{\longrightarrow} p' \parallel q'}$ if $n_1 \mid n_2 = n \in Sig(E).Comm^*$,
  - $\dfrac{p \overset{n_1(t_1,\ldots,t_m)}{\longrightarrow} \surd \quad q \overset{n_2(t_1,\ldots,t_m)}{\longrightarrow} q'}{p \parallel q \overset{n(t_1,\ldots,t_m)}{\longrightarrow} q'}$ if $n_1 \mid n_2 = n \in Sig(E).Comm^*$,
  - $\dfrac{p \overset{n_1(t_1,\ldots,t_m)}{\longrightarrow} p' \quad q \overset{n_2(t_1,\ldots,t_m)}{\longrightarrow} \surd}{p \parallel q \overset{n(t_1,\ldots,t_m)}{\longrightarrow} p'}$ if $n_1 \mid n_2 = n \in Sig(E).Comm^*$,
  - $\dfrac{p \overset{n_1(t_1,\ldots,t_m)}{\longrightarrow} \surd \quad q \overset{n_2(t_1,\ldots,t_m)}{\longrightarrow} \surd}{p \parallel q \overset{n(t_1,\ldots,t_m)}{\longrightarrow} \surd}$ if $n_1 \mid n_2 = n \in Sig(E).Comm^*$.
- $\dfrac{p \xrightarrow{l} p'}{p \mathbin{\lfloor\!\lfloor} q \xrightarrow{l} p' \parallel q}$,
  - $\dfrac{p \xrightarrow{l} \surd}{p \mathbin{\lfloor\!\lfloor} q \xrightarrow{l} q}$.
- $\dfrac{p \overset{n_1(t_1,\ldots,t_m)}{\longrightarrow} p' \quad q \overset{n_2(t_1,\ldots,t_m)}{\longrightarrow} q'}{p \mid q \overset{n(t_1,\ldots,t_m)}{\longrightarrow} p' \parallel q'}$ if $n_1 \mid n_2 = n \in Sig(E).Comm^*$,
  - $\dfrac{p \overset{n_1(t_1,\ldots,t_m)}{\longrightarrow} \surd \quad q \overset{n_2(t_1,\ldots,t_m)}{\longrightarrow} q'}{p \mid q \overset{n(t_1,\ldots,t_m)}{\longrightarrow} q'}$ if $n_1 \mid n_2 = n \in Sig(E).Comm^*$,
  - $\dfrac{p \overset{n_1(t_1,\ldots,t_m)}{\longrightarrow} p' \quad q \overset{n_2(t_1,\ldots,t_m)}{\longrightarrow} \surd}{p \mid q \overset{n(t_1,\ldots,t_m)}{\longrightarrow} p'}$ if $n_1 \mid n_2 = n \in Sig(E).Comm^*$,
  - $\dfrac{p \overset{n_1(t_1,\ldots,t_m)}{\longrightarrow} \surd \quad q \overset{n_2(t_1,\ldots,t_m)}{\longrightarrow} \surd}{p \mid q \overset{n(t_1,\ldots,t_m)}{\longrightarrow} \surd}$ if $n_1 \mid n_2 = n \in Sig(E).Comm^*$.

- $$\frac{p \stackrel{l}{\longrightarrow} p'}{\tau(\{n_1,...,n_k\},p) \stackrel{l}{\longrightarrow} \tau(\{n_1,...,n_k\},p')}$$
  if $l \equiv n(t_1,...,t_m)$ and $n \not\equiv n_i$ for all $1 \leq i \leq k$, or $l \equiv \tau$,

  - $$\frac{p \stackrel{l}{\longrightarrow} \surd}{\tau(\{n_1,...,n_k\},p) \stackrel{l}{\longrightarrow} \surd}$$
    if $l \equiv n(t_1,...,t_m)$ and $n \not\equiv n_i$ for all $1 \leq i \leq k$, or $l \equiv \tau$,

  - $$\frac{p \stackrel{n(t_1,...,t_m)}{\longrightarrow} p'}{\tau(\{n_1,...,n_k\},p) \stackrel{\tau}{\longrightarrow} \tau(\{n_1,...,n_k\},p')}$$
    if $n \equiv n_i$ for some $1 \leq i \leq k$,

  - $$\frac{p \stackrel{n(t_1,...,t_m)}{\longrightarrow} \surd}{\tau(\{n_1,...,n_k\},p) \stackrel{\tau}{\longrightarrow} \surd}$$ if $n \equiv n_i$ for some $1 \leq i \leq k$.

- $$\frac{p \stackrel{l}{\longrightarrow} p'}{\rho(\{n_1 \to n_1',...,n_k \to n_k'\},p) \stackrel{l}{\longrightarrow} \rho(\{n_1 \to n_1',...,n_k \to n_k'\},p')}$$
  if $l \equiv n(t_1,...,t_m)$ and $n \not\equiv n_i$ for all $1 \leq i \leq k$, or $l \equiv \tau$,

  - $$\frac{p \stackrel{l}{\longrightarrow} \surd}{\rho(\{n_1 \to n_1',...,n_k \to n_k'\},p) \stackrel{l}{\longrightarrow} \surd}$$
    if $l \equiv n(t_1,...,t_m)$ and $n \not\equiv n_i$ for all $1 \leq i \leq k$, or $l \equiv \tau$,

  - $$\frac{p \stackrel{n(t_1,...,t_m)}{\longrightarrow} p'}{\rho(\{n_1 \to n_1',...,n_k \to n_k'\},p) \stackrel{n'(t_1,...,t_m)}{\longrightarrow} \rho(\{n_1 \to n_1',...,n_k \to n_k'\},p')}$$
    if $n \equiv n_i$ and $n' \equiv n_i'$ for some $1 \leq i \leq k$,

  - $$\frac{p \stackrel{n(t_1,...,t_m)}{\longrightarrow} \surd}{\rho(\{n_1 \to n_1',...,n_k \to n_k'\},p) \stackrel{n'(t_1,...,t_m)}{\longrightarrow} \surd}$$
    if $n \equiv n_i$ and $n' \equiv n_i'$ for some $1 \leq i \leq k$.

- $$\frac{p \stackrel{l}{\longrightarrow} p'}{\partial(\{n_1,...,n_k\},p) \stackrel{l}{\longrightarrow} \partial(\{n_1,...,n_k\},p')}$$
  if $l \equiv n(t_1,...,t_m)$ and $n \not\equiv n_i$ for all $1 \leq i \leq k$, or $l \equiv \tau$,

  - $$\frac{p \stackrel{l}{\longrightarrow} \surd}{\partial(\{n_1,...,n_k\},p) \stackrel{l}{\longrightarrow} \surd}$$
    if $l \equiv n(t_1,...,t_m)$ and $n \not\equiv n_i$ for all $1 \leq i \leq k$, or $l \equiv \tau$.

- $$\frac{\sigma(P) \xrightarrow{l} p'}{\Sigma(x:S,P) \xrightarrow{l} p'}$$
where σ is a substitution over $Sig(E)$ and $\{\langle x:S\rangle\}$ such that $\sigma(\langle x:S\rangle) = t$ for some *data-term* t that is SSC w.r.t. $Sig(E)$ and $\emptyset$,

- $$\frac{\sigma(P) \xrightarrow{l} \surd}{\Sigma(x:S,P) \xrightarrow{l} \surd}$$
where σ is a substitution over $Sig(E)$ and $\{\langle x:S\rangle\}$ such that $\sigma(\langle x:S\rangle) = t$ for some *data-term* t that is SSC w.r.t. $Sig(E)$ and $\emptyset$.

According to the convention in 2.12 we often write $\mathcal{A}(\mathbb{A}, r, p)$ instead of $\mathcal{A}(\mathbb{A}, r, p$ **from** $E)$. Again, the following lemma serves as a justification for our definition.

Lemma 5.12. *Let E be a well-formed specification, $\mathbb{A}$ be a minimal model of E that is boolean preserving and r a representation function of E and $\mathbb{A}$. Consider a process-expression p that is SSC w.r.t. $Sig(E)$ and $\emptyset$ and let $(S, L, \longrightarrow, s) \stackrel{\text{def}}{=} \mathcal{A}(\mathbb{A}, r, p)$. If for some sequence of labels $l_1, ..., l_m$ it holds that $p \xrightarrow{l_1} \ldots \xrightarrow{l_m} p'$, then either $p' \equiv \surd$ or p' is SSC w.r.t. $Sig(E)$ and $\emptyset$.*

We feel that our operational semantics is somewhat ad hoc; we can easily provide an alternative that is also satisfactory in the sense that for each *process-expression* the generated transition system is strongly bisimilar with that generated by the rules above. Therefore, we generally consider transition systems modulo strong bisimulation equivalence. This means that the operational semantics for μCRL as given in this document has only a *referential* meaning, and any generated transition system is therefore called a *referential transition system*. A consequence of this view is that for the generation of transition systems for a μCRL-*process-expression* an operational semantics generating a smaller number of states can be used.

Definition 5.13. Let $\mathcal{A}_1 = (S_1, L_1, \longrightarrow_1, s_1)$ and $\mathcal{A}_2 = (S_2, L_2, \longrightarrow_2, s_2)$ be two transition systems. We say that $\mathcal{A}_1$ and $\mathcal{A}_2$ are bisimilar, notation $\mathcal{A}_1 \leftrightarrow \mathcal{A}_2$, iff there is a relation $R \subseteq S_1 \times S_2$ such that

- $(s_1, s_2) \in R$,
- for each pair $(t_1, t_2) \in R$:
  - $t_1 \xrightarrow{a}_1 t_1' \Rightarrow \exists t_2'\ t_2 \xrightarrow{a}_2 t_2'$ and $(t_1', t_2') \in R$,
  - $t_2 \xrightarrow{a}_2 t_2' \Rightarrow \exists t_1'\ t_1 \xrightarrow{a}_1 t_1'$ and $(t_1', t_2') \in R$.

Let E be a well-formed *specification*, $\mathbb{A}$ a minimal boolean preserving model of E, and r a representation function of E and $\mathbb{A}$. For two μCRL-*process-expressions* p and q that are SSC w.r.t. $Sig(E)$ and $\emptyset$, we write

p **from** $E \leftrightarrow_{\mathbb{A},r} q$ **from** E

iff $\mathcal{A}(\mathbb{A}, r, p\ \mathbf{from}\ E) \leftrightarrow \mathcal{A}(\mathbb{A}, r, q\ \mathbf{from}\ E)$.

The following lemma allows us to write $\leftrightarrow_{\mathbb{A}}$ instead of $\leftrightarrow_{\mathbb{A},r}$. Moreover, it gives us a useful property of bisimulation, i.e. that it is a congruence for all process operators. Note that according to our own convention we do not explicitly say where p and q stem from as they can only come from E.

Lemma 5.14. *Let E be a specification, $\mathbb{A}$ a minimal, boolean preserving model of E and p, q process-expressions that are SSC w.r.t. E and $\emptyset$.*

- *If $p \leftrightarrow_{\mathbb{A},r} q$ for some representation function r of E and $\mathbb{A}$, then $p \leftrightarrow_{\mathbb{A},r'} q$ for each representation function r' of E and $\mathbb{A}$.*
- *For all representation functions of E and $\mathbb{A}$, $\leftrightarrow_{\mathbb{A},r}$ is a congruence for all μCRL operators working on process-expressions.*

6 Effective μCRL-specifications

In order to provide a process language with tools, such as for instance a simulator, it is very important that the language has a computable operational semantics, i.e. it is decidable what the next (finite number of) steps of a process are. This is not at all the case for μCRL. Due to the undecidability of data equivalence, the use of possibly unguarded recursion and infinite sums, the next step relation need not be enumerable. We deal with this situation by restricting μCRL to *effective* μCRL. In effective μCRL data equivalence is decidable, only finite sums are allowed and recursion must be guarded. For effective μCRL the next step relation is indeed decidable.

6.1 Semi complete rewriting systems

For the data we require that the rewriting system is semi-complete (= weakly terminating and confluent) [26]. This implies that data equivalence between closed terms is decidable. Moreover, this is (in some sense) not too restrictive: every data type for which data equivalence is decidable, can be specified by a complete (= strongly terminating and confluent) term rewriting system [6]. As a complete term rewriting system is also semi-complete, all decidable data types can be expressed in effective μCRL.

We first define all required rewrite relations.

Definition 6.1. Let E be a well-formed *specification*. We define the *elementary rewrite relation* $\longrightarrow^e_E$ by:

$$\longrightarrow^e_E \stackrel{\text{def}}{=} \{\sigma(u) \longrightarrow \sigma(u') \mid u = u' \in R \text{ with } \langle R, \mathcal{V}\rangle \in \mathit{rewrites}(E), \sigma \text{ is a substitution over } \mathit{Sig}(E) \text{ and } \mathcal{V} \text{ such that } \mathit{Var}_{\mathit{Sig}(E),\mathcal{V}}(\sigma(u)) = \emptyset\}.$$

The one-step reduction relation $\longrightarrow_E$ is inductively defined by:

- $u \longrightarrow u' \in \longrightarrow_E$ if $u \longrightarrow u' \in \longrightarrow^e_E$.

- $n(t_1, ..., t_m) \longrightarrow n(t'_1, ..., t'_m) \in \longrightarrow_E$ if for some $1 \leq i \leq m$
  - $t_i \longrightarrow t'_i \in \longrightarrow_E$,
  - for $j \neq i$ it holds that $t_j \equiv t'_j$ and $n(t_1, ..., t_m)$ is SSC w.r.t. $Sig(E)$ and $\emptyset$.

The *reduction relation* $\twoheadrightarrow_E$ is the reflexive and transitive closure of $\longrightarrow_E$. We write $t \longrightarrow_E u$ and $t \twoheadrightarrow_E u$ for $t \longrightarrow u \in \longrightarrow_E$ and $t \twoheadrightarrow u \in \twoheadrightarrow_E$, respectively.

The following lemma is meant to reassure ourselves that the definitions of the rewrite relations are correct. Moreover, it gives a basic but useful property.

Lemma 6.2. *Let E be a well-formed specification. Let t be a data-term that is SSC w.r.t. $Sig(E)$ and $\emptyset$. If $t \twoheadrightarrow_E t'$, then t' is also SSC w.r.t. $Sig(E)$ and $\emptyset$.*

With these rewrite relations it is easy to define confluence and termination.

Definition 6.3. Let E be a well-formed *specification*. E is *data-confluent* iff for *data-terms* t, t' and t'' that are SSC w.r.t. $Sig(E)$ and $\emptyset$ it holds that:

$$\left.\begin{array}{l} t \twoheadrightarrow_E t' \\ t \twoheadrightarrow_E t'' \end{array}\right\} \text{ implies that there is a } \textit{data-term } t''' \text{ such that } \left\{\begin{array}{l} t' \twoheadrightarrow_E t''' \\ t'' \twoheadrightarrow_E t'''. \end{array}\right.$$

A *data-term* t that is SSC w.r.t. $Sig(E)$ and $\emptyset$ is a *normal form* if for no *data-term* u it holds that $t \longrightarrow_E u$. E is *data-terminating* if for each *data-term* t that is SSC w.r.t. $Sig(E)$ and $\emptyset$ there is some normal form t'' such that $t \twoheadrightarrow_E t''$. E is *data-semi-complete* if E is data-confluent and data-terminating.

The following lemma states that in μCRL we can find a unique normal form for each *data-term* that can be obtained from a well-formed *specification*.

Lemma 6.4. *Let E be a well-formed specification that is data-semi-complete. For any data-term t that is SSC with respect to $Sig(E)$ and $\emptyset$, there is a unique data-term $N_E(t)$ satisfying*

$t \twoheadrightarrow_E N_E(t)$ *and* $N_E(t)$ *is a normal form.*

$N_E(t)$ is called the normal form of t and there is an algorithm to find $N_E(t)$ for each data-term t that is SSC w.r.t. $Sig(E)$ and $\emptyset$.

Effective μCRL is based on the following algebra of normal forms.

Definition 6.5. Let E be a well-formed *specification* that is data-semi-complete. The $Sig(E)$-algebra $\mathbb{A}_{N_E}$ of normal forms is defined by:

- for each name $S \in Sig(E).Sort$ there is a domain $D(\mathbb{A}_{N_E}, S) \stackrel{\text{def}}{=} \{N_E(t) \mid sort_{Sig(E),\emptyset}(t) = S$ and t is a *data-term* that is SSC w.r.t. $Sig(E)$ and $\emptyset\}$,
- $C(\mathbb{A}_{N_E}, n) \stackrel{\text{def}}{=} N_E(n)$ provided $n :\rightarrow S \in Sig(E).Fun$,

- $F(\mathbb{A}_{N_E}, n : S_1 \times ... \times S_m) = f$ where the function f is defined by:

$$f(t_1, ..., t_m) = N_E(n(t_1, ..., t_m))$$

with $t_i \in D(\mathbb{A}_{N_E}, S_i)$ for $1 \leq i \leq m$ provided $n : S_1 \times ... \times S_m \to S \in Sig(E).Fun$.

Note that in $\mathbb{A}_{N_E}$ it is easy to determine that $T \neq F$. It is however undecidable that the sort **Bool** has at most two elements. We must use the *finite sort tool* of section 6.5 to determine this. Often the algebra $\mathbb{A}_{N_E}$ is called the *canonical term algebra* of E.

6.2 Finite sums

If a μCRL specification contains infinite sums, then the operational behaviour is not finitely branching anymore. Consider for instance the behaviour of the following process:

X	**from**	**sort**	**Bool**
		func	$T, F :\to$ **Bool**
		sort	Nat
		func	$0 : Nat$
			$succ : Nat \to Nat$
		act	$a : Nat$
		proc	$X = \sum(x : Nat, a(x))$

The process X can perform an $a(m)$ step for each natural number m. We judge an infinitely branching operational behaviour undesirable and therefore exclude sums over infinite sorts from effective μCRL.

Definition 6.6. Let E be a well-formed *specification* and let $\mathbb{A}$ be a model of E. We say that E has *finite sums* w.r.t. $\mathbb{A}$ iff for each occurrence $\Sigma(x : S, p)$ in E the set $D(\mathbb{A}, S)$ is finite.

6.3 Guarded recursive specifications

Also unguarded recursion may lead to an infinitely branching operational behaviour. Consider for instance the following example:

X	**from**	**sort**	**Bool**
		func	$T, F :\to$ **Bool**
		act	a
		proc	$X = X \cdot a + a$

The *process-expression* $X \cdot a$ can perform an a step to any *process-expression* a^m ($m \geq 1$) where a^m is the sequential composition of m a's. Therefore, we also exclude unguarded recursion from effective μCRL.

In the next definition it is said what a guarded μCRL specification is in very general terms.

Definition 6.7. Let E be a well-formed *specification* and $\mathbb{A}$ be a model of E that is boolean preserving. Let p be a *process-expression* of the form n or $n(t_1, ..., t_m)$ for some *name* n that is SSC w.r.t. $Sig(E)$ and $\emptyset$. Let q be a *process-expression* that is SSC w.r.t. $Sig(E)$ and $\emptyset$. We say that p is *guarded* w.r.t. $\mathbb{A}$ in q iff

- $q \equiv q_1 + q_2$, $q \equiv q_1 \parallel q_2$ or $q \equiv q_1 \,|\, q_2$, and p is guarded w.r.t. $\mathbb{A}$ in q_1 and q_2,
- $q \equiv q_1 \triangleleft c \triangleright q_2$ and either $\mathbb{A} \models c = T$ and p is guarded w.r.t. $\mathbb{A}$ in q_1, or $\mathbb{A} \models c = F$ and p is guarded w.r.t. $\mathbb{A}$ in q_2,
- $q \equiv q_1 \cdot q_2$, $q \equiv q_1 \mathbin{\|\!\_} q_2$, $q \equiv \partial(\{n_1, ..., n_m\}, q_1)$, $q \equiv \tau(\{n_1, ..., n_m\}, q_1)$, $q \equiv \rho(\{n_1 \to n_1', ..., n_m \to n_m'\}, q_1)$ or $q \equiv (q_1)$ and p is guarded w.r.t. $\mathbb{A}$ in q_1,
- $q \equiv \Sigma(x : S, q_1)$ and p is guarded w.r.t. $\mathbb{A}$ in $\sigma(q_1)$ for any substitution σ over $Sig(E)$ and $\{\langle x : S\rangle\}$,
- $q \equiv \tau$ or $q \equiv \delta$,
- $q \equiv n'$ for a *name* n' and $p \not\equiv n'$ or
- $q \equiv n'(u_1, ..., u_{m'})$ for a *basic-expression* $n'(u_1, ..., u_{m'})$ and $n \not\equiv n'$, $m \neq m'$ or $[\![u_i]\!]_{\mathbb{A}} \neq [\![t_i]\!]_{\mathbb{A}}$ for some $1 \leq i \leq m$.

If p is not guarded w.r.t. $\mathbb{A}$ in q we say that p appears *unguarded* w.r.t. $\mathbb{A}$ in q.

Definition 6.8. Let E be a well-formed *specification* and $\mathbb{A}$ be a model of E that is boolean preserving. The *Process Name Dependency Graph* of E and $\mathbb{A}$, notation $PNDG(E, \mathbb{A})$, is constructed as follows:

- for each $n = p \in Sig(E).Proc$, n is a node of $PNDG(E, \mathbb{A})$,
- for each $n(x_1 : S_1, ..., x_m : S_m) = p \in Sig(E).Proc$ and *data-terms* $t_1, ..., t_m$ that are SSC w.r.t. $Sig(E)$ and $\emptyset$ such that $sort_{Sig(E),\emptyset}(t_i) = S_i$ $(1 \leq i \leq m)$, $n(t_1, ..., t_m)$ is a node of $PNDG(E, \mathbb{A})$,
- if n is a node of $PNDG(E, \mathbb{A})$ and $n = p \in Sig(E).Proc$, then there is an edge

  $$n \longrightarrow q$$

  for a node $q \in PNDG(E, \mathbb{A})$ iff q is unguarded w.r.t. $\mathbb{A}$ in p,
- if $n(x_1 : sort_{Sig(E),\emptyset}(t_1), ..., x_m : sort_{Sig(E),\emptyset}(t_m)) = p \in Sig(E).Proc$ and $n(t_1, ..., t_m)$ is a node of $PNDG(E, \mathbb{A})$, then there is an edge

  $$n(t_1, ..., t_m) \longrightarrow q$$

  for a node $q \in PNDG(E, \mathbb{A})$ iff q is unguarded w.r.t. $\mathbb{A}$ in $\sigma(p)$ where σ is the substitution over $Sig(E)$ and $\{\langle x_i : sort_{Sig(E),\emptyset}(t_i)\rangle \mid 1 \leq i \leq m\}$ defined by

  $$\sigma(\langle x_i : sort_{Sig(E),\emptyset}(t_i)\rangle) = t_i.$$

Definition 6.9. Let E be a well-formed *specification* and $\mathbb{A}$ be a model of E that is boolean preserving. We say that E is ***guarded*** w.r.t. $\mathbb{A}$ iff $PNDG(E, \mathbb{A})$ is well founded, i.e. does not contain an infinite path.

6.4 Effective μCRL-specifications

Here we define the operational semantics of effective μCRL by combining all definitions given above.

Definition 6.10. Let E be a *specification*. We call E an *effective μCRL specification* or for short an *effective specification* iff

- E is well-formed,
- E is data-semi-complete,
- E has finite sums w.r.t. $\mathbb{A}_{N_E}$,
- E is guarded w.r.t. $\mathbb{A}_{N_E}$.

Definition 6.11. Let E be an effective μCRL *specification*. Let p be a *process-expression* that is SSC w.r.t. $Sig(E)$ and $\emptyset$. The behaviour of p is the transition system

$$\mathcal{A}(\mathbb{A}_{N_E}, r, p \textbf{ from } E)$$

where the representation function r of E and $\mathbb{A}_{N_E}$ is the identity.

In effective μCRL data equivalence is indeed decidable and the operational behaviour is finitely branching and computable:

Theorem 6.12. *Let E be an effective μCRL specification, $(S, L, \longrightarrow, s) = \mathcal{A}(\mathbb{A}_{N_E}, r, p)$ for some data-term p that is SSC w.r.t. $Sig(E)$ and $\emptyset$, and let r be the identity. Then*

- *for each pair of data-terms t_1, t_2 that are SSC w.r.t. $Sig(E)$ and $\emptyset$:*

  $$t_1 =_E t_2 \text{ is decidable,}$$

- *for each process-expression p' that is SSC w.r.t. $Sig(E)$ and $\emptyset$:*

  $$\{\langle a, p''\rangle \mid p' \stackrel{a}{\longrightarrow} p''\}$$

  is finite and effectively computable. Moreover, its cardinality is also effectively computable from E and p.

The second point of the previous theorem says that $\mathcal{A}(\mathbb{A}_{N_E}, r, p \textbf{ from } E)$ is a *computable* transition system. In a recursion theoretic setting a *computable* transition system is defined as follows: let $\mathcal{A} = (S, L, \longrightarrow, s_0)$ be a transition system with S and L sets of natural numbers and $s_0 \in S$ is represented by 0. We say that $\mathcal{A}$ is a *computable* transition system iff $\longrightarrow$ is represented by a *total* recursive function ϕ that maps each number in S to (a coding of) a finite set of pairs $\{\langle l, s' \rangle \mid s \xrightarrow{l} s'\}$.

6.5 Proving μCRL-specifications effective

In general it is not decidable whether a μCRL *specification* is effective. But there are many tools available that can prove the effectiveness for quite large classes of *specifications*. These tools provide, given a specification, a 'yes' or a 'don't know' answer.

Definition 6.13. Let $\mathcal{E}$ be the set of all well-formed *specifications*. A *data-semi-completeness tool*, notation DC, a *finite-sort tool*, notation FS, and a *guardedness tool*, notation GD, are all decidable predicates over $\mathcal{E}$, i.e. $DC \subseteq \mathcal{E}, FS \subseteq \mathcal{N} \times \mathcal{E}, GD \subseteq \mathcal{E}$.

A tool is called *sound* if each claim of a certain property it makes about a well-formed *specification* is correct. In the definition of a sound finite-sort tool and a sound guardedness tool we assume that specifications are data-semi-complete because we expect that this is a minimal requirement for these tools to operate.

Definition 6.14. A data-semi-completeness tool DC is called *sound* iff for each *specification* E that is well-formed:

> if $DC(E)$ holds, then E is data-semi-complete.

A finite-sort tool FS is called *sound* iff for each *name* n and *specification* E that is well-formed and data-semi-complete:

> if $FS(n, E)$ holds, then $n \in Sig(E).Sort$ and $D(\mathbb{A}_{N_E}, n)$ is a finite set.

A guardedness tool GD is called *sound* iff for each *specification* E that is well-formed and data-semi-complete:

> if $GD(E)$ holds, then E is guarded w.r.t. $\mathbb{A}_{N_E}$.

Sometimes a tool needs auxiliary information per *specification* to perform its task. In this case such a tool may work on a tuple containing a specification and a finite amount of such information. There is no prescribed format for this information, and it may vary from tool to tool. If a tool requires auxiliary information, then the soundness of the tool may not depend on this information. In this case the definition of soundness is modified as follows (the definition is only given for DC, the other cases can be defined likewise):

Definition 6.15. A data-semi-completeness tool DC requiring auxiliary information, is called ***sound*** iff for each well-formed ***specification*** E and each instance of auxiliary information $\mathcal{I}$:

if $DC(E, \mathcal{I})$ holds, then E is data-semi-complete.

This definition guarantees that even with incorrect auxiliary information DC always produces correct answers. DC has to be ***robust***.

Below we describe some techniques for constructing sound tools, except in those cases where techniques are provided in the literature. As time proceeds, more and more powerful techniques will appear. In order to incorporate these technological advancements in μCRL, the techniques mentioned here are only possible candidates for sound tools. They may be replaced by others, as long as these also lead to sound tools.

There are many techniques for proving termination and confluence (see HUET and OPPEN [24] and DERSHOWITZ [10] for termination, NEWMAN [32] for confluence if termination has been shown and KLOP [26] for an overview). Therefore we will not go into details here.

The problem whether a sort has a finite number of elements [5] is undecidable and as far as we know no general techniques have been developed to prove that a sort has only a finite number of elements in a minimal algebra.

We present a possible approach that can only be applied to a restricted case: let E be a ***specification*** in $\mathcal{E}$ such that $DC(E)$ for some sound data-semi-completeness tool DC and assume that we are interested in the finiteness of sorts $S_1, ..., S_k$ occurring in E. Let F be the set of all functions specified in E that have as target sort one of the sorts S_i $(1 \leq i \leq k)$. We assume that their parameter sorts also originate from $S_1, ..., S_k$. As auxiliary information we use finite sets $\mathcal{I}_i$ of (closed) ***data-terms*** that ought to represent all elements of sort S_i.

We compute for each function $f \in F$ (with target sort S_j) and for all arguments in the sets $\mathcal{I}_i$ of appropriate sorts, whether application of f leads to a ***data-term*** equivalent to one of the elements of $\mathcal{I}_j$. This can be done as we assume that $DC(E)$ holds. If this is successful, then obviously the sorts $S_1, ..., S_k$ have a finite number of elements.

Also the question whether a ***specification*** is guarded is undecidable. Still very good results can be obtained when guardedness is checked abstracting from the data parameters of process names. This is done by the following function HV. Its first argument contains the ***process-expression*** that is being searched for unguarded occurrences of ***names*** of processes and its second argument guarantees that the bodies of ***process-declarations*** are not searched twice.

Definition 6.16. Let E be a well-formed ***specification*** and let $\mathcal{V}$ be a set of variables over $Sig(E)$. A ***process-type*** is an expression $\langle n : S_1 \times ... \times S_m \rangle$ for some $m \geq 0$ with n a ***name*** and $S_1, ..., S_m$ ***names***. The function HV maps pairs of a ***process-expression*** and a set of ***process-types*** to sets of ***process-types***.

- $HV(\delta, PT) \stackrel{\text{def}}{=} \emptyset$.
- $HV(p_1 + p_2, PT) = HV(p_1 \triangleleft c \triangleright p_2, PT) = HV(p_1 \parallel p_2, PT) = HV(p_1 \mid p_2, PT) \stackrel{\text{def}}{=} HV(p_1, PT) \cup HV(p_2, PT)$.

- $HV(p_1 \cdot p_2, PT) = HV(p_1 \,\|\!\_\, p_2, PT) = HV(\partial(\{n_1, ..., n_m\}, p_1), PT) =$ $HV(\tau(\{n_1, ..., n_m\}, p_1), PT) = HV(\rho(\{n_1 \to n'_1, ..., n_m \to n'_m\}, p_1), PT)$ $= HV(\Sigma(x : S, p_1), PT) \stackrel{\text{def}}{=} HV(p_1, PT)$.

- $HV(n(t_1, ..., t_m), PT) \stackrel{\text{def}}{=}$

  - $\{\langle n : sort_{Sig(E),\mathcal{V}}(t_1) \times ... \times sort_{Sig(E),\mathcal{V}}(t_m)\rangle\}$ if $\langle n : sort_{Sig(E),\mathcal{V}}(t_1) \times ... \times sort_{Sig(E),\mathcal{V}}(t_m)\rangle \in PT$.
  - $HV(p, PT \cup \{\langle n : sort_{Sig(E),\mathcal{V}}(t_1) \times ... \times sort_{Sig(E),\mathcal{V}}(t_m)\rangle\}) \cup$ $\{\langle n : sort_{Sig(E),\mathcal{V}}(t_1) \times ... \times sort_{Sig(E),\mathcal{V}}(t_m)\rangle\}$ if $\langle n : sort_{Sig(E),\mathcal{V}}(t_1) \times ... \times sort_{Sig(E),\mathcal{V}}(t_m)\rangle \notin PT$ and $n(x_1 : sort_{Sig(E),\mathcal{V}}(t_1), ..., x_m : sort_{Sig(E),\mathcal{V}}(t_m)) = p \in Sig(E).Proc$ for some *names* $x_1, ..., x_m$.

- $HV(n, PT) \stackrel{\text{def}}{=}$

  - $\{\langle n :\rangle\}$ if $\langle n :\rangle \in PT$,
  - $HV(p, PT \cup \{\langle n :\rangle\}) \cup \{\langle n :\rangle\}$ if $\langle n :\rangle \notin PT$ and $n = p \in Sig(E).Proc$.

- $HV((p), PT) \stackrel{\text{def}}{=} HV(p, PT)$.

Theorem 6.17. *Let E be a well-formed specification. If for each process-decl $n(x_1 : S_1, ..., x_m : S_m) = p \in Sig(E).Proc$ it holds that $\langle n : S_1 \times ... \times S_m\rangle \notin HV(p, \emptyset)$ and for each process-decl $n = p \in Sig(E).Proc$ $n \notin HV(p, \emptyset)$, then E is guarded.*

Appendix An SDF-syntax for μCRL

We present an SDF-syntax for μCRL [20] which serves two purposes. It provides a syntax that does not employ special characters and, using it as input for the ASF+SDF-system, it yields an interactive editor for μCRL-specifications (see eg. [21]). The ASF+SDF system is also used to provide a well-formedness checker [27].

According to the convention in SDF we write syntactical categories with a capital and keywords with small letters. The first LAYOUT rule says that spaces (' '), tabs (`\t`) and newlines (`\n`) may be used to generate some attractive layout and are not part of the μCRL specification itself. The second LAYOUT rule says that lines starting with a `%`-sign followed by zero or more non-newline characters (`~[\n]*`) followed by a newline (`\n`) must be taken as comments and are therefore also not a part of the μCRL syntax.

In this syntax *names* are arbitrary strings over `a-z`, `A-Z` and `0-9` except that keywords are not *names*. In the context free syntax most items are self-explanatory. The symbol `+` stands for one or more and `*` for zero or more occurrences. For instance { `Name ","`}+ is a list of one or more *names* separated by commas.

The phrase **right** means that an operator is right-associative and **assoc** means that an operator is associative. The phrase **bracket** says that the defined construct is not an operator, but just a way to disambiguate the construction of a syntax tree. Instead of δ, ∂, τ and ρ we write **delta**, **encap**, **tau**, **hide** and **rename**. These keywords are taken from PSF [30].

The priorities say that '.' has highest and + has lowest priority on *process-expressions*.

```
exports
 sorts Name
       Name-list
       X-name-list
       Space-name-list
       Sort-specification
       Function-specification
       Function-decl
       Rewrite-specification
       Variable-decl-section
       Variable-decl
       Data-term
       Rewrite-rules-section
       Rewrite-rule
       Process-exp
       Renaming-decl
       Single-variable-decl
       Process-specification
       Process-decl
       Action-specification
       Action-decl
       Comm-specification
       Comm-decl
       Specification

lexical syntax
       [ \t\n]                                      -> LAYOUT
       "%" ~[\n]* "\n"                              -> LAYOUT
       [a-zA-Z0-9]*                                 -> Name
context-free syntax
       { Name ","}+                                 -> Name-list
       { Name "#"}+                                 -> X-name-list
         Name+                                      -> Space-name-list
       sort Space-name-list                         -> Sort-specification
       func Function-decl+                          -> Function-specification
       Name-list ":" X-name-list "->" Name          -> Function-decl
       Name-list ":" "->" Name                      -> Function-decl

       Variable-decl-section
               Rewrite-rules-section                -> Rewrite-specification
       var Variable-decl+                           -> Variable-decl-section
                                                    -> Variable-decl-section
```

```
Name-list ":" Name                          -> Variable-decl
Name                                        -> Data-term
Name "(" { Data-term "," }+ ")"             -> Data-term
rew Rewrite-rule+                           -> Rewrite-rules-section
Name "(" { Data-term "," }+ ")"
                  "=" Data-term             -> Rewrite-rule
Name "=" Data-term                          -> Rewrite-rule

Process-exp "+" Process-exp                 -> Process-exp right
Process-exp "||" Process-exp                -> Process-exp right
Process-exp "||_" Process-exp               -> Process-exp
Process-exp "|"  Process-exp                -> Process-exp right
Process-exp "<|" Data-term "|>"
           Process-exp                      -> Process-exp
Process-exp "." Process-exp                 -> Process-exp right
delta                                       -> Process-exp
tau                                         -> Process-exp
encap "(" "{" Name-list "}" ","
            Process-exp ")"                 -> Process-exp
hide "(" "{" Name-list "}" ","
            Process-exp ")"                 -> Process-exp
rename "(" "{" { Renaming-decl "," }+
          "}" "," Process-exp ")"           -> Process-exp
sum "(" Single-variable-decl ","
               Process-exp ")"              -> Process-exp
Name "(" { Data-term "," }+ ")"             -> Process-exp
Name                                        -> Process-exp
"(" Process-exp ")"                         -> Process-exp bracket

Name "->" Name                              -> Renaming-decl
Name ":" Name                               -> Single-variable-decl
proc Process-decl+                          -> Process-specification
Name "(" { Single-variable-decl
         "," }+ ")" "=" Process-exp         -> Process-decl
Name "=" Process-exp                        -> Process-decl

act Action-decl+                            -> Action-specification
Name-list ":" X-name-list                   -> Action-decl
Name                                        -> Action-decl

comm Comm-decl+                             -> Comm-specification
Name "|" Name "=" Name                      -> Comm-decl

Sort-specification                          -> Specification
Function-specification                      -> Specification
Rewrite-specification                       -> Specification
Action-specification                        -> Specification
Comm-specification                          -> Specification
Process-specification                       -> Specification
Specification Specification                 -> Specification  assoc
```

```
priorities
       "+" < { "||", "|", "||_"} < "<|" "|>" < "."
```

As an example we provide a μCRL-specification of an alternating bit protocol. This is almost exactly the protocol as described in [3] to which we also refer for an explanation.

```
sort    Bool
func    T,F:->Bool

sort    D
func    d1,d2,d3 : -> D

sort    error
func    e        : -> error

sort    bit
func    0,1      : -> bit
        invert   : bit -> bit

rew     invert(1)=0
        invert(0)=1

act     r1,s4    : D
        s2,r2,c2 : D#bit
        s3,r3,c3 : D#bit
        s3,r3,c3 : error
        s5,r5,c5 : bit
        s6,r6,c6 : bit
        s6,r6,c6 : error

comm    r2|s2 = c2
        r3|s3 = c3
        r5|s5 = c5
        r6|s6 = c6

proc    S             = S(0).S(1).S
        S(n:bit)      = sum(d:D,r1(d).S(d,n))
        S(d:D,n:bit) = s2(d,n).((r6(invert(n))+r6(e)).S(d,n)+r6(n))

        R             = R(1).R(0).R
        R(n:bit)      = (sum(d:D,r3(d,n))+r3(e)).s5(n).R(n)+
                          sum(d:D,r3(d,invert(n)).s4(d).s5(invert(n)))

        K             = sum(d:D,sum(n:bit,r2(d,n).(tau.s3(d,n)+
                                                        tau.s3(e)))).K

        L             = sum(n:bit,r5(n).(tau.s6(n)+tau.s6(e))).L
```

```
ABP            = hide({c2,c3,c5,c6},
                      encap({r2,r3,r5,r6,s2,s3,s5,s6},S||R||K||L))
```

References

[1] D.J. Andrews, J.F. Groote, and C.A. Middelburg, editors. *Proceedings of the International Workshop on Semantics of Specification Languages.* Workshops in Computing, Springer-Verlag, 1994.

[2] J.C.M. Baeten and J.A. Bergstra. Process algebra with signals and conditions. In M. Broy, editor, *Programming and Mathematical Methods, Proceedings Summer School Marktoberdorf 1991*, pages 273–323. Springer-Verlag, 1992. NATO ASI Series F88.

[3] J.C.M. Baeten and W.P. Weijland. *Process Algebra.* Cambridge Tracts in Theoretical Computer Science 18. Cambridge University Press, 1990.

[4] J.A. Bergstra and J.W. Klop. Process algebra for synchronous communication. *Information and Control*, 60(1/3):109–137, 1984.

[5] J.A. Bergstra and J.V. Tucker. A characterisation of computable data types by means of a finite equational specification method. In J.W. de Bakker and J. van Leeuwen, editors, *Proceedings 7th ICALP,* Noorwijkerhout, volume 85 of *Lecture Notes in Computer Science*, pages 76–90. Springer-Verlag, July 1980.

[6] J.A. Bergstra and J.V. Tucker. The completeness of the algebraic specification methods for computable data types. *Information and Control*, 12:186–200, 1982.

[7] M.A. Bezem and J.F. Groote. Invariants in process algebra with data. Technical Report 98, Logic Group Preprint Series, Utrecht University, 1993. To appear in Proceedings Concur'94, Uppsala.

[8] M.A. Bezem and J.F. Groote. A correctness proof of a one bit sliding window protocol in μCRL. Technical Report 99, Logic Group Preprint Series, Utrecht University, 1993. To appear in the Computer Journal Volume 37(4), 1994.

[9] CCITT Working Party X/1. *Recommendation Z.100 (SDL)*, 1987.

[10] N. Dershowitz. Computing with rewrite systems. *Information and Computation*, 65:122–157, 1985.

[11] H. Ehrig and B. Mahr. *Fundamentals of algebraic specifications I*, volume 6 of *EATCS Monographs on Theoretical Computer Science.* Springer-Verlag, 1985.

[12] J.F. Groote. *Process Algebra and Structured Operational Semantics.* PhD thesis, University of Amsterdam, 1991.

[13] J.F. Groote and H. Korver. A correctness proof of the bakery protocol in μCRL. In this volume: A. Ponse, C Verhoef and S.F.M. van Vlijmen, editors. *Proceedings of ACP94*. Workshops in Computing, Springer-Verlag, 1994.

[14] J.F. Groote and J.C. van de Pol. A bounded retransmission protocol for large data packets. A case study in computer checked verification. Technical Report 100, Logic Group Preprint Series, Utrecht University, 1993.

[15] J.F. Groote and A. Ponse. The syntax and semantics of μCRL. Report CS-R9076, CWI, Amsterdam, 1990.

[16] J.F. Groote and A. Ponse. Proof theory for μCRL. Report CS-R9138, CWI, 1991.

[17] J.F. Groote and A. Ponse. Proof theory for μCRL: a language for processes with data. In Andrews et al. [1], pages 231–250.

[18] J.F. Groote and J. van Wamel. Algebraic Data Types and Induction in μCRL. Report P9409, University of Amsterdam, Amsterdam, 1994.

[19] J.F. Groote and F.W. Vaandrager. Structured operational semantics and bisimulation as a congruence. *Information and Computation*, 100(2):202–260, October 1992.

[20] J. Heering, P.R.H. Hendriks, P. Klint, and J. Rekers. The syntax definition formalism SDF – reference manual –. *ACM SIGPLAN Notices*, 24(11):43–75, 1989.

[21] P.R.H. Hendriks. *Implementation of Modular Algebraic Specifications.* PhD thesis, University of Amsterdam, 1991. To appear.

[22] C.A.R. Hoare. *Communicating Sequential Processes.* Prentice-Hall International, Englewood Cliffs, 1985.

[23] C.A.R. Hoare, I.J. Hayes, He Jifeng, C.C. Morgan, A.W. Roscoe, J.W. Sanders, I.H. Sorensen, J.M. Spivey, and B.A. Sufrin. Laws of programming. *Communications of the ACM*, 30(8):672–686, August 1987.

[24] G. Huet and D.D. Oppen. Equations and rewrite rules: A survey. In R. Book, editor, *Formal Language Theory: Perspectives and Open Problems*, pages 349–405. Academic Press, 1980.

[25] ISO. *Information processing systems – open systems interconnection – LOTOS – a formal description technique based on the temporal ordering of observational behaviour* ISO/TC97/SC21/N DIS8807, 1987.

[26] J.W. Klop. Term rewriting systems. In S. Abramsky, D.M. Gabbay, and T.S.E. Maibaum, eds., *Handbook of Logic in Computer Science, Volume 2*, pages 2–108. Oxford University Press, 1992.

[27] H. Korver. Private communications, 1991.

[28] H. Korver. Protocol Verification in μCRL. PhD. Thesis, University of Amsterdam, Amsterdam, The Netherlands. 1994.

[29] H. Korver and J. Springintveld. A computer-checked verification of Milner's scheduler. In M. Hagiya and J.C. Mitchel, editors. *Proceedings of the 2nd International Symposium on Theoretical Aspects of Computer Software, TACS '94,* Sendai, Japan, pages 161–178, volume 789 of *Lecture Notes in Computer Science.* Springer-Verlag, 1994.

[30] S. Mauw and G.J. Veltink. A process specification formalism. *Fundamenta Informaticae,* XIII:85–139, 1990.

[31] R. Milner. *A Calculus of Communicating Systems,* volume 92 of *Lecture Notes in Computer Science.* Springer-Verlag, 1980.

[32] M.H.A. Newman. On theories with a combinatorial definition of equivalence. *Annals of Mathematics,* 43(2):223–243, 1942.

[33] G.D. Plotkin. An operational semantics for CSP. In D. Bjørner, editor, *Proceedings IFIP TC2 Working Conference on Formal Description of Programming Concepts – II,* Garmisch, pages 199–225, Amsterdam, 1983. North-Holland.

[34] A. Ponse. Computable processes and bisimulation equivalence. Report CS-R9207, CWI, Amsterdam, January 1992.

[35] M.P.A. Sellink. Verifying process algebra proofs in type theory. In Andrews et al. [1], pages 314–338.

[36] SPECS-semantics. *Definition of MR and CRL Version 2.1,* 1990.

A Correctness Proof of the Bakery Protocol in μCRL

Jan Friso Groote
Department of Philosophy, Utrecht University
Heidelberglaan 8, 3584 CS Utrecht, The Netherlands

Henri Korver
Department of Software Technology, CWI
Kruislaan 413, 1098 SJ Amsterdam, The Netherlands

Abstract

A specification of a bakery protocol is given in μCRL. We provide a simple correctness criterion for the protocol. Then the protocol is proven correct using a proof system that has been developed for μCRL. The proof primarily consists of algebraic manipulations based on specifications of abstract data types and elementary rules and axioms from process algebra.

1 Introduction

The main purpose of this paper is to show that μCRL, or in more general terms process algebra with abstract data types, offers a framework for verifications of processes. This is done by the verification of a bakery protocol, which is a non trivial protocol with unbounded state space. Neither process algebra nor data type theory seems to form a suitable vehicle for the verification of this protocol on their own, showing that the verification capacities of μCRL go beyond those found in both its constituents. Actually, this observation has been confirmed by the verification of a number of other 'difficult' protocols (see e.g. [BG93a, GP93, KS94]). Process algebra in its basic form does not include processes that are parametrised with data: parameterised sums, conditionals, parametrised actions, etc., and very importantly induction over these parameters. All these are essential in the verification given in this paper.

Our work structurally differs from the more conventional 'assertional' verification techniques (see [Apt81, Apt84, CM88]). These are mainly based on data and do not often allow for algebraic reductions of processes. In particular, simple and elegant correctness identities such as given in section 3 cannot be formulated.

There are two other points that deserve mention. First, the proof system of μCRL has been defined in such way that it allows for automatic proof checking [Sel93, BG93b, GP93, KS94]. This is important, as a minor mistake in a program or a protocol may have disastrous impacts. And actually, we have so often detected 'oversights' in calculations that we may expect that also the proof in this paper is not completely flawless. The only way to systematically increase the reliability of proofs is by having these automatically checked using a computer tool. This of course does not decrease the value of this paper, because finding a proof remains the essential step in a verification.

The other point is about the proof in this paper. Although initially the proof was not easy to find due to the large number of possible proof strategies, the resulting proof follows a reasonable and straightforward line of thought. This is promising, because we think that if we get more skill and experience in doing calculations such as given in this paper, most communication protocols can be verified in μCRL by a fixed selection of standard strategies.

2 The specification of the Bakery Protocol

We describe a simple system that captures a well-known protocol that has been used over the centuries, especially in bakery shops, and prove its correctness in the proof system for μCRL [GP94b, GP94a]. We assume that the reader is familiar with μCRL which is a straightforward combination of process algebra [BW90] and abstract data types [EM85]. But see Appendix B for the axiom system of μCRL.

The Bakery Protocol derives its name from the well-known situation in a busy bakery where customers pick a number when entering the shop in order to guarantee that they are served in proper order. The system basically consists of n 1-place buffers (see *Buf* in Figure 1), that may each contain a customer waiting to be served. Before waiting, each customer picks a sequence number (which are distributed modulo n) indicating when it is his turn. This is modeled by the in-sequencer *INS* in Figure 1. A customer is served when his number matches that of the baker, modeled by the out-sequencer *OUTS* in Figure 1. The system is supposed to work on a first come first served basis, i.e. it should behave like a queue. We take the existence of basic data types, which are in this case booleans and natural numbers, for granted. These data types are specified in appendix A. We also need modulo calculations, e.g. for specifying the in-sequencer *INS* and out-sequencer *OUTS*. For this purpose $+_n$ is introduced, which is addition modulo n. Its specification can also be found in appendix A.

The customers are supposed to be given by a (non empty) sort D. Sort D contains a bottom element $d_\perp$ for denoting undefined data elements. We have a sort $queue_d$ that consists of queues of customers (see appendix A for its specification). In order to attach a number to a customer the data type *Pair* is introduced together with a pairing $\langle \_, \_ \rangle$ and an equality function *eq*. We do not completely obey the syntax of μCRL to increase readability, e.g. by using infix notation and omitting $\cdot$ for sequential composition.

sort *Pair*
func $\langle , \rangle : D \times nat \to Pair$
$\quad eq : Pair \times Pair \to \mathbf{Bool}$
var $d, e : D$
$\quad n, m : nat$
rew $eq(\langle d, n\rangle, \langle e, m\rangle) = eq(d, e)$ and $eq(n, m)$

We also need queues which can contain pairs. Therefore, the data type $queue_p$ is introduced in appendix A.

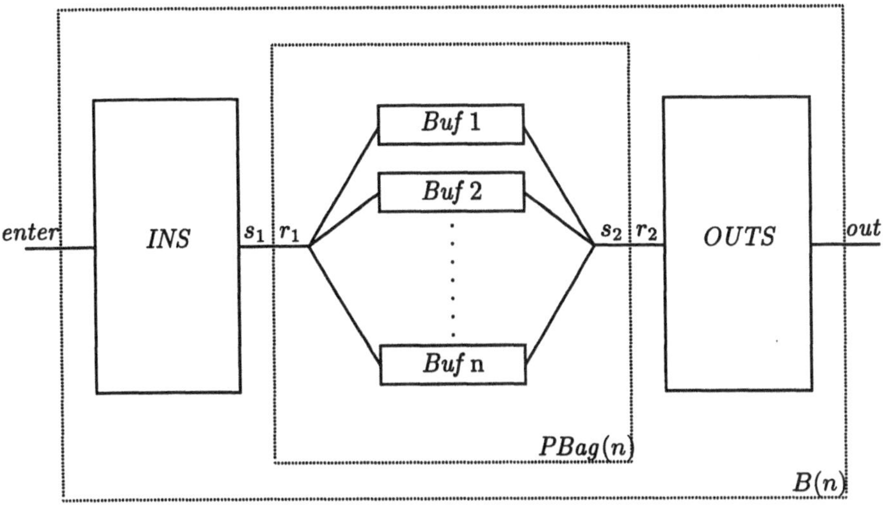

Figure 1: The Bakery Protocol.

A buffer process that can contain a customer with a ticket is straightforwardly specified as follows.

act $r_1, s_2 : Pair$
proc $Buf = \sum_{p:Pair}(r_1(p) \cdot s_2(p)) \cdot Buf$

A customer with a ticket, modeled by the pair $\langle d, i\rangle$, can enter the buffer at gate r_1 and leave it at gate s_2.

By putting n of these buffers in parallel, we model that n customers can wait in the shop. As this is the behaviour of a bag which is essentially described by processes, we call this process *PBag*, derived from 'Process bag'.

proc $PBag(n : nat) = \delta \triangleleft eq(n, 0) \triangleright (Buf \parallel PBag(n-1))$

Note the way in which *PBag* has been recursively defined, e.g. $PBag(1) = \delta \parallel Buf$ which exactly corresponds to our intuition because $\delta \parallel Buf = Buf$.[1]

The process $INS(n, i)$ assigns a successive number modulo n to each customer. The number i represents the first number that is assigned. A customer enters at the entrance of the bakery, represented by the action $enter(d)$. With a number he walks into the shop, which is modelled by $s_1(\langle d, i\rangle)$. The fact that he directly enters a place in a buffer is modelled by a communication between s_1 and r_1. The process $OUTS(n, i)$ selects the customer to be served. In this case i represents the first number that will be served. Entering *OUTS* is modelled via the r_2 gate that must communicate with gate s_2. After being served the customer leaves the counter via *out*.

[1]In general, the identity $\delta \parallel x = x$ does not hold, e.g. consider the counter example $\delta \parallel a = a\delta \neq a$. However, the identity $\delta \parallel Buf = Buf$ does hold because *Buf* is a non-terminating process.

$$\begin{array}{ll} \textbf{act} & enter, out : D \\ & s_1, r_2, c_1, c_2 : Pair \\ \textbf{proc} & INS(n, i{:}nat) = \sum_{d:D}(enter(d)\, s_1(\langle d, i\rangle))\, INS(n, i +_n 1) \\ & OUTS(n, i{:}nat) = \sum_{d:D}(r_2(\langle d, i\rangle)\, out(d))\, OUTS(n, i +_n 1) \end{array}$$

The whole bakery $B(n)$ is given by:

$$\begin{array}{ll} \textbf{comm} & r_1 \,|\, s_1 = c_1,\ r_2 \,|\, s_2 = c_2 \\ \textbf{proc} & B(n{:}nat) = \tau_I(\partial_H(INS(n, 0) \parallel PBag(n) \parallel OUTS(n, 0))) \end{array}$$

where $I = \{c_1, c_2\}$ and $H = \{r_1, s_1, r_2, s_2\}$.

3 The correctness criterion for the Bakery Protocol

The Bakery Protocol $B(n)$ is supposed to work as a bounded queue of size $n + 2$; there can be n customers waiting in the buffers, one can be busy obtaining a number and one can already be selected to be served. The 'standard' specification of a process $Q(n)$ modelling a queue of size n containing elements of D is:

$$\begin{array}{ll} \textbf{proc} & Q(n : nat, b : queue_d) = \\ & \qquad \sum_{d:D}(enter(d) \cdot Q(n, in(d, b))) \lhd size(b) < n \rhd \delta\ + \\ & \qquad out(toe(b)) \cdot Q(n, untoe(b)) \lhd size(b) > 0 \rhd \delta \\ & \\ & Q(n : nat) = Q(n, \varnothing_d) \end{array}$$

With this specification of a queue the correctness of the Bakery Protocol is stated as follows.

$$\boxed{n > 0 \rightarrow B(n) = Q(n + 2)}$$

The condition $n > 0$ is necessary to guarantee that there is at least one buffer place for a customer to wait. Otherwise, as is easy to see, no customer can reach the counter.

4 Basic lemmas for μCRL

In this section, we present a number of elementary lemmas that are used in the verification of the Bakery Protocol. These lemmas are interesting in their own right as it is very likely that they are necessary in almost every verification in μCRL.

In this section, we assume that the reader is familiar with the following conventions about open terms and variables. The letters $d, e, \ldots$ stand for *data* variables. The symbols $t, t_1, t_2, \ldots$ stand for data terms (possibly containing variables). The symbols $b, b_1, b_2, \ldots$ stand for data variables of sort **Bool** and the symbols $c, c_1, c_2, \ldots$ stand for data terms of sort **Bool**. The letters $x, y, z, \ldots$ stand for *process* variables. The symbols $p, p_1, p_2, \ldots$ stand for process terms.

The following lemma expresses that μCRL supports the Excluded Middle Principle.

Lemma 4.1 (*Excluded Middle Principle*). *Let ϕ and ψ be two arbitrary property-formulas over a given μCRL specification.*

1. *If $\phi \to \psi$ and $\neg\phi \to \psi$ then ψ.*
2. *If $b = \mathsf{t} \to \psi$ and $b = \mathsf{f} \to \psi$ then ψ.*

Proof.

1. See [GP94a].
2. Immediate by 4.1.1 and BOOL2.

□

Lemma 4.1.2 is an instance of the Excluded Middle Principle which is often applied in μCRL proofs, for instance to prove the conditional identities in the next lemma.

Lemma 4.2.

1. $x \triangleleft b \triangleright x = x$,
2. $x + x \triangleleft b \triangleright \delta = x$,
3. $(x + y) \triangleleft b \triangleright z = x \triangleleft b \triangleright z + y \triangleleft b \triangleright z$,
4. $(x + y) \triangleleft b \triangleright y = x \triangleleft b \triangleright \delta + y$,
5. $(x \triangleleft b \triangleright \delta) \mathbin{\lfloor\!\lfloor} y = (x \mathbin{\lfloor\!\lfloor} y) \triangleleft b \triangleright \delta$,
6. $x \mid (y \triangleleft b \triangleright \delta) = (x \mid y) \triangleleft b \triangleright \delta$,
7. $p[e/d] = p \triangleleft eq(d,e) \triangleright \delta + p[e/d]$, *provided that* $eq(d,e) \to d = e$,
8. $a(d) \cdot x \mid b(e) \cdot y = (a(d) \cdot x \mid b(e) \cdot y) \triangleleft eq(d,e) \triangleright \delta$, *provided that* $d = e \to eq(d,e)$.

These identities are frequently used in the verification of the Bakery Protocol and μCRL verifications in general.

Proof.

1–7. Easy with axioms COND1, COND2 and Lemma 4.1.2.

8. By COND1, COND2, CF and Lemma 4.1.2.

□

The following lemma presents a rule which is derived from the SUM axioms. This rule appears to be a powerful tool to eliminate sum expressions in μCRL calculations. We first introduce an auxiliary proposition, which enables us to identify processes via summand inclusion.

Proposition 4.3. *Let $p \subseteq q$ be a shorthand for $q = q + p$ expressing that p is a summand of q. Then, we have:*

$$p \subseteq q \wedge q \subseteq p \to p = q.$$

Proof. $q = q + p \stackrel{\text{A1}}{=} p + q = p$. □

Sometimes it is more convenient to reason with summands instead of equations.

Lemma 4.4 (*Sum Elimination*). *Assume that there is an equality function eq such that $eq(d,e) = \mathsf{t} \leftrightarrow d = e$. Then*

$$\sum_{d:D}(p \triangleleft eq(d,e) \triangleright \delta) = p[e/d].$$

Proof. By showing summand inclusion in both directions (Proposition 4.3).

$\supseteq$: Consider the following obvious identity

$$\sum_{d:D}(p \triangleleft eq(d,e) \triangleright \delta) \stackrel{\text{SUM3}}{=} \sum_{d:D}(p \triangleleft eq(d,e) \triangleright \delta) + p \triangleleft eq(d,e) \triangleright \delta \qquad (1)$$

By applying the substitution $[e/d]$ to equation (1) we obtain

$$\sum_{d:D}(p \triangleleft eq(d,e) \triangleright \delta) = \sum_{d:D}(p \triangleleft eq(d,e) \triangleright \delta) + p[e/d] \triangleleft eq(e,e) \triangleright \delta \qquad (2)$$

Note that the substitution does not affect both sum constructs above because in μCRL substitutions do not change bound variables (see [GP94a]). By applying axiom COND1 to the second summand on the right hand side of equation (2), we get

$$\sum_{d:D}(p \triangleleft eq(d,e) \triangleright \delta) = \sum_{d:D}(p \triangleleft eq(d,e) \triangleright \delta) + p[e/d]. \qquad (3)$$

$\subseteq$: By the calculation:

$$\begin{aligned} p[e/d] &\stackrel{\text{SUM1}}{=} \textstyle\sum_{d:D}(p[e/d]) \\ &\stackrel{4.2.7}{=} \textstyle\sum_{d:D}(p \triangleleft eq(d,e) \triangleright \delta + p[e/d]) \\ &\stackrel{\text{SUM4}}{=} \textstyle\sum_{d:D}(p \triangleleft eq(d,e) \triangleright \delta) + \sum_{d:D}(p[e/d]) \\ &\stackrel{\text{SUM1}}{=} \textstyle\sum_{d:D}(p \triangleleft eq(d,e) \triangleright \delta) + p[e/d]. \end{aligned}$$

□

In the next lemma, we generalise axiom CM3 with a conditional construct and a sum operator.

Lemma 4.5 (*Left Merge with SUM and COND*). *We assume that the variable d does not occur free in term q.*

1. $\sum_{d:D}(a(d) \cdot p) \mathbin{⌊\!⌊} q = \sum_{d:D}(a(d) \cdot (p \parallel q))$
2. $\sum_{d:D}(a(d) \cdot p \triangleleft c \triangleright \delta) \mathbin{⌊\!⌊} q = \sum_{d:D}(a(d) \cdot (p \parallel q) \triangleleft c \triangleright \delta)$

Proof.

1. By the following calculation

$$\begin{aligned} &\textstyle\sum_{d:D}(a(d)\cdot p) \mathbin{\lfloor\!\lfloor} q \\ \overset{\text{SUM6}}{=}\quad &\textstyle\sum_{d:D}(a(d)\cdot p \mathbin{\lfloor\!\lfloor} q) \\ \overset{\text{CM3}}{=}\quad &\textstyle\sum_{d:D}(a(d)\cdot (p \parallel q)) \end{aligned}$$

2. By the following calculation

$$\begin{aligned} &\textstyle\sum_{d:D}(a(d)\cdot p \lhd c \rhd \delta) \mathbin{\lfloor\!\lfloor} q \\ \overset{\text{SUM6}}{=}\quad &\textstyle\sum_{d:D}(a(d)\cdot p \lhd c \rhd \delta \mathbin{\lfloor\!\lfloor} q) \\ \overset{4.2.5}{=}\quad &\textstyle\sum_{d:D}(a(d)\cdot p \mathbin{\lfloor\!\lfloor} q \lhd c \rhd \delta) \\ \overset{\text{CM3}}{=}\quad &\textstyle\sum_{d:D}(a(d)\cdot (p \parallel q) \lhd c \rhd \delta) \end{aligned}$$

□

There are two remarks about the lemma above. At first, note that we can avoid the restriction that d is not allowed to occur free in q by renaming it with axiom SUM2. So, the restriction is just a formality and no generality is lost. Secondly, note that in stating properties about sum expressions as in 4.5.2, we often use a boolean term c instead of a boolean variable b to be as general as possible. Otherwise b can never be substituted by a term containing a variable d, as substitution may not create bound variables.

Next, we generalise axiom CM7 with a conditional construct and a sum operator.

Lemma 4.6 (*Communication with SUM and COND*). *Let $d, e : D$ and $d' : D'$ be variables, and let $t : D$ a term. Assume there is an equality function eq such that $eq(d,e) = \mathsf{t} \leftrightarrow d = e$, and variable d does not occur free in terms q and c_2. Variable d' does not occur free in p, c and c_1.*

1. $\sum_{d:D}(a(d)\cdot p) \mid b(e)\cdot q = (a(e) \mid b(e))\cdot (p[e/d] \parallel q)$
2. $\sum_{d:D}(a(d)\cdot p \lhd c \rhd \delta) \mid b(e)\cdot q = (a(e) \mid b(e))\cdot (p[e/d] \parallel q) \lhd c[e/d] \rhd \delta$
3. $\sum_{d:D}(a(d)\cdot p \lhd c \rhd \delta) \mid \sum_{d':D'}(b(t)\cdot q) = \sum_{d':D'}((a(t) \mid b(t))\cdot (p[t/d] \parallel q) \lhd c[t/d] \rhd \delta)$
4. $\sum_{d:D}(a(d)\cdot p \lhd c_1 \rhd \delta) \mid \sum_{d':D'}(b(t)\cdot q \lhd c_2 \rhd \delta) = \sum_{d':D'}((a(t) \mid b(t))\cdot (p[t/d] \parallel q) \lhd c_1[t/d] \text{ and } c_2 \rhd \delta)$

Proof.

1. By the following calculation

$$\begin{aligned} &\textstyle\sum_{d:D}(a(d)\cdot p) \mid b(e)\cdot q \\ \overset{\text{SUM7}}{=}\quad &\textstyle\sum_{d:D}(a(d)\cdot p \mid b(e)\cdot q) \\ \overset{4.2.8}{=}\quad &\textstyle\sum_{d:D}((a(d)\cdot p \mid b(e)\cdot q) \lhd eq(d,e) \rhd \delta) \\ \overset{\text{SumEl.}}{=}\quad &a(e)\cdot p[e/d] \mid b(e)\cdot q \\ \overset{\text{CM7}}{=}\quad &(a(e) \mid b(e))\cdot (p[e/d] \parallel q) \end{aligned}$$

Note that the assumption $eq(d,e) = \mathsf{t} \leftrightarrow d = e$ is needed for application of lemmas 4.2.8 and 4.4 (Sum Elimination) in the calculation above.

2. By the following calculation

$$\begin{aligned}
&\textstyle\sum_{d:D}(a(d)\cdot p \lhd c \rhd \delta) \mid b(e)\cdot q\\
\overset{\text{SUM7}}{=}\quad &\textstyle\sum_{d:D}((a(d)\cdot p \lhd c \rhd \delta) \mid b(e)\cdot q)\\
\overset{4.2.6}{=}\quad &\textstyle\sum_{d:D}((a(d)\cdot p \mid b(e)\cdot q) \lhd c \rhd \delta)\\
\overset{4.2.8}{=}\quad &\textstyle\sum_{d:D}(((a(d)\cdot p \mid b(e)\cdot q) \lhd eq(d,e) \rhd \delta) \lhd c \rhd \delta)\\
\overset{\text{SumEl.}}{=}\quad &(a(e)\cdot p[e/d] \mid b(e)\cdot q) \lhd c \rhd \delta\\
\overset{\text{CM7}}{=}\quad &(a(e) \mid b(e))\cdot (p[e/d] \parallel q) \lhd c \rhd \delta
\end{aligned}$$

Then by the substitution rule SUB (see [GP94a]) we have

$$\sum_{d:D}(a(d)\cdot p \lhd c \rhd \delta) \mid b(e)\cdot q = (a(e) \mid b(e))\cdot (p[e/d] \parallel q) \lhd c[e/d] \rhd \delta$$

as the substitution $[e/d]$ does not effect the right-hand side of this equation.

3. By the following calculation

$$\begin{aligned}
&\textstyle\sum_{d:D}(a(d)\cdot p \lhd c \rhd \delta) \mid \sum_{d':D'}(b(t)\cdot q)\\
\overset{\text{SUM7}}{=}\quad &\textstyle\sum_{d':D'}(\sum_{d:D}(a(d)\cdot p \lhd c \rhd \delta) \mid (b(t)\cdot q))\\
\overset{4.6.2}{=}\quad &\textstyle\sum_{d':D'}((a(t) \mid b(t))\cdot (p[t/d] \parallel q) \lhd c[t/d] \rhd \delta)
\end{aligned}$$

4. In a similar way as the proof of 4.6.3.

□

In the proofs given above we often needed an equality function *eq* for comparing elements of data type D. In order to have the rather desirable property that $eq(d,e) \leftrightarrow d = e$, one can extend the specification of D with a selector function *if* and four axioms. This is formulated in Lemma 4.7. The axioms are due to Jan Bergstra.

$$\begin{array}{ll}
\textbf{sort} & D\\
\textbf{func} & \ldots\\
 & eq : D \times D \to \textbf{Bool}\\
 & if : \textbf{Bool} \times D \times D \to D\\
\textbf{var} & d, e : D\\
\textbf{rew} & \ldots\\
 & if(\mathsf{t}, d, e) = d\\
 & if(\mathsf{f}, d, e) = e\\
 & eq(d,d) = \mathsf{t}\\
 & if(eq(d,e), d, e) = e
\end{array}$$

Lemma 4.7. *Let d, e be variables of sort D, and let eq be the equality function specified above. Then, $eq(d,e) \leftrightarrow d = e$.*

5 The correctness proof of the Bakery Protocol

In this section we prove that the Bakery Protocol $B(n)$ indeed satisfies the criterion as stated in section 3. The proof transforms the process style description in two steps into a data style description. First, we use the fact that $PBag(n)$ behaves as a 'standard' bounded bag which is usually described by a data type bag. This fact has already be proven in μCRL (see [Kor94]). Then we show that this bounded bag combined with the in-sequencer *INS* and the out-sequencer *OUTS* is equal to the bounded queue described above.

Part I: $PBag(n)$ is a bounded bag

The standard behaviour of a bounded bag is specified as follows:

$$\begin{aligned}\textbf{proc } DBag(n: nat, b: queue_p) = \\ &\textstyle\sum_{p:Pair}(r_1(p)\cdot DBag(n, in(p,b))) \lhd size(b) < n \rhd \delta\ + \\ &\textstyle\sum_{p:Pair}(s_2(p)\cdot DBag(n, rem(p,b)) \lhd test(p,b) \rhd \delta)\end{aligned}$$

The D in *DBag* refers to 'Data'. Although the data type $queue_p$ is actually a queue, it has been extended with functions *test* and *rem* (*remove*) which makes it possible to use it as a bag.

The next theorem taken from [Kor94] says that $PBag(n)$ behaves like a bounded bag of size n.

Theorem 5.1. $PBag(n) = DBag(n, \varnothing_p)$.

Proof. By induction on n and RSP (see [Kor94]). □

Part II: $B(n)$ is a queue

The second part of showing the Bakery Protocol correct consists of proving $INS(i,n)$ and $OUTS(i,n)$ in combination with the $DBag(n)$ equal to a bounded queue.

The essential observation in our proof is to distinguish the following four situations:

0 No customer is busy getting a ticket and no customer is being served by the baker.

1 No customer is busy getting a ticket but there is a customer being served by the baker.

2 A customer is busy getting a ticket and no customer is being served by the baker.

3 A customer is busy getting a ticket and another customer is being served by the baker.

In order to calculate with the four situations above we make these explicit as processes.

Imagine the ideal situation towards which we are working, namely a queue b of customers. How are these customers distributed over the bakery in situation 3? The customer that entered the queue first is being served by the baker. So, $toe(b)$ is in $OUTS$, ready to leave the bakery. The person that entered the queue last is still picking a number. So $hd(b)$ is in INS. All other persons in the queue are waiting with a ticket in $PBag$. They are assigned consecutive numbers which is modelled by the function $number(i, n, b)$. It makes a queue of pairs ($queue_p$) out of a queue of customers $b : queue_d$ by taking the elements in b and number it from the end to the start of b, starting with i, modulo n. Below the four situations are described as processes $CQ_0, \ldots, CQ_3$. The number n indicates the size of the queue (the actual size of the queue is $n+2$), b is the queue of customers and i is the number on the ticket of the first customer. Note that the behaviour of CQ_j is chosen to be δ if $i \geq n$ or the size of the queue b is too small or too large. In these cases the values of i, n and b do not matter. For instance in case of CQ_3 if $size(b) < 2$, then there cannot be customers in both INS and $OUTS$, which does not conform to the intention of CQ_3.

$$
\begin{array}{l}
\textbf{proc } CQ_0(n : nat, i : nat, b : queue_d) = \\
\quad \tau_{\{c_1,c_2\}}(\partial_{\{r_1,s_1,r_2,s_2\}}(\\
\quad\quad INS(n, i +_n size(b)) \parallel \\
\quad\quad DBag(n, number(i, n, b)) \parallel \\
\quad\quad OUTS(n, i))) \\
\quad \triangleleft size(b) \leq n \text{ and } i < n \triangleright \delta
\end{array}
$$

$$
\begin{array}{l}
CQ_1(n : nat, i : nat, b : queue_d) = \\
\quad \tau_{\{c_1,c_2\}}(\partial_{\{r_1,s_1,r_2,s_2\}}(\\
\quad\quad INS(n, i +_n size(b)) \parallel \\
\quad\quad DBag(n, number(i +_n 1, n, untoe(b))) \parallel \\
\quad\quad out(toe(b))\, OUTS(n, i +_n 1))) \\
\quad \triangleleft 0 < size(b) \leq n+1 \text{ and } i < n \triangleright \delta
\end{array}
$$

$$
\begin{array}{l}
CQ_2(n : nat, i : nat, b : queue_d) = \\
\quad \tau_{\{c_1,c_2\}}(\partial_{\{r_1,s_1,r_2,s_2\}}(\\
\quad\quad s_1(\langle hd(b), i +_n size(tl(b))\rangle)\, INS(n, i +_n size(b)) \parallel \\
\quad\quad DBag(n, number(i, n, tl(b))) \parallel \\
\quad\quad OUTS(n, i))) \\
\quad \triangleleft 0 < size(b) \leq n+1 \text{ and } i < n \triangleright \delta
\end{array}
$$

$$
\begin{array}{l}
CQ_3(n : nat, i : nat, b : queue_d) = \\
\quad \tau_{\{c_1,c_2\}}(\partial_{\{r_1,s_1,r_2,s_2\}}(\\
\quad\quad s_1(\langle hd(b), i +_n size(tl(b))\rangle)\, INS(n, i +_n size(b)) \parallel \\
\quad\quad DBag(n, number(i +_n 1, n, tl(untoe(b)))) \parallel \\
\quad\quad out(toe(b))\, OUTS(n, i +_n 1))) \\
\quad \triangleleft 1 < size(b) \leq n+2 \text{ and } i < n \triangleright \delta
\end{array}
$$

Note that obviously $CQ_0(n, 0, \emptyset_d) = B(n)$ if $n > 0$.

Now let us consider say $CQ_0(n, i, b)$ and pose the question what the behaviour of CQ_0 would be. The process $CQ_0(n, i, b)$ can perform an action $enter(d)$ and arrive in the situation $CQ_2(n, i, in(d, b))$, namely the situation where customer d is busy picking a ticket. If $size(b) > 0$, there is a customer in $CQ_0(n, i, b)$ that can become served by the baker. So, via an internal step $CQ_0(n, i, b)$ becomes $CQ_1(n, i, b)$. This analysis can be made in all four cases. In other words, CQ_j substituted for G_j should satisfy the equations below. Indeed this is confirmed by Theorem 5.2.3.

$$
\begin{aligned}
&\textbf{proc } G_0(n:nat, i:nat, b:queue_d) = \\
&\qquad \textstyle\sum_{d:D}(enter(d)\,G_2(n,i,in(d,b))) \triangleleft size(b) \leq n \text{ and } i < n \triangleright \delta + \\
&\qquad \tau\, G_1(n,i,b) \triangleleft 0 < size(b) \leq n \text{ and } i < n \triangleright \delta
\end{aligned}
$$

$$
\begin{aligned}
&G_1(n:nat, i:nat, b:queue_d) = \\
&\qquad \textstyle\sum_{d:D}(enter(d)\,G_3(n,i,in(d,b))) \\
&\qquad\qquad\qquad \triangleleft 0 < size(b) \leq n+1 \text{ and } i<n \triangleright \delta + \\
&\qquad out(toe(b))\,G_0(n, i +_n 1, untoe(b)) \\
&\qquad\qquad\qquad \triangleleft 0 < size(b) \leq n+1 \text{ and } i < n \triangleright \delta
\end{aligned}
$$

$$
\begin{aligned}
&G_2(n:nat, i:nat, b:queue_d) = \\
&\qquad \tau\, G_0(n,i,b) \triangleleft 0 < size(b) \leq n \text{ and } i < n \triangleright \delta + \\
&\qquad \tau\, G_3(n,i,b) \triangleleft 1 < size(b) \leq n+1 \text{ and } i < n \triangleright \delta
\end{aligned}
$$

$$
\begin{aligned}
&G_3(n:nat, i:nat, b:queue_d) = \\
&\qquad \tau\, G_1(n,i,b) \triangleleft 1 < size(b) \leq n+1 \text{ and } i < n \triangleright \delta + \\
&\qquad out(toe(b))\,G_2(n, i +_n 1, untoe(b)) \\
&\qquad\qquad\qquad \triangleleft 1 < size(b) \leq n+2 \text{ and } i < n \triangleright \delta
\end{aligned}
$$

Now it is tempting to state that the queue $Q(n, b)$ is a solution of $G_0(n, i, b)$. But this can not easily be shown. The most important reason is that Q must perform τ-steps in a rather irregular way in order to be a solution. We model this by defining the following four processes Q_j ($j = 0, 1, 2, 3$). Obviously, $Q_0(n, 0, \varnothing_d)$ is equal to $Q(n + 2)$. $Q_j(n, i, b)$ is also a solution for $G_j(n, i, b)$ from which it follows that $Q_j(n, i, b) = CQ(n, i, b)$. Combination of these facts leads to the correctness of the protocol.

$$
\begin{aligned}
&\textbf{proc } Q_0(n:nat, i:nat, b:queue_d) = \\
&\qquad (Q(n+2,b) \triangleleft empty(b) \triangleright \tau\, Q(n+2,b)) \\
&\qquad \triangleleft size(b) \leq n \text{ and } i < n \triangleright \delta
\end{aligned}
$$

$$
\begin{aligned}
&Q_1(n:nat, i:nat, b:queue_d) = \\
&\qquad Q(n+2,b) \triangleleft 0 < size(b) \leq n+1 \text{ and } i < n \triangleright \delta
\end{aligned}
$$

$$
\begin{aligned}
&Q_2(n:nat, i:nat, b:queue_d) = \\
&\qquad \tau\, Q(n+2,b) \triangleleft 0 < size(b) \leq n+1 \text{ and } i < n \triangleright \delta
\end{aligned}
$$

$$
\begin{aligned}
&Q_3(n:nat, i:nat, b:queue_d) = \\
&\qquad (Q(n+2,b) \triangleleft eq(size(b), n+2) \triangleright \tau\, Q(n+2,b)) \\
&\qquad\qquad\qquad \triangleleft 1 < size(b) \leq n+2 \text{ and } i < n \triangleright \delta
\end{aligned}
$$

Theorem 5.2. *Let $i, n : nat$, $b : queue_d$.*

1. $n > 0 \to B(n) = CQ_0(n, 0, \varnothing_d)$,

2. $n > 0 \rightarrow Q(n+2) = Q_0(n, 0, \varnothing_d)$,

3. $n > 0 \rightarrow CQ_j(n,i,b) = Q_j(n,i,b)$ *for* $j = 0,1,2,3$,

4. $n > 0 \rightarrow Q(n+2) = B(n)$.

Proof.

1. $$\begin{array}{rl} B(n) = & \tau_I\partial_H(INS(n,0) \parallel PBag(n) \parallel OUTS(n,0)) \\ \overset{5.1}{=} & \tau_I\partial_H(INS(n,0) \parallel DBag(n,\varnothing_p) \parallel OUTS(n,0)) \\ = & CQ_0(n,0,\varnothing_d) \end{array}$$

2. Immediate using the definitions of $Q(n{:}nat)$ and $Q_0(n, i{:}nat, b{:}queue_d)$.

3. We show that both $CQ_j(n,i,b)$ and $Q_j(n,i,b)$ are solutions for the equations defining $G_j(n,i,b)$. As $G_j(n,i,b)$ is guarded (see appendix B) this immediately implies that $CQ_j(n,i,b) = Q_j(n,i,b)$ ($j = 0,1,2,3$). First we show that the processes $CQ_j(n,i,b)$ satisfy the equations for G_j and then we do the same for $Q_j(n,i,b)$. In each case the proof consists of a straightforward expansion using Theorem 4.5 and 4.6 and the applications of some lemmas about data given in appendix A.

$$\begin{array}{ll}
\multicolumn{2}{l}{CQ_0(n,i,b)} \\
= & \tau_{\{c_1,c_2\}}(\partial_{\{r_1,s_1,r_2,s_2\}}(\\
& \quad INS(n, i +_n size(b)) \parallel \\
& \quad DBag(n, number(i,n,b)) \parallel OUTS(n,i))) \\
& \triangleleft size(b) \leq n \text{ and } i < n \triangleright \delta \\[2ex]
\overset{4.5}{=} & \sum_{d:D}(enter(d) \\
& \quad \tau_{\{c_1,c_2\}}(\partial_{\{r_1,s_1,r_2,s_2\}}(\\
& \quad\quad s_1(\langle d, i +_n size(b)\rangle)\, INS(n, i +_n size(b) +_n 1) \parallel \\
& \quad\quad DBag(n, number(i,n,b)) \parallel OUTS(n,i)))) \\
& \triangleleft size(b) \leq n \text{ and } i < n \triangleright \delta \\
+ & \tau_{\{c_1,c_2\}}(\partial_{\{r_1,s_1,r_2,s_2\}}(\\
& \quad (\sum_{p:Pair}(s_2(p)\, DBag(n, rem(p, number(i,n,b)))) \\
& \quad\quad\quad \triangleleft test(p, number(i,n,b)) \triangleright \delta) \parallel \\
& \quad \sum_{d:D}(r_2(\langle d,i\rangle)\, out(d))\, OUTS(n, i +_n 1)) \mathbin{\lfloor\!\lfloor} \\
& \quad INS(n, i +_n size(b)))) \\
& \triangleleft size(b) \leq n \text{ and } i < n \triangleright \delta \\[2ex]
\overset{4.6.3}{=} & \sum_{d:D}(enter(d)\, CQ_2(n,i,in(d,b))) \triangleleft size(b) \leq n \text{ and } i < n \triangleright \delta \\
+ & \tau_{\{c_1,c_2\}}(\partial_{\{r_1,s_1,r_2,s_2\}}(\\
& \quad \sum_{d:D}(c_1(\langle d,i\rangle) \\
& \quad\quad (DBag(n, rem(\langle d,i\rangle, number(i,n,b))) \parallel \\
& \quad\quad out(d)\, OUTS(n, i +_n 1)) \\
& \quad\quad \triangleleft test(\langle d,i\rangle, number(i,n,b)) \triangleright \delta)) \mathbin{\lfloor\!\lfloor} \\
& \quad INS(n, i +_n size(b)))) \\
& \triangleleft size(b) \leq n \text{ and } i < n \triangleright \delta
\end{array}$$

$$\stackrel{\text{A.6.1, Sum El.}}{=} \begin{array}{l} \sum_{d:D}(enter(d)\, CQ_2(n,i,in(d,b))) \lhd size(b) \leq n \text{ and } i < n \rhd \delta \\ + \quad \tau_{\{c_1,c_2\}}(\partial_{\{r_1,s_1,r_2,s_2\}}(\\ \quad\quad (c_1(\langle toe(b), i\rangle) \\ \quad\quad\quad (DBag(n, rem(\langle toe(b), i\rangle, number(i,n,b)))) \parallel \\ \quad\quad\quad out(toe(b))\, OUTS(n, i +_n 1)) \,\|\!\!\_ \\ \quad\quad INS(n, i +_n size(b)))) \\ \quad \lhd 0 < size(b) \leq n \text{ and } i < n \rhd \delta \end{array}$$

$$\stackrel{\text{A.6.3,TI2}}{=} \begin{array}{l} \sum_{d:D}(enter(d)\, CQ_2(n,i,in(d,b))) \lhd size(b) \leq n \text{ and } i < n \rhd \delta \\ + \quad \tau \cdot \tau_{\{c_1,c_2\}}(\partial_{\{r_1,s_1,r_2,s_2\}}(\\ \quad\quad INS(n, i +_n size(b)) \parallel \\ \quad\quad DBag(n, number(i +_n 1, n, untoe(b))) \parallel \\ \quad\quad out(toe(b))\, OUTS(n, i +_n 1))) \\ \quad \lhd 0 < size(b) \leq n \text{ and } i < n \rhd \delta \end{array}$$

$$\begin{array}{ll} = & \sum_{d:D}(enter(d)\, CQ_2(n,i,in(d,b))) \lhd size(b) \leq n \text{ and } i < n \rhd \delta \\ + & \tau\, CQ_1(n,i,b) \lhd 0 < size(b) \leq n \text{ and } i < n \rhd \delta \end{array}$$

The calculations showing that $CQ_j(n,i,b)$ satisfy the equation defining G_1 are analogous to the ones given above. They are omitted here but can be found in the full version [GK92].

We continue by showing that the processes $CQ_j(n,i,b)$ also satisfy the equation for G_2.

$$CQ_2(n,i,b)$$

$$= \begin{array}{l} \tau_{\{c_1,c_2\}}(\partial_{\{r_1,s_1,r_2,s_2\}}(\\ \quad s_1(\langle hd(b), i +_n size(tl(b))\rangle)\, INS(n, i +_n size(b)) \parallel \\ \quad DBag(n, number(i,n,tl(b))) \parallel OUTS(n,i))) \\ \lhd 0 < size(b) \leq n+1 \text{ and } i < n \rhd \delta \end{array}$$

$$\stackrel{\text{4.6.1, 4.6.3}}{=} \begin{array}{l} \tau_{\{c_1,c_2\}}(\partial_{\{r_1,s_1,r_2,s_2\}}(\\ \quad (c_1(\langle hd(b), i +_n size(tl(b))\rangle) \\ \quad\quad (INS(n, i +_n size(b)) \parallel \\ \quad\quad DBag(n, in(\langle hd(b), i +_n size(tl(b))\rangle, \\ \quad\quad\quad\quad\quad number(i,n,tl(b))))) \\ \quad \lhd size(b) \leq n \rhd \delta) \,\|\!\!\_\, OUTS(n,i))) \\ \lhd 0 < size(b) \leq n+1 \text{ and } i < n \rhd \delta \\ + \quad \tau_{\{c_1,c_2\}}(\partial_{\{r_1,s_1,r_2,s_2\}}(\\ \quad \sum_{d:D}(c_2(\langle d,i\rangle) \\ \quad\quad (DBag(n, rem(\langle d,i\rangle, number(i,n,tl(b)))) \parallel \\ \quad\quad out(d)\, OUTS(n, i +_n 1)) \\ \quad \lhd test(\langle d,i\rangle, number(i,n,tl(b))) \rhd \delta) \,\|\!\!\_ \\ \quad s_1(\langle hd(b), i +_n size(tl(b))\rangle)\, INS(n, i +_n size(b)))) \\ \lhd 0 < size(b) \leq n+1 \text{ and } i < n \rhd \delta \end{array}$$

$$\stackrel{\text{TI2, A.6.7}}{=} \begin{array}{l} \tau\, CQ_0(n,i,b) \lhd 0 < size(b) \leq n \text{ and } i < n \rhd \delta \\ + \quad \tau_{\{c_1,c_2\}}(\partial_{\{r_1,s_1,r_2,s_2\}}(\end{array}$$

$$\begin{array}{l}\sum_{d:D}(c_2(\langle d,i\rangle)\\ \quad (DBag(n, rem(\langle d,i\rangle, number(i,n,tl(b)))) \parallel\\ \quad\ out(d)\ OUTS(n, i +_n 1))\\ \triangleleft test(\langle d,i\rangle, number(i,n,tl(b))) \triangleright \delta) \lfloor\!\lfloor\\ s_1(\langle hd(b), i +_n size(tl(b))\rangle)\ INS(n, i +_n size(b))))\\ \triangleleft 0 < size(b) \leq n+1 \text{ and } i < n \triangleright \delta\end{array}$$

$$\begin{array}{rl}\stackrel{\text{A.6.2, Sum El., TI2}}{=} & \\ & \tau\, CQ_0(n,i,b) \triangleleft 0 < size(b) \leq n \text{ and } i < n \triangleright \delta\\ + & \tau \cdot \tau_{\{c_1,c_2\}}(\partial_{\{r_1,s_1,r_2,s_2\}}(\\ & \quad s_1(\langle hd(b), i +_n size(tl(b))\rangle)\ INS(n, i +_n size(b)) \parallel\\ & \quad DBag(n, rem(\langle toe(b), i\rangle, number(i,n,tl(b)))) \parallel\\ & \quad out(toe(b))\ OUTS(n, i +_n 1)))\\ & \triangleleft 1 < size(b) \leq n+1 \text{ and } i < n \triangleright \delta\end{array}$$

$$\begin{array}{rl}\stackrel{\text{A.6.8}}{=} & \tau\, CQ_0(n,i,b) \triangleleft 0 < size(b) \leq n \text{ and } i < n \triangleright \delta\\ + & \tau\, CQ_3(n,i,b) \triangleleft 1 < size(b) \leq n+1 \text{ and } i < n \triangleright \delta\end{array}$$

That $CQ_j(n,i,b)$ even satisfy the equation for G_3 can be shown in a similar way as above. These calculations are omitted here but can be found in [GK92].

Now we show that the processes $Q_j(n,i,b)$ are solutions for the equations for $G_0, \ldots, G_3$. It is worth noting that the only place where the τ-laws are used is below.

$$\begin{array}{rl}Q_0(n,i,b) & \\ = & (Q(n+2,b) \triangleleft empty(b) \triangleright \tau\, Q(n+2,b))\\ & \triangleleft size(b) \leq n \text{ and } i < n \triangleright \delta\\ \stackrel{\text{T2}}{=} & (\sum_{d:D}(enter(d)\, Q(n+2, in(d,b))) \triangleleft empty(b) \triangleright\\ & \quad \sum_{d:D}(enter(d)\, Q(n+2, in(d,b)) + \tau\, Q(n+2,b)))\\ & \triangleleft size(b) \leq n \text{ and } i < n \triangleright \delta\\ = & (\sum_{d:D}(enter(d)\, Q(n+2, in(d,b)) \triangleleft size(b) \leq n \triangleright \delta) +\\ & \tau\, Q_1(n,i,b) \triangleleft size(b) > 0 \triangleright \delta) \triangleleft empty(b) \triangleright\\ & \quad (\sum_{d:D}(enter(d)\, Q(n+2, in(d,b)) \triangleleft size(b) \leq n \triangleright \delta) +\\ & \tau\, Q_1(n,i,b) \triangleleft size(b) > 0 \triangleright \delta) \triangleleft size(b) \leq n \text{ and } i < n \triangleright \delta\\ \stackrel{\text{4.2.1, T1}}{=} & \sum_{d:D}(enter(d)\, Q_2(n,i,in(d,b))) \triangleleft size(b) \leq n \text{ and } i < n \triangleright \delta\\ + & \tau Q_1(n,i,b) \triangleleft 0 < size(b) \leq n \text{ and } i < n \triangleright \delta\end{array}$$

The processes $Q_j(n,i,b)$ are also a solution of the defining equation for G_1.

$$\begin{array}{rl}Q_1(n,i,b) & \\ = & Q(n+2,b) \triangleleft 0 < size(b) \leq n+1 \text{ and } i < n \triangleright \delta\\ = & \sum_{d:D}(enter(d)\, Q(n+2, in(d,b)))\\ & \qquad\qquad \triangleleft 0 < size(b) \leq n+1 \text{ and } i < n \triangleright \delta\\ + & out(toe(b))\, Q(n+2, untoe(b)) \triangleleft 0 < size(b) \leq n \text{ and } i < n \triangleright \delta\\ \stackrel{\text{T1}}{=} & \sum_{d:D}(enter(d)\, Q_3(n,i,in(d,b)))\\ & \qquad\qquad \triangleleft 0 < size(b) \leq n+1 \text{ and } i < n \triangleright \delta\end{array}$$

$$+ \quad out(toe(b))\, Q_0(n, i +_n 1, untoe(b)) \\ \lhd 0 < size(b) \leq n+1 \text{ and } i < n \rhd \delta$$

Note that we need that $n > 0$ in the second step below to show that the processes $Q_j(n, i, b)$ are a solution for the equation for G_2.

$$\begin{array}{ll} & Q_2(n,i,b) \\ = & \tau\, Q(n+2,b) \lhd 0 < size(b) \leq n+1 \text{ and } i < n \rhd \delta \\ = & \tau\, Q(n+2,b) \lhd 0 < size(b) \leq n \text{ and } i < n \rhd \delta \\ + & \tau\, Q(n+2,b) \lhd 1 < size(b) \leq n+1 \text{ and } i < n \rhd \delta \\ \overset{\text{T1}}{=} & \tau\, Q_0(n,i,b) \lhd 0 < size(b) \leq n \text{ and } i < n \rhd \delta \\ + & \tau\, Q_3(n,i,b) \lhd 1 < size(b) \leq n+1 \text{ and } i < n \rhd \delta \end{array}$$

And last, we show that $CQ_j(n, i, b)$ also satisfies the equation for Q_3.

$$\begin{array}{ll} & Q_3(n,i,b) \\ = & (Q(n+2,b) \lhd eq(size(b), n+2) \rhd \tau\, Q(n+2,b)) \\ & \lhd 1 < size(b) \leq n+2 \text{ and } i < n \rhd \delta \\ \overset{\text{T2}}{=} & (out(toe(b))\, Q(n+2, untoe(b)) \lhd eq(size(b), n+2) \rhd \\ & \qquad (\tau\, Q(n+2,b) + out(toe(b))\, Q(n+2, untoe(b)))) \\ & \lhd 1 < size(b) \leq n+2 \text{ and } i < n \rhd \delta \\ \overset{4.2.4,\, A.3.5,\, A.1.1}{=} & \\ & \tau\, Q(n+2,b) \lhd 1 < size(b) < n+2 \text{ and } i < n \rhd \delta \\ + & out(toe(b))\, Q(n+2, untoe(b)) \\ & \qquad \lhd 1 < size(b) \leq n+2 \text{ and } i < n \rhd \delta \\ \overset{\text{T1}}{=} & \tau\, Q_1(n,i,b) \lhd 1 < size(b) < n+2 \text{ and } i < n \rhd \delta \\ + & out(toe(b))\, Q_2(n, i +_n 1, untoe(b)) \\ & \qquad \lhd 1 < size(b) \leq n+2 \text{ and } i < n \rhd \delta \end{array}$$

4. Now, given the calculations above, this proof is straightforward:

$$Q(n+2) \overset{5.2.2}{=} Q_0(n, 0, \varnothing_d) \overset{5.2.3}{=} CQ_0(n, 0, \varnothing_d) \overset{5.2.1}{=} B(n).$$

□

Acknowledgements

We would like to thank Jos Baeten, Jan Bergstra, Wan Fokkink, Gert-Jan Kamsteeg, Alban Ponse and Frits Vaandrager for their comments on earlier drafts of this paper. In particular, we are very grateful to Jos van Wamel for spotting dozens of bugs in, and helpful comments on our verification of the Bakery Protocol.

A Elementary data types

In this appendix, we specify the data types that are used in the specification and in proof of the Bakery Protocol. Furthermore, the properties of the data types needed for the verification of the Bakery are presented as data laws together with their proofs.

A.1 About booleans

The predefined booleans extended with their well-known connectives *not, and, or*[2] are often used in the Bakery proof. The (rewrite) rules for the added connectives are consistent with the predefined rules BOOL1 and BOOL2.

$$
\begin{array}{ll}
\textbf{sort} & \textbf{Bool} \\
\textbf{func} & \mathsf{t}, \mathsf{f} :\to \textbf{Bool} \\
 & \mathrm{not} : \textbf{Bool} \to \textbf{Bool} \\
 & \mathrm{and} : \textbf{Bool} \times \textbf{Bool} \to \textbf{Bool} \\
 & \mathrm{or} : \textbf{Bool} \times \textbf{Bool} \to \textbf{Bool} \\
\textbf{var} & b, b_1, b_2, b_3 : \textbf{Bool} \\
\textbf{rew} & \mathrm{not}(\mathsf{t}) = \mathsf{f} \\
 & \mathrm{not}(\mathsf{f}) = \mathsf{t} \\
 & \mathsf{t} \text{ and } b = b \\
 & \mathsf{f} \text{ and } b = \mathsf{f} \\
 & \mathsf{t} \text{ or } b = \mathsf{t} \\
 & \mathsf{f} \text{ or } b = b
\end{array}
$$

Lemma A.1.

1. $p \triangleleft b \triangleright q = q \triangleleft \mathrm{not}(b) \triangleright p$,
2. $p \triangleleft b_1 \text{ or } b_2 \triangleright \delta = p \triangleleft b_1 \triangleright \delta + p \triangleleft b_2 \triangleright \delta$,
3. $b_1 \text{ and } b_1 = b_1$,
4. $b_1 \text{ or } b_1 = b_1$,
5. $b_1 \text{ and } b_2 = b_2 \text{ and } b_1$.

Proof. By COND1, COND2 and Lemma 4.1. □

A.2 About natural numbers

The natural numbers represented by sort *nat* play an important role in both the specification and the verification of the Bakery Protocol. Below the operators on natural numbers $0, P$ (Predecessor), $S, +, -$ (monus), $\geq, \leq, <, >, \mathit{if}, \mathit{eq}$ used in this paper are specified. (We will use infix notation wherever we find it convenient to do so.)

$$
\begin{array}{ll}
\textbf{sort} & \mathit{nat} \\
\textbf{func} & 0 :\to \mathit{nat} \\
 & S, P : \mathit{nat} \to \mathit{nat} \\
 & +, -, : \mathit{nat} \times \mathit{nat} \to \mathit{nat} \\
 & \mathit{eq}, \geq, \leq, <, > : \mathit{nat} \times \mathit{nat} \to \textbf{Bool} \\
 & \mathit{if} : \textbf{Bool} \times \mathit{nat} \times \mathit{nat} \to \mathit{nat} \\
\textbf{var} & n, m, z : \mathit{nat} \\
\textbf{rew} & P(0) = 0 \\
 & P(S(n)) = n \\
 & n + 0 = n \\
 & n + S(m) = S(n + m)
\end{array}
$$

[2]The symbols $\neg, \vee, \wedge, \rightarrow, \leftrightarrow$ are reserved in the proof system of μCRL as operators for connecting properties.

$$n - 0 = n$$
$$n - S(m) = P(n - m)$$
$$eq(n, n) = \mathsf{t}$$
$$if(eq(n, m), n, m) = m$$
$$n \geq 0 = \mathsf{t}$$
$$0 \geq S(n) = \mathsf{f}$$
$$S(n) \geq S(m) = n \geq m$$
$$n \leq m = m \geq n$$
$$n > m = n \geq S(m)$$
$$n < m = S(n) \leq m$$
$$if(\mathsf{t}, n, m) = n$$
$$if(\mathsf{f}, n, m) = m$$

The *if* function given above will be used for specifying modulo arithmetic in section A.3.

Notation A.2. We write $n \leq m$ for $n \leq m = \mathsf{t}$. Idem for $\geq$,$>$ and $<$. We write $eq(n, m)$ for $eq(n, m) = \mathsf{t}$. We write 1 for $S(0)$ and 2 for $S(S(0))$.

Lemma A.3.

1. $eq(n, m) = \mathsf{t} \leftrightarrow n = m$,
2. $n = m \vee \neg(n = m)$,
3. $n + (m + z) = (n + m) + z$,
4. $n + m = m + n$,
5. $n \geq m = not(n < m)$,
6. *etc.*

Proof. By (nested) induction on *nat* (see [TD88]). □

A.3 About modulo arithmetic

On top of the natural numbers *nat* we specify the *mod* operator and the $+_m$ operator (addition modulo m) as follows.

func $\mathrm{mod} : nat \times nat \to nat$
$+ : nat \times nat \times nat \to nat$
var $n, n', m : nat$
rew $n \bmod 0 = n$
$n \bmod m = if(n \geq m, (n - m) \bmod m, n)$
$n +_m n' = (n + n') \bmod m$

Lemma A.4.

1. $n_1 +_m n_2 = n_2 +_m n_1$,
2. $(n_1 +_m n_2) +_m n_3 = n_1 +_m (n_2 +_m n_3)$,
3. $n_2 > 0 \to (n_1 \bmod n_2) < n_2$,
4. $(n_1 \bmod n_2) \bmod n_2 = n_1 \bmod n_2$.

A.4 About data queues

In this section, we specify the data type $queue_d$ which can contain elements of a set D of customers. D may be finite or infinite, but it should at least contain a bottom element $d_\perp$ for denoting an undefined data element. We assume that D is equipped with an equality function $eq : D \times D \to \mathbf{Bool}$ that has the desired property $eq(d,e) = \mathrm{t} \leftrightarrow d = e$.

$$
\begin{array}{ll}
\textbf{sort} & D \\
\textbf{func} & d_\perp : D \\
 & d_1, \ldots, d_n : D \\
 & eq : D \times D \to \mathbf{Bool}
\end{array}
$$

The specification of $queue_d$ is given below.

$$
\begin{array}{ll}
\textbf{sort} & queue_d \\
\textbf{func} & \varnothing_d : \to queue_d \\
 & in, rem : D \times queue_d \to queue_d \\
 & test : D \times queue_d \to \mathbf{Bool} \\
 & size : queue_d \to nat \\
 & hd, toe : queue_d \to D \\
 & tl, untoe : queue_d \to queue_d \\
 & if : \mathbf{Bool} \times queue_d \times queue_d \to queue_d \\
 & empty : queue_d \to \mathbf{Bool} \\
\textbf{var} & d, e : D \\
 & b, c : queue_d \\
\textbf{rew} & test(d, \varnothing_d) = \mathrm{f} \\
 & test(d, in(e,b)) = if(eq(d,e), \mathrm{t}, test(d,b)) \\
 & rem(d, \varnothing_d) = \varnothing_d \\
 & rem(d, in(e,b)) = if(eq(d,e), b, in(e, rem(d,b))) \\
 & size(\varnothing_d) = 0 \\
 & size(in(d,b)) = S(size(b)) \\
 & hd(\varnothing_d) = d_\perp \\
 & hd(in(d,b)) = d \\
 & toe(\varnothing_d) = d_\perp \\
 & toe(in(d, \varnothing_d)) = d \\
 & toe(in(d, in(e,b))) = toe(in(e,b)) \\
 & tl(\varnothing_d) = \varnothing_d \\
 & tl(in(d,b)) = b \\
 & untoe(\varnothing_d) = \varnothing_d \\
 & untoe(in(d, \varnothing_d)) = \varnothing_d \\
 & untoe(in(d, in(e,b))) = in(d, untoe(in(e,b))) \\
 & empty(b) = eq(size(b), 0)
\end{array}
$$

Lemma A.5.

1. $size(untoe(b)) = size(b) - 1$,
2. $size(tl(b)) = size(b) - 1$,
3. $\neg b = \varnothing_d \to in(hd(b), tl(b)) = b$,
4. $size(b) \leq 0 \leftrightarrow b = \varnothing_d$,

5. $\neg b = \varnothing_d \to test(d, tl(b))$ or $eq(d, hd(b)) = test(d, b)$,
6. $rem(hd(b), b) = tl(b)$,
7. $size(b) \leq 0 \leftrightarrow b = \varnothing_p$,
8. $b = \varnothing_p \to size(b) \leq n$.

The function *number* is specified as follows.

sort	$queue_p$
func	$\varnothing_p :\to queue_p$
	$number : nat \times nat \times queue_d \to queue_p$
rew	$number(i, n, \varnothing_d) = \varnothing_p$
	$number(i, n, in(d, b)) = in(\langle d, i +_n size(b)\rangle, number(i, n, b))$

Here $queue_p$ is exactly the same data type as $queue_d$ except that it contains elements of sort *Pair* instead of sort D.

Lemma A.6.

1. $size(b) \leq n \to$
   $test(\langle d, i\rangle, number(i, n, b)) = (eq(d, toe(b))$ and $size(b) > 0)$,
2. $size(b) \leq n \to$
   $test(\langle d, i\rangle, number(i, n, b)) = (size(b) \geq 1$ and $eq(d, toe(b)))$,
3. $size(b) \leq n \to$
   $rem(\langle toe(b), i\rangle, number(i, n, b)) = number(i +_n 1, n, untoe(b))$,
4. $size(b) > 0 \to size(tl(b)) < n = size(b) \leq n$,
5. $size(b) > 0 \to$
   $size(number(i +_n 1, n, tl(untoe(b)))) < n = size(b) < n + 2$,
6. $size(number(i, n, b)) = size(b)$,
7. $size(b) \leq n \to$
   $in(\langle hd(b), i +_n size(tl(b))\rangle, number(i, n, tl(b))) = number(i, n, b)$,
8. $1 < size(b) \leq n \to$
   $rm(\langle toe(b), i\rangle, number(i, n, tl(b))) = number(i +_n 1, n, tl(untoe(b)))$.

B An overview of the proof theory for μCRL

B.1 The proof system

In [GP94a] a proof system has been given which allows to prove identities about processes with data. Table 1 lists the axioms of ACP in μCRL, followed by the axioms of Standard Concurrency in Table 2 and the axioms for hiding in Table 3. For an explanation of these axioms we refer to [BW90], except for the following points. In the tables x, y are process variables and p, q are process terms in which the variable d may occur. The letters $t_1, \ldots, t_n$ stand for data terms, and $\overline{t}$ for a sequence of data terms where ϵ is the empty sequence. The symbols a, b represent δ, τ or range over (declared) actions $n(\overline{t})$, where $n(\overline{t})$ represents n if $\overline{t} = \epsilon$. $\tilde{\gamma}$ is the pre-communication function such that $\tilde{\gamma}(n_1, n_2) = n_3$ if a rule **comm** $n_1 \,|\, n_2 = n_3$ appears in the μCRL specification.

A1	$x + y = y + x$	CF	$n(\bar{t}) \mid m(\bar{t}) = \begin{cases} \gamma(n,m)(\bar{t}) & \text{if } \gamma(n,m)\downarrow \\ \delta & \text{otherwise} \end{cases}$
A2	$x + (y + z) = (x + y) + z$		
A3	$x + x = x$		
A4	$(x + y) \cdot z = x \cdot z + y \cdot z$		
A5	$(x \cdot y) \cdot z = x \cdot (y \cdot z)$		
A6	$x + \delta = x$		
A7	$\delta \cdot x = \delta$	CD1	$\delta \mid x = \delta$
		CD2	$x \mid \delta = \delta$
CM1	$x \parallel y = x \mathbin{⌊\!⌊} y + y \mathbin{⌊\!⌊} x + x \mid y$	CT1	$\tau \mid x = \delta$
CM2	$a \mathbin{⌊\!⌊} x = a \cdot x$	CT2	$x \mid \tau = \delta$
CM3	$a \cdot x \mathbin{⌊\!⌊} y = a \cdot (x \parallel y)$		
CM4	$(x + y) \mathbin{⌊\!⌊} z = x \mathbin{⌊\!⌊} z + y \mathbin{⌊\!⌊} z$	DD	$\partial_H(\delta) = \delta$
CM5	$a \cdot x \mid b = (a \mid b) \cdot x$	DT	$\partial_H(\tau) = \tau$
CM6	$a \mid b \cdot x = (a \mid b) \cdot x$	D1	$\partial_H(n(\bar{t})) = n(\bar{t})$ if $n \notin H$
CM7	$a \cdot x \mid b \cdot y = (a \mid b) \cdot (x \parallel y)$	D2	$\partial_H(n(\bar{t})) = \delta$ if $n \in H$
CM8	$(x + y) \mid z = x \mid z + y \mid z$	D3	$\partial_H(x + y) = \partial_H(x) + \partial_H(y)$
CM9	$x \mid (y + z) = x \mid y + x \mid z$	D4	$\partial_H(x \cdot y) = \partial_H(x) \cdot \partial_H(y)$

Table 1: The axioms of ACP in μCRL.

Otherwise $\tilde{\gamma}(n_1, n_2) = \delta$. γ is the symmetrical closure of $\tilde{\gamma}$. We write $\gamma(n, m)\downarrow$ if γ is defined on n and m.

Tables 4, 5 and 6 lists the typical μCRL axioms and rules for interaction between data and processes. The axioms for summation are denoted by SUM, the axioms for the conditional by COND and the rules for the booleans by BOOL.

Beside the axioms and rules mentioned above, μCRL incorporates two other important proof principles. First, it supports an principle for induction not only on data but also on data in processes. The second principle is RSP (Recursive Specification Principle) taken from [BW90] extended to processes with data. Informally, it says that each guarded recursive specification has at most one solution.

$(x \mathbin{⌊\!⌊} y) \mathbin{⌊\!⌊} z = x \mathbin{⌊\!⌊} (y \parallel z)$	$(x \mid y) \mid z = x \mid (y \mid z)$
$x \parallel \delta = x\delta$	$x \mid (y \mathbin{⌊\!⌊} z) = (x \mid y) \mathbin{⌊\!⌊} z$
$x \mid y = y \mid x$	$x \mid (y \mid z) = \delta$ Handshaking

Table 2: Axioms of Standard Concurrency (SC).

B.2 Adding τ-laws to the proof system

The proof system as presented above is considered as a kernel and does not yet contain axioms for τ. In this section, we extend the proof theory with axioms

TID	$\tau_I(\delta) = \delta$	
TIT	$\tau_I(\tau) = \tau$	
TI1	$\tau_I(n(\bar{t})) = n(\bar{t})$	if $n \notin I$
TI2	$\tau_I(n(\bar{t})) = \tau$	if $n \in I$
TI3	$\tau_I(x + y) = \tau_I(x) + \tau_I(y)$	
TI4	$\tau_I(x \cdot y) = \tau_I(x) \cdot \tau_I(y)$	

Table 3: Axioms for abstraction.

SUM1	$\sum_{d:D}(p) = p$	if d not free in p
SUM2	$\sum_{d:D}(p) = \sum_{e:D}(p[e/d])$	if e not free in p
SUM3	$\sum_{d:D}(p) = \sum_{d:D}(p) + p$	
SUM4	$\sum_{d:D}(p_1 + p_2) = \sum_{d:D}(p_1) + \sum_{d:D}(p_2)$	
SUM5	$\sum_{d:D}(p_1 \cdot p_2) = \sum_{d:D}(p_1) \cdot p_2$	if d not free in p_2
SUM6	$\sum_{d:D}(p_1 \mathbin{\lfloor\!\lfloor} p_2) = \sum_{d:D}(p_1) \mathbin{\lfloor\!\lfloor} p_2$	if d not free in p_2
SUM7	$\sum_{d:D}(p_1 \mid p_2) = \sum_{d:D}(p_1) \mid p_2$	if d not free in p_2
SUM8	$\sum_{d:D}(\partial_H(p)) = \partial_H(\sum_{d:D}(p))$	
SUM9	$\sum_{d:D}(\tau_I(p)) = \tau_I(\sum_{d:D}(p))$	
SUM11	$\dfrac{\begin{matrix}\mathcal{D}\\ p_1 = p_2\end{matrix}}{\sum_{d:D}(p_1) = \sum_{d:D}(p_2)}$	provided d not free in the assumptions of $\mathcal{D}$

Table 4: Axioms for summation.

COND1	$x \lhd \mathsf{t} \rhd y$	$=$	x
COND2	$x \lhd \mathsf{f} \rhd y$	$=$	y

Table 5: Axioms for the conditional construct.

BOOL1	$\neg(\mathsf{t} = \mathsf{f})$
BOOL2	$\neg(b = \mathsf{t}) \rightarrow b = \mathsf{f}$

Table 6: Axioms for **Bool**.

for τ as we need these axioms in the verification of the Bakery Protocol. One can add the τ-laws of Table 7 taken from MILNER [Mil89] to the proof system. These axioms correspond to the well-known *observation equivalence*. These

T1	$x\,\tau$	$=$	x
T2	$\tau\,x$	$=$	$\tau\,x + x$
T3	$a\,(\tau\,x + y)$	$=$	$a\,(\tau\,x + y) + a\,x$

Table 7: τ-laws for observation equivalence.

axioms can be added to the proof system under the restriction that the a and b in the ACP axioms of μCRL do not range over τ. Otherwise, we are able to derive inconsistent identities (see [BW90], page 165). The axioms in Table 8 model the interaction between τ and the other operators (see [BW90]).

TM1	$\tau \mathbin{⌊\!⌊} x$	$=$	τx
TM2	$\tau x \mathbin{⌊\!⌊} y$	$=$	$\tau(x \parallel y)$
TC3	$\tau x \mid y$	$=$	$x \mid y$
TC4	$x \mid \tau y$	$=$	$x \mid y$

Table 8: Completing τ-laws for observation equivalence.

Another point is that the axiom $(x \mid y) \mathbin{⌊\!⌊} z = x \mid (y \mathbin{⌊\!⌊} z)$ of standard concurrency is not consistent in the context of the τ-laws given in Table 7. It must be replaced by the weaker axiom $(x \mid ay) \mathbin{⌊\!⌊} z = x \mid (ay \mathbin{⌊\!⌊} z)$.

The RSP rule mentioned above is restricted to guarded systems of process-equations. A definition for guardedness that works in a setting of Milner's observation equivalence can be given as follows:

Definition B.1 (*Guardedness of G*). A term p is a *guard* iff:

- $p \equiv \delta$,
- $p \equiv n(t_1, \ldots, t_n)$ or $p \equiv n$ and n is an action label,
- $p \equiv q_1 \circ q_2$ with $\circ \in \{+, \lhd t \rhd\}$ and q_1 and q_2 are guards,
- $p \equiv q_1 \circ q_2$ with $\circ \in \{\cdot, \parallel, \mathbin{⌊\!⌊}\}$ and q_1 or q_2 are guards,
- $p \equiv q_1 \mid q_2$,
- $p \equiv \sum_{x:S}(q_1)$ and q_1 is a guard,
- $p \equiv \partial_{nl}(q_1)$ with nl being a list of names, and q_1 is a guard.

Let G be a system of process-equations and let N be the left-hand side of one of the equations of G. We say that N is *guarded* in r, where r is a subterm of one of the right-hand sides of G, iff

- $r \equiv q_1 \circ q_2$ with $\circ \in \{+, \parallel, \mid, \lhd t \rhd\}$, and N is guarded in q_1 and q_2,

- $r \equiv q_1 \circ q_2$ with $\circ \in \{\cdot, \|\!\lfloor\}$ and N is guarded in q_1, and q_1 is a *guard* or N is guarded in q_2,
- $r \equiv \sum_{x:S}(q_1)$ and N is guarded in q_1,
- $r \equiv \partial_{nl}(q_1)$ with nl being a list of names, and N is guarded in q_1,
- $r \equiv \delta$ or $r \equiv \tau$,
- $r \equiv n'$ for a *name* n' and $N \not\equiv n'$,
- $r \equiv n'(u_1, \ldots, u_{m'})$ and $N \not\equiv n'(x_{i1}, \ldots, x_{im_i})$.

If N is not guarded in r we say that N appears *unguarded* in r.

The *Identifier Dependency Graph* of G, notation *IDG*(G), is constructed as follows:

- each left-hand side of the equations of G is a node,
- if N is a node of *IDG*(G) and $N = r \in G$, then there is an edge $N \rightarrow N'$ for any node N' that appears unguarded in r.

We call G *guarded* iff

- *IDG*(G) is well founded, i.e. does not contain an infinite path, and
- none of the right-hand sides of G has a subterm of the form $\tau_{nl}(q)$, where nl is a list of names.

□

References

[AGM94] D.J. Andrews, J.F. Groote, and C.A. Middelburg, editors. *Proceedings of the International Workshop on Semantics of Specification Languages.* Workshops in Computing, Springer-Verlag, 1994.

[Apt81] K.R. Apt. Ten years of Hoare's logic, a survey, part I. *ACM Transactions on Programming Languages and Systems*, 3(4):431–483, 1981.

[Apt84] K.R. Apt. Ten years of Hoare's logic, a survey, part II: Nondeterminism. *Theoretical Computer Science*, 28:83–109, 1984.

[BW90] J.C.M. Baeten and W.P. Weijland. *Process Algebra.* Cambridge Tracts in Theoretical Computer Science 18. Cambridge University Press, 1990.

[BG93a] M.A. Bezem and J.F. Groote. A correctness proof of a one bit sliding window protocol in μCRL. Technical Report 99, Logic Group Preprint Series, Utrecht University, 1993. To appear in the Computer Journal, Volume 37(4), 1994.

[BG93b] M.A. Bezem and J.F. Groote. A formal verification of the alternating bit protocol in the calculus of constructions. Technical Report 88, Logic Group Preprint Series, Utrecht University, March 1993.

[CM88] K.M. Chandy and J. Misra. *Parallel Program Design. A Foundation.* Addison-Wesley, 1988.

[EM85] H. Ehrig and B. Mahr. *Fundamentals of algebraic specifications I*, volume 6 of *EATCS Monographs on Theoretical Computer Science.* Springer-Verlag, 1985.

[GK92] J.F. Groote and H. Korver. A correctness proof of the bakery protocol in μCRL. Logic Group Preprint Series 80, Dept. of Philosophy, Utrecht University, October 1992.

[GP93] J.F. Groote and J.C. van de Pol. A bounded retransmission protocol for large data packets. A case study in computer checked verification. Technical Report 100, Logic Group Preprint Series, Utrecht University, 1993.

[GP94a] J.F. Groote and A. Ponse. Proof theory for μCRL: a language for processes with data. In D.J. Andrews, e.a. [AGM94], pages 232–251. Full version is available as CWI Report CS-R9138, Amsterdam, The Netherlands, August 1991.

[GP94b] J.F. Groote and A. Ponse. The syntax and semantics of μCRL. In this volume: A. Ponse, C Verhoef and S.F.M. van Vlijmen, editors. *Proceedings of ACP94.* Workshops in Computing, Springer-Verlag, 1994.

[GW89] R.J. van Glabbeek and W.P. Weijland. Branching time and abstraction in bisimulation semantics (extended abstract). In G.X. Ritter, editor, *Information Processing 89*, pages 613–618. North-Holland, 1989. Full version available as Report CS-R9120, CWI, Amsterdam, 1991.

[Kor94] H. Korver. *Protocol Verification in μCRL.* PhD thesis, University of Amsterdam, 1994.

[KS94] H. Korver and J. Springintveld. A computer-checked verification of Milner's scheduler. In M. Hagiya and J.C. Mitchel, editors. *Proceedings of the 2nd International Symposium on Theoretical Aspects of Computer Software, TACS '94*, Sendai, Japan, pages 161–178, volume 789 of *Lecture Notes in Computer Science.* Springer-Verlag, 1994.

[Mil89] R. Milner. *Communication and Concurrency.* Prentice-Hall International, Englewood Cliffs, 1989.

[Sel93] M.P.A. Sellink. Verifying process algebra proofs in type theory. In Andrews et al. [AGM94], pages 315–339.

[TD88] A.S. Troelstra and D. van Dalen. *Constructivism in Mathematics, An Introduction (vol I).* North-Holland, 1988.

Inductive Proofs with Sets, and some Applications in Process Algebra

Jos van Wamel *
Department of Mathematics and Computer Science
University of Amsterdam
The Netherlands

Abstract

We formalise proofs by structural induction on sets, specified as an algebraic data type. The core idea is that the standard scheme for constructor induction can be used for the derivation of alternative schemes. This way, considerable flexibility can be obtained for proofs in the inductive theory for the sets. For more general purposes we formulate a rule for 'hybrid' induction, which allows a variety of induction schemes.

A number of examples is provided, most of which contain well-known and intuitively clear facts about sets that are nevertheless not always easy to prove. We also use our material on sets for proving some interesting properties of two special processes that have sets as a parameter, specified in μCRL style, representing generalised alternative and parallel compositions of processes. We moreover demonstrate how our generalised parallel composition can be used for a specification of broadcasting.

1 Introduction

The aim of this paper is to formalise proofs by structural induction on sets, specified as an algebraic data type. We hope to demonstrate that 'everything you always wanted to know about sets' can be be proved in an elegant, straightforward manner, once the right induction schemes are available. Another thing we wish to establish is that process algebras that use algebraic data types can benefit from knowledge that primarily concerns data. The specific formalism we use throughout this paper is the process specification language μCRL, and its proof theory [6, 7]. For studying sets as a data type, however, we could have chosen any other formalism based on algebraic specification theory as well.

Data types in μCRL are algebraically specified in the standard way, using sorts, functions and axioms (see e.g. [11, 14]). Process specification in μCRL is based on the process algebra ACP (Algebra of Communicating Processes, see e.g. [2, 3]), which is particularly suited for the formal description of concurrent, communicating systems.

The axiomatic style of specifying abstract data types makes it possible to derive properties of data with the inference rules of equational logic, and moreover term rewriting techniques can be applied. The proof theory of μCRL can

*Partly supported by the Foundation for Computer Science in the Netherlands (SION) and by the Netherlands Organisation for Scientific Research (NWO).

be regarded as a logical framework, defined over a language of *property formulas*, which contains identities between data and identities between processes, linked with the propositional connectives $\wedge, \vee, \neg, \rightarrow, \leftrightarrow$. The proof system also supports proofs by *constructor induction*, which is a well-known technique for proving properties of data by structural induction on terms, with an induction rule.

As a result of this purely algebraic character, verifications in μCRL lend themselves very well for automated proof checking. As examples of such work we mention [1, 10]. For a brief overview of μCRL and its proof theory, and a survey of verifications of concurrent systems in it we refer to the paper [8]. The derivation and use of alternative induction schemes as we use it originates from [9], where a large collection of standard data types and structures was investigated. This approach was in its turn inspired by [13], where a scheme for *double induction* on natural numbers was derived.

Preliminaries. As to the property language of μCRL, we use the convention that $=$ binds stronger than any of the logical operators, that $\neg$ binds stronger than any of the binary logical operators, and that $\wedge, \vee$ bind stronger than $\rightarrow, \leftrightarrow$. Variables are implicitly universally quantified. For readability, we do not always strictly adhere to the syntax of μCRL.

In this paper we assume an algebraic specification of a sort **Bool**, with constants t and f, satisfying $\neg(\mathsf{t} = \mathsf{f})$. We also assume a unary function *not* and binary functions *and* and *or* with obvious meanings. Also a specification of a sort *Nat* is assumed, which consists of a constant 0, a unary function S (successor), and addition $+$. We often write $x := t$ for the substitution of term t for variable x.

As a consequence of the induction rule in μCRL, two very useful and common induction schemes can be formulated for these sorts. Let ϕ, ψ be property formulas, b : **Bool**, and m, n : *Nat*, then

(Case Distinction)	$\phi(\mathsf{t}) \wedge \phi(\mathsf{f}) \rightarrow \phi(b)$,
(Standard Induction)	$\psi(0) \wedge (\psi(m) \rightarrow \psi(Sm)) \rightarrow \psi(n)$.

Both induction schemes are used in this paper, which is further organised as follows. In Section 2 an algebraic specification of sets is given. There we formulate the 'standard induction scheme' for sets, and we derive a number of induction schemes that can make life a lot easier. In Section 3 a rule for 'hybrid' induction on sets is provided, and in Section 4 we demonstrate how sets interact with processes by means of an example with generalised sums and merges. We also give an example which shows that a broadcast process can be modelled with a generalised merge.

Acknowledgements. Thanks to Jan Bergstra, Willem Jan Fokkink, Henri Korver and Alban Ponse for helpful comments. Special thanks go also to Henri for drawing my attention to the generalised merge, which was used in the specification and verification of Milner's famous scheduler [12]. In his paper [10] he had a hard job handling this special process operator.

2 Algebraic specification of sets

Sets with elements of some arbitrary sort D are specified in a standard fashion. [1] We do not need any specific assumptions on D, except that there has to be a function $eq : D \times D \to \mathbf{Bool}$ satisfying $d = e \leftrightarrow eq(d,e) = \mathsf{t}$ for $d, e : D$. The empty set is represented by the constant s_0, and, not in accordance with the definition of μCRL, we use infix operators $\cup$, $\cap$, and $\setminus$ for representing the standard functions on sets.

sort	Set	
cons	$s_0 :\to Set$	
	$in : D \times Set \to Set$	
func	$rem : D \times Set \to Set$	
	$\cup, \cap, \setminus : Set \times Set \to Set$	
	$if : \mathbf{Bool} \times Set \times Set \to Set$	
	$test : D \times Set \to \mathbf{Bool}$	
var	$d, e : D$	
	$s, t : Set$	
rew	$if(\mathsf{t}, s, t) = s$	Con1
	$if(\mathsf{f}, s, t) = t$	Con2
	$test(d, s_0) = \mathsf{f}$	Test1
	$test(d, in(e, s)) = or(eq(d,e), test(d,s))$	Test2
	$in(d, in(d, s)) = in(d, s)$	Sets1
	$in(d, in(e, s)) = in(e, in(d, s))$	Sets2
	$rem(d, s_0) = s_0$	Sets3
	$rem(d, in(e, s)) = if(eq(d,e), rem(d,s), in(e, rem(d,s)))$	Sets4
	$s_0 \cup s = s$	Sets5
	$in(d, s) \cup t = in(d, s \cup t)$	Sets6
	$s_0 \cap s = s_0$	Sets7
	$in(d, s) \cap t = if(test(d,t), in(d, s \cap t), s \cap t)$	Sets8
	$s \setminus s_0 = s$	Sets9
	$s \setminus in(d, t) = rem(d, s \setminus t)$	Sets10

The characteristic properties of sets are captured by axiom Sets1, which implies that every element in a set occurs only once, and axiom Sets2, according to which there is no specific order in which the elements in a set are arranged. Typical consequences of these axioms are respectively idempotency, and commutativity of the operators $\cup$ and $\cap$.

The standard scheme for constructor induction on sets is given below.

Definition 1 (Standard Set Induction). A proof of a property formula $\phi(s)$, by standard set induction on $s : Set$, is the deduction

$$\frac{\phi(s_0) \quad \phi(t) \to \phi(in(d,t))}{\phi(s)}$$

where $d : D$ and $t : Set$.

[1] The presentation of the axioms, and therefore the naming, differs from [9].

Proposition 2. *Proofs by standard set induction are correct.*

Proof. A proof that anything derived with the above scheme is valid in the inductive theory, i.e., for all closed terms, is standard. First it has to be shown that the standard scheme follows from the induction rule of μCRL, parameterised with the set $\{s_0 :\to Set, in : D \times Set \to Set\}$, and then it has to be proved from the above specification that every closed term of sort *Set* can be represented as either s_0 or $in(e_0, u_0)$ for some $e_0 : D$ and $u_0 : Set$, i.e., that the section **cons** indeed specifies a *constructor set*. Finally the soundness of the above scheme follows from the soundness of the induction rule. □

We start off with some facts about the most basic operations on sets. These property formulas are used later on in proofs.

Proposition 3 (Simple Set Identities). *For all $s, t : Set$ and $d, e : D$ we have:*

1. $in(d, rem(d, s)) = in(d, s)$,
2. $rem(d, rem(d, s)) = rem(d, s)$,
3. $rem(d, rem(e, s)) = rem(e, rem(d, s))$,
4. $test(d, s) = \mathsf{t} \to in(d, s) = s$,
5. $test(d, s) = \mathsf{f} \to rem(d, s) = s$,
6. $test(d, s \cup t) = or(test(d, s), test(d, t))$,
7. $test(d, s \cap t) = and(test(d, s), test(d, t))$,
8. $rem(d, s \cup t) = rem(d, s) \cup rem(d, t)$,
9. $rem(d, s \cap t) = rem(d, s) \cap rem(d, t)$,
10. $test(d, rem(e, s)) = \mathsf{t} \to in(e, rem(d, s)) = rem(d, in(e, s))$.

Proof. By induction on s. As an example we prove 1:
We have $in(d, rem(d, s_0)) \stackrel{\text{Sets3}}{=} in(d, s_0)$.
If the i.h. is $in(d, rem(d, t)) = in(d, t)$ then if $eq(d, e) = \mathsf{t}$

$$\begin{aligned} in(d, rem(d, in(e,t))) &\stackrel{\text{Sets4,Con1}}{=} in(d, rem(d,t)) \stackrel{\text{i.h.}}{=} in(d,t) \\ &\stackrel{\text{Sets1}}{=} in(d, in(d,t)) = in(d, in(e,t)). \end{aligned}$$

If $eq(d, e) = \mathsf{f}$ then

$$\begin{aligned} in(d, rem(d, in(e,t))) &\stackrel{\text{Sets4,Con2}}{=} in(d, in(e, rem(d,t))) \\ &\stackrel{\text{Sets2}}{=} in(e, in(d, rem(d,t))) \\ &\stackrel{\text{i.h.}}{=} in(e, in(d,t)) \stackrel{\text{Sets2}}{=} in(d, in(e,t)). \end{aligned}$$

So $in(d, rem(d, t)) = in(d, t) \to in(d, rem(d, in(e, t))) = in(d, in(e, t))$. Now we may apply standard set induction, and we conclude Proposition 3.1.

□

Note the correspondence between the *rem* function (propositions 3.2 and 3.3), and the *in* function (axioms Sets1 and Sets2).

At this stage we can already formulate an alternative induction scheme, which is a variant of the standard one.

Definition 4 (Special Set Induction 1). A proof of a property formula $\phi(s)$, by special set induction 1 on $s : Set$, is the deduction

$$\frac{\phi(s_0) \quad \phi(rem(d,t)) \rightarrow \phi(in(d,t))}{\phi(s)}$$

where $d : D$ and $t : Set$.

Proposition 5. *Proofs by special set induction with the above scheme are correct.*

Proof. We prove that this special induction scheme is equivalent to the standard induction scheme in Definition 1, i.e. that

$$\phi(s_0) \wedge (\phi(s) \rightarrow \phi(in(d,s))) \longleftrightarrow \phi(s_0) \wedge (\phi(rem(d,t)) \rightarrow \phi(in(d,t))).$$

$\rightarrow$: Substitute $rem(d,t)$ for s and apply Proposition 3.1.
$\leftarrow$: If $test(d,t) = \mathsf{f}$ then according to Proposition 3.5 $rem(d,t) = t$, so the premise of special scheme implies the premise of the standard scheme. If $test(d,t) = \mathsf{t}$ then with Proposition 3.4 we find $\phi(t) \leftrightarrow \phi(in(d,t))$, so $\phi(s) \rightarrow \phi(in(d,s))$ follows trivially. □

There are a number of well-known properties of operations on sets, such as commutativity, associativity, idempotency, absorption of $\cup$ and $\cap$, and distribution laws, such as

$$s \cap (t \cup u) = (s \cap t) \cup (s \cap u), \qquad (t \cap u) \setminus s = (t \setminus s) \cap (u \setminus s).$$

All these laws can be derived using the presented material and a number of other basic property formulas. Of course one also needs the scheme for standard induction on sets, as well as properties of the booleans.

The following step we take is the introduction of some familiar boolean functions on sets. We axiomatise equality of sets since it can be useful to have equality as a predicate *eq*, e.g. for use in the conditional construct *if*. Moreover we define subset predicates $\subset$ and $\subseteq$, which also uses *eq*.

func	$eq, \subset, \subseteq : Set \times Set \rightarrow \mathbf{Bool}$	
rew	$eq(s_0, s_0) = \mathsf{t}$	Eq1
	$eq(s_0, in(d,s)) = \mathsf{f}$	Eq2
	$eq(in(d,s),t) = and(test(d,t), eq(rem(d,s), rem(d,t)))$	Eq3
	$s_0 \subset s_0 = \mathsf{f}$	Sub1
	$s_0 \subset in(d,s) = \mathsf{t}$	Sub2
	$in(d,s) \subset t = and(test(d,t), rem(d,s) \subset rem(d,t))$	Sub3
	$s \subseteq t = or(eq(s,t), s \subset t)$	Sub4

For proving property formulas with sets that contain the above boolean functions, a useful new induction scheme can be introduced. The reason for this becomes clear when looking at the specification, and in particular at the axioms Eq3 and Sub3. E.g. the definition of $eq(in(d,s),t)$ contains the term $eq(rem(d,s), rem(d,t))$ in the r.h.s., so in order to prove a property formula $\phi(eq(v,w))$, it would be very helpful to have $\phi(eq(rem(d,s), rem(d,t)))$ as induction hypothesis. This is possible according to the following special induction scheme, which allows function applications on two variables. Note that axiom Sub3 has a similar inductive structure as axiom Eq3, and that axiom Sub4 will give no problems once the functions eq and $\subset$ are studied.

Definition 6 (Special Set Induction 2). Let $f : D \times Set \to Set$ be a function. A proof of a property formula $\phi(v,w)$, by special set induction 2 on $v : Set$, is the deduction

$$\frac{\phi(s_0,t) \quad \phi(rem(d,s), f(d,t)) \to \phi(in(d,s),t)}{\phi(v,w)}$$

where $d : D$ and $s,t,w : Set$.

Proposition 7. *Proofs by special set induction with the above scheme are correct.*

Proof. If we assume $\phi(rem(d,s), f(d,t))$ we conclude from the premise of the above scheme that $\phi(in(d,s),t)$. If also $test(d,s) = \mathsf{t}$ we have according to Proposition 3.4 that $in(d,s) = s$, and consequently $\phi(s,t)$. Therefore, we may assume

$$(i) \quad \phi(s_0,t),$$
$$(ii) \quad test(d,s) = \mathsf{t} \wedge \phi(rem(d,s), f(d,t)) \to \phi(s,t).$$

The next step is to prove that for all closed terms $v_0, w_0 : Set$ the property formula $\phi(v_0, w_0)$ can be derived using (i) and (ii). First we define an auxiliary function $g : Set \to Set$:

$$(iii) \quad g(s_0) \stackrel{\text{def}}{=} w_0,$$
$$(iv) \quad g(in(d,u)) \stackrel{\text{def}}{=} if(test(d,v_0), f(d,g(u)), g(u)),$$

where $u : Set$.
Let $\Phi(d,u) \equiv (test(d,v_0) = \mathsf{t} \to (\phi(v_0 \setminus in(d,u), g(in(d,u))) \to \phi(v_0 \setminus u, g(u))))$.
If $test(d,u) = \mathsf{t}$ then $u = in(d,u)$. It therefore easily follows that

$$(v) \quad test(d,u) = \mathsf{t} \to \Phi(d,u).$$

If $test(d,u) = \mathsf{f}$ and $test(d,v_0) = \mathsf{t}$ then clearly $test(d, v_0 \setminus u) = \mathsf{t}$, so we find from (ii), after substitution of $v_0 \setminus u$ for s and $g(u)$ for t, that
$\phi(rem(d, v_0 \setminus u), f(d,g(u))) \to \phi(v_0 \setminus u, g(u))$. After application of (iv) and axiom Sets10 we find

$$(vi) \quad test(d,u) = \mathsf{f} \to \Phi(d,u).$$

Combining (v) and (vi) yields $\Phi(d,u)$.

Let $\Psi(u) \equiv (\phi(v_0 \setminus u, g(u)) \rightarrow \phi(v_0, w_0))$. We prove $\Psi(u)$ by standard set induction on u. $\Psi(s_0)$ follows trivially using (iii). We now have to prove that $\Psi(in(e,x))$ holds, where $e : D$ and $x : Set$. The i.h. is $\Psi(x)$.
If $test(e, v_0) = \mathsf{t}$ then it follows from $\Phi(e,x)$ that $\phi(v_0 \setminus in(e,x), g(in(e,x))) \rightarrow \phi(v_0 \setminus x, g(x))$. Combining this with the i.h. yields

$$(vii) \quad test(e, v_0) = \mathsf{t} \rightarrow \Psi(in(e,x)).$$

If $test(e, v_0) = \mathsf{f}$ then $v_0 \setminus u = v_0 \setminus in(e,u)$, so using (iv) it follows from the i.h. that

$$(viii) \quad test(e, v_0) = \mathsf{f} \rightarrow \Psi(in(e,x)).$$

So it follows from (vii) and $(viii)$ that $\Psi(in(e,x))$, and hence by induction $\Psi(u)$. Finally, using (i), we conclude from $\Psi(v_0)$ that $\phi(v_0, w_0)$. □

The following proposition states that derivable equality of sets is equivalent to equality as axiomatised by the *eq* function. In the proof of this proposition we demonstrate the use of the first and second special induction scheme in detail.

Proposition 8 (Set Equality). *For all $s, t : Set$ we have:*

$$s = t \leftrightarrow eq(s,t) = \mathsf{t}.$$

Proof. By induction on s. Let $d : D$ and $u : Set$.
$\rightarrow$: We use the first scheme for special induction. If $s := s_0$ the result follows easily, and if $s = t$ and $s := in(d,u)$ then $eq(s,t) = eq(in(d,u), in(d,u))$. Now we can first apply the axioms Eq3 and Test2, and subsequently the i.h.
$\leftarrow$: This time we use the second scheme for special induction. Let $\phi(s,t) \equiv (eq(s,t) = \mathsf{t} \rightarrow s = t)$. If $s := s_0$ then $\phi(s,t)$ follows easily by standard induction on t. Now we have to prove $\phi(in(d,u), t)$. The i.h. is $\phi(rem(d,u), rem(d,t))$. Two cases are distinguished:
If $test(d,t) = \mathsf{t}$ then it follows from $eq(in(d,u),t) = \mathsf{t}$ after application of axiom Eq3 that $eq(rem(d,u), rem(d,t)) = \mathsf{t}$. From the i.h. we conclude $rem(d,u) = rem(d,t)$, as well as $in(d, rem(d,u)) = in(d, rem(d,t))$. According to Proposition 3.1 we also have $in(d,u) = in(d,t)$. Since $test(d,t) = \mathsf{t}$ we can apply Proposition 3.4, and we conclude that $in(d,t) = t$. Combining these results yields

$$(i) \quad test(d,t) = \mathsf{t} \rightarrow (\phi(rem(d,u), rem(d,t)) \rightarrow \phi(in(d,u),t)).$$

If $test(d,t) = \mathsf{f}$ then according to axiom Eq3 $eq(in(d,u),t) = \mathsf{f}$. So we can easily derive a contradiction. It follows that

$$(ii) \quad test(d,t) = \mathsf{f} \rightarrow (\phi(rem(d,u), rem(d,t)) \rightarrow \phi(in(d,u),t)).$$

Finally we combine (i) and (ii) with $\phi(s_0, t)$, and we conclude $\phi(s,t)$. □

It is also not hard to prove property formulas that contain the relational operators $\subset$ or $\subseteq$. As an example we give the following two identities.

Proposition 9 (Relational Set Identities). *For all $s, t : Set$ we have:*

1. $s \subset s = \mathsf{f}$,
2. $s \cap t \subseteq t = \mathsf{t}$.

Proof. By induction on s. Let $d : D$. Without proof we state the following intermediate results:

(*i*) $s_0 \subseteq s = \mathsf{t}$,
(*ii*) $in(d,s) \subseteq in(d,t) = rem(d,s) \subseteq rem(d,t)$.

1. We use the first scheme for special induction. Axiom Sub1 says that $s_0 \subset s_0 = \mathsf{f}$, and the i.h. is $rem(d,t) \subset rem(d,t) = \mathsf{f}$. With the axioms Sub3 and Test2 we find that $in(d,t) \subset in(d,t) = rem(d,t) \subset rem(d,t)$. It follows immediately from the i.h. that $in(d,t) \subset in(d,t) = \mathsf{f}$, so we conclude $s \subset s = \mathsf{f}$.

2. We use the second scheme for special induction.

   If $s := s_0$ then $s \cap t \subseteq t = s_0 \cap t \subseteq t \overset{\text{Sets7}}{=} s_0 \subseteq t \overset{i}{=} \mathsf{t}$.

   We take $rem(d,u) \cap rem(d,t) \subseteq rem(d,t) = \mathsf{t}$ as i.h.
   If $test(d,t) = \mathsf{t}$ then

$$\begin{aligned} in(d,u) \cap t \subseteq t &\overset{\text{Sets8}}{=} in(d, u \cap t) \subseteq t \\ &\overset{3.4}{=} in(d, u \cap t) \subseteq in(d,t) \\ &\overset{ii}{=} rem(d, u \cap t) \subseteq rem(d,t) \\ &\overset{3.9}{=} rem(d,u) \cap rem(d,t) \subseteq rem(d,t). \end{aligned}$$

   If $test(d,t) = \mathsf{f}$ then

$$\begin{aligned} in(d,u) \cap t \subseteq t &\overset{3.1}{=} in(d, rem(d,u)) \cap t \subseteq t \\ &\overset{\text{Sets8}}{=} rem(d,u) \cap t \subseteq t \\ &\overset{3.5}{=} rem(d,u) \cap rem(d,t) \subseteq rem(d,t). \end{aligned}$$

   It follows from the i.h. that $in(d,u) \cap t \subseteq t = \mathsf{t}$, so we finally may conclude $s \cap t \subseteq t = \mathsf{t}$. □

We finish our specification of sets with two simple functions. The *size* function is needed for a 'neat' proof of the correctness of the third special induction scheme; it represents the number of elements in a set. The predicate *empty* is defined for cosmetic reasons. It is used in Section 4.

func	$size : Set \to Nat$	
	$empty : Set \to \mathbf{Bool}$	
rew	$size(s_0) = 0$	Size1
	$size(in(d,s)) = size(rem(d,s)) + 1$	Size2
	$empty(s) = eq(size(s), 0)$	Size3

Definition 10 (Special Set Induction 3). A proof of a property formula $\phi(s)$, by special set induction 3 on $s : Set$, is the deduction

$$\frac{\phi(s_0) \quad test(d,t) = \mathsf{t} \wedge \phi(rem(d,t)) \to \phi(t)}{\phi(s)}$$

where $d : D$ and $t : Set$.

Proposition 11. *Proofs by special set induction with the above scheme are correct.*

Proof. We assume that the premise of this induction scheme is satisfied, so that we have

$$(i) \quad \phi(s_0),$$
$$(ii) \quad test(d,t) = \mathsf{t} \wedge \phi(rem(d,t)) \rightarrow \phi(t).$$

The following results are needed:

$$(iii) \quad size(t) = 0 \rightarrow t = s_0,$$
$$(iv) \quad test(d,t) = \mathsf{t} \wedge size(t) = Sn \rightarrow size(rem(d,t)) = n,$$

where $n : Nat$. Property formula (iii) is easily proved by standard set induction on t, and (iv) follows with the help of Proposition 3.4. If $test(d,t) = \mathsf{t}$ then $Sn = size(t) = size(in(d,t))$, which equals $size(rem(d,t)) + 1$ by axiom Size2.

Let $\Phi(s,m) \equiv (size(s) = m \rightarrow \phi(s))$, where $m : Nat$. We prove $\Phi(s,m)$ by standard induction on m. $\Phi(s,0)$ follows easily from (i) and (iii). The i.h. is $\Phi(s,n)$. If $test(d,t) = \mathsf{t} \wedge size(t) = Sn$ then it follows from (iv) and the i.h. that $\phi(rem(d,t))$. It follows from (ii) that $\phi(t)$, so we have proved that $\Phi(s,n) \rightarrow (test(d,t) = \mathsf{t} \wedge size(t) = Sn \rightarrow \phi(t))$. It follows by standard set induction on t that $\Phi(s,n) \rightarrow (size(t) = Sn \rightarrow \phi(t))$, and therefore that $\Phi(s,n) \rightarrow \Phi(s,Sn)$, and hence by induction we may conclude $\Phi(s,m)$. Finally we can substitute $size(s)$ for m, so that we find $\phi(s)$. □

Sometimes it is possible to specify a function $f : Set \rightarrow D$ that selects a particular element from a set. E.g. when D is linearly ordered a function can be specified that picks the maximum or minimum element from a set. We incorporate this idea in the following induction scheme.

Definition 12 (Special Set Induction 4). Let $f : Set \rightarrow D$ be a function with the property $eq(s, s_0) = \mathsf{f} \rightarrow test(f(s), s) = \mathsf{t}$. A proof of a property formula $\phi(s)$, by special set induction 4 on $s : Set$, is the deduction

$$\frac{\phi(s_0) \quad \phi(rem(f(t),t)) \rightarrow \phi(t)}{\phi(s)}$$

where $t : Set$.

Proposition 13. *Proofs by special set induction with the above scheme are correct.*

Proof. Follows from the previous induction scheme. □

3 A rule for hybrid set induction

Thus far we gave induction schemes that concerned variables of a single sort. The careful reader may have noticed that correctness proofs of these schemes could be provided within the formal framework of μCRL, extended with some second order arguments, i.e. we had to reason about property formulas, using variables ϕ, ψ ranging over some universe of property formulas. Moreover, the proof we gave for Proposition 7 is not finite.

One can imagine that in many cases our schemes do not suffice for providing proofs on a more or less formal basis. As an example we have Proposition 21.1. In general, induction schemes where more than one variable varies in the induction step, see e.g. Definition 6, will be hard to prove on a formal basis. If we allow some 'meta-theoretical' reasoning, however, we can easily obtain a considerable generalisation of the standard scheme in the form of a rule.

We refer to the below rule as "hybrid" because it allows simultaneous reasoning on the structure of data terms and process terms. Here it is maybe interesting to note that μCRL has no rules for structural induction on process terms, although many useful results in process algebra have to be derived this way. We also remark that the need for hybrid induction once again illustrates the close interaction between data and processes.

Definition 14 (Hybrid Set Induction). Let $x, y_1, ..., y_n$ be data variables typed by $x, y_1 : S_1, y_2 : S_2, ..., y_n : S_n$ and let $f : S_1 \times ... \times S_n \rightarrow S_1$ be a function, or let $x, y_1, ..., y_n$ be process variables and f a process operator. A proof of a property formula $\phi(s, x)$, by hybrid set induction on $s : Set$, is the deduction

$$\frac{\phi(s_0, y_1) \quad \phi(rem(d,t), f(y_1, ..., y_n)) \rightarrow \phi(in(d,t), y_1)}{\phi(s, x)}$$

where $d : D$ and $t : Set$.

Proposition 15. *Proofs by hybrid set induction are correct.*

Proof. We prove $\phi(u_0, x)$ from the above clauses for all closed $u_0 : Set$. If $u_0 = s_0$ then this follows immediately. If $u_0 = in(e_0, v_0)$, where $e_0 : D$ and $v_0 : Set$, then if we prove $\phi(rem(e_0, v_0), f(y_1, ..., y_n))$ we are ready. Clearly, after another $size(rem(e_0, v_0))$ applications of the rem function we can obtain the empty set. Therefore, it suffices to prove $\phi(s_0, f^{size(u_0)}(y_1, ..., y_n))$ if u_0 is not the empty set (we use $f^{n+2}(y_1, ..., y_n)$ to denote $f(f^{n+1}(y_1, ..., y_n), y'_2, ..., y'_n))$. This follows immediately from the property formula $\phi(s_0, y_1)$. □

Note that if we adopt the rule for hybrid set induction, the first two schemes for special set induction in the previous section become superfluous.

4 Inductive proofs with sets in process algebra

In order to demonstrate the use of our induction schemes for sets in a different context, we study two special processes that can be regarded generalisations of the alternative composition, and parallel composition from process algebra. Moreover we demonstrate how our generalised parallel composition can be used for modelling a broadcast process.

4.1 Process algebra

We first recall the system PA_δ (Process Algebra with δ) from [2]. The signature of PA_δ has a set A of *atomic actions*, with typical elements $a, b, ...$, as a parameter. The special constant δ, or *deadlock*, represents the process that 'blocks'. We also have the binary operator $+$ for alternative composition, and $\cdot$ for sequential composition available. Concurrency is described by the operators $\|$ for *merge* or parallel composition, and the auxiliary operator $\lfloor\!\lfloor$ for *left-merge*. In term formation, brackets and variables from a set $x, y, z, ...$ are used. Letters $p, q, r, ...$ denote process terms, and brackets are omitted according to the convention that $\cdot$ binds stronger than any other operator, and that $+$ binds weaker than any other operator.

The axioms in Table 1 represent the axiom system PA_δ. The operator $+$ is commutative, associative and idempotent (A1–A3). The operator $\cdot$ right distributes over $+$ and is associative (A4, A5). Furthermore δ behaves as the neutral element for $+$ (A6), and absorbs terms at the right side of $\cdot$ (A7). The operator $\|$ is defined in terms of the operator $\lfloor\!\lfloor$ (M1), and the axioms M2–M3 define the interaction between $\lfloor\!\lfloor$ and a, $a \cdot x$, and $x + y$, respectively.

(A1)	$x + y = y + x$	(M1)	$x \parallel y = x \lfloor\!\lfloor y + y \lfloor\!\lfloor x$
(A2)	$x + (y + z) = (x + y) + z$	(M2)	$a \lfloor\!\lfloor x = a \cdot x$
(A3)	$x + x = x$	(M3)	$a \cdot x \lfloor\!\lfloor y = a \cdot (x \parallel y)$
(A4)	$(x + y) \cdot z = x \cdot z + y \cdot z$	(M4)	$(x + y) \lfloor\!\lfloor z = x \lfloor\!\lfloor z + y \lfloor\!\lfloor z$
(A5)	$(x \cdot y) \cdot z = x \cdot (y \cdot z)$		
(A6)	$x + \delta = x$		
(A7)	$\delta \cdot x = \delta$		

Table 1: The axioms of PA_δ, where $a \in A \cup \{\delta\}$.

Associativity and commutativity of $+$ is denoted with AC, and associativity and commutativity of $\|$ is denoted with SC (Standard Concurrency, which is derivable for regular process expressions).

From μCRL we use the conditional construct and the Σ operator, which denotes a generalisation of $+$, e.g. the expression $\Sigma_{d:D}(P(d))$ denotes the (possibly infinite) sum of processes P, parameterised with elements from the sort D. The necessary axioms are given in Table 2. The binding strength of $. \lhd b \rhd .$ is taken between $+$ and the other operators of PA_δ. If we prove identities with conditionals using case distinction, we usually annotate this with CD.

4.2 Generalised sums

We define a generalised sum process S_P, not to be confused with the Σ operator. Then we prove some properties of this process, and demonstrate how it can be defined in terms of Σ.

Definition 16 (Generalised Sums). Let $P(d : D)$ be a given process, and $f : \mathit{Set} \to D$ a function with the property $eq(s, s_0) = \mathsf{f} \to \mathit{test}(f(s), s) = \mathsf{t}$. A

(Cond1)	$x \lhd \mathsf{t} \rhd y = x$
(Cond2)	$x \lhd \mathsf{f} \rhd y = y$
(Sum1)	$\Sigma_{d:D}(q) = q$
(Sum4)	$\Sigma_{d:D}(p + q) = \Sigma_{d:D}(p) + \Sigma_{d:D}(q)$
(Sum5)	$\Sigma_{d:D}(p \cdot q) = \Sigma_{d:D}(p) \cdot q$
(Sum6)	$\Sigma_{d:D}(p \mathbin{\|\!\_} q) = \Sigma_{d:D}(p) \mathbin{\|\!\_} q$
	provided d is not free in q in Sum1, Sum5 and Sum6

Table 2: Axioms for the conditional construct and Σ.

generalised sum for the process $P(d : D)$ over a set s is

$$S_P(s : Set) \stackrel{\text{def}}{=} \delta \lhd empty(s) \rhd (P(f(s)) + S_P(rem(f(s), s))).$$

For example, if D represents the natural numbers and $f(s)$ the minimum in s then it is easy to derive $S_P(in(2, in(5, in(1, s_0)))) = P(1) + (P(2) + P(5))$.

Proposition 17 (Sets and Sums). *Let $d : D$ and $s, t : Set$. For generalised sums for a process $P(d : D)$ we have:*

1. $S_P(s_0) = \delta$,
2. $S_P(in(d, s)) = P(d) + S_P(rem(d, s))$,
3. $S_P(s \cup t) = S_P(s) + S_P(t)$.

Proof.

1. Immediately from the definition of S_P.
2. By induction on s. We use the scheme in Definition 12.

   Using the definition of S_P it easily follows that

   $$S_P(in(d, s_0)) = P(d) + S_P(rem(d, s_0)).$$

   The i.h. is $S_P(in(d, rem(f(s), s))) = P(d) + S_P(rem(d, rem(f(s), s)))$, and we use the following intermediate results. If $eq(f(in(d, s)), d) = \mathsf{f}$ then

   (i) $f(in(d, s)) = f(s)$,
   (ii) $f(in(d, s)) = f(rem(d, s))$,

   which follows easily from the properties of f and the *test* function.

   If $eq(f(in(d, s)), d) = \mathsf{t}$ then Proposition 17.2 follows directly from the definition of S_P.

   If $eq(f(in(d, s)), d) = \mathsf{f}$ then we derive

$$\begin{array}{ll} S_P(in(d,s)) & \\ = & P(f(in(d,s))) + S_P(rem(f(in(d,s)), in(d,s))) \\ \stackrel{\text{Sets4}}{=} & P(f(in(d,s))) + S_P(in(d, rem(f(in(d,s)), s))) \\ \stackrel{i,ii}{=} & P(f(rem(d,s))) + S_P(in(d, rem(f(s), s))) \\ \stackrel{\text{i.h.}}{=} & P(f(rem(d,s))) + (P(d) + S_P(rem(d, rem(f(s), s)))) \\ \stackrel{3.3}{=} & P(f(rem(d,s))) + (P(d) + S_P(rem(f(s), rem(d,s)))) \\ \stackrel{i,ii,\text{AC}}{=} & P(d) + \\ & (P(f(rem(d,s))) + S_P(rem(f(rem(d,s)), rem(d,s)))) \\ = & P(d) + S_P(rem(d,s)). \end{array}$$

3. Also by induction on s. This time we use the scheme in Definition 6.

If $s := s_0$ then $S_P(s \cup t) \stackrel{\text{Sets5}}{=} S_P(t) \stackrel{\text{A6,AC}}{=} \delta + S_P(t) \stackrel{17.1}{=} S_P(s) + S_P(t)$.

The i.h. is $S_P(rem(d,s) \cup rem(d,t)) = S_P(rem(d,s)) + S_P(rem(d,t))$.

If $test(d,t) = \mathsf{t}$ then

$$\begin{array}{lll} S_P(in(d,s) \cup t) & \stackrel{\text{Sets6}}{=} & S_P(in(d, s \cup t)) \\ & \stackrel{17.2}{=} & P(d) + S_P(rem(d, s \cup t)) \\ & \stackrel{3.8}{=} & P(d) + S_P(rem(d,s) \cup rem(d,t)) \\ & \stackrel{\text{i.h.}}{=} & P(d) + (S_P(rem(d,s)) + S_P(rem(d,t))) \\ & \stackrel{\text{A3,AC}}{=} & (P(d) + S_P(rem(d,s))) + \\ & & (P(d) + S_P(rem(d,t))) \\ & \stackrel{17.2}{=} & S_P(in(d,s)) + S_P(in(d,t)) \\ & \stackrel{3.4}{=} & S_P(in(d,s)) + S_P(t). \end{array}$$

If $test(d,t) = \mathsf{f}$ we simply follow a different route after application of the i.h.:

$$\begin{array}{lll} S_P(in(d,s) \cup t) & = & \ldots \\ & \stackrel{\text{i.h.}}{=} & P(d) + (S_P(rem(d,s)) + S_P(rem(d,t))) \\ & \stackrel{\text{AC}}{=} & (P(d) + S_P(rem(d,s))) + S_P(rem(d,t)) \\ & \stackrel{17.2}{=} & S_P(in(d,s)) + S_P(rem(d,t)) \\ & \stackrel{3.5}{=} & S_P(in(d,s)) + S_P(t). \end{array}$$

□

The first two identities of the above proposition imply that we can argue about S_P without using f. By definition of constructor sets we know that any process term $S_P(t)$ can be rewritten by either the first or the second identity. The third identity shows how set composition interacts with the generalised sum process. This correspondence is due to the fact that both + and ∪ are idempotent, e.g.

$$S_P(s) = S_P(s \cup s) \overset{17.3}{=} S_P(s) + S_P(s) \overset{A3}{=} S_P(s).$$

The above identities can be also be proven correct using properties of Σ, which would allow here for easier proofs. (This is not the case, however, with the generalised merge.) Using the axioms for Σ, it can be proved that $S_P(s)$ is a summand of $\Sigma_{d:D}(P(d))$. If s contains all elements from D then they are even equal.

Proposition 18 (Sums and Σ). *Let* $s : Set$. *For generalised sums for a process* $P(d : D)$ *we have:*

$$S_P(s) = \Sigma_{d:D}(P(d) \lhd test(d, s) \rhd \delta).$$

Proof. We need some basic identities for Σ:

$$\begin{array}{ll} (i) & \Sigma_{d:D}(p \lhd test(d, s_0) \rhd \delta) = \delta, \\ (ii) & \Sigma_{d:D}(p \lhd eq(d, e) \rhd \delta) = p[e/d], \\ (iii) & \Sigma_{d:D}(p \lhd test(d, in(e, s)) \rhd \delta) \;= \\ & \qquad \Sigma_{d:D}(p \lhd test(d, rem(e, s)) \rhd \delta) + p[e/d], \end{array}$$

where $p[e/d]$ denotes the process term p with $e : D$ substituted for d. Identity (i) follows using axiom Test1 and axiom Sum1, (ii) is proved in [5], and (iii) is proved as follows:

$$\begin{array}{ll} & \Sigma_{d:D}(p \lhd test(d, in(e, s)) \rhd \delta) \\ \overset{3.1,\text{Test2}}{=} & \Sigma_{d:D}(p \lhd or(eq(d, e), test(d, rem(e, s))) \rhd \delta) \\ \overset{\text{CD}}{=} & \Sigma_{d:D}(p \lhd test(d, rem(e, s)) \rhd \delta + p \lhd eq(e, s) \rhd \delta) \\ \overset{\text{Sum4},ii}{=} & \Sigma_{d:D}(p \lhd test(d, rem(e, s)) \rhd \delta) + p[e/d]. \end{array}$$

Proposition 18 now follows by induction on s, using the first scheme for special set induction. If $s := s_0$ the result follows trivially from (i), and for $s := in(e, t)$ then use (iii) and apply the i.h. □

4.3 Generalised merges

The generalised merge M_P as defined below is inspired by [12], where it was used for the specification and verification of 'Milner's scheduler'; a simple communication network with a large amount of states, and therefore often used for testing the performance of tools.

Definition 19 (Generalised Merges). Let $P(d : D)$ be a given process, and $f : Set \to D$ a function with the property $eq(s, s_0) = \mathsf{f} \to test(f(s), s) = \mathsf{t}$. A generalised merge for the process $P(d : D)$ over a set s is

$$\begin{array}{l} M_P(s : Set) \overset{\text{def}}{=} \delta \lhd empty(s) \rhd \\ \quad (P(f(s)) \lhd empty(rem(f(s), s)) \rhd P(f(s)) \parallel M_P(rem(f(s), s))). \end{array}$$

For example, if D represents the natural numbers and $f(s)$ the minimum in s then it is easy to derive $M_P(in(5, in(2, in(5, in(1, s_0))))) = P(1) \parallel (P(2) \parallel P(5))$.

Proposition 20 (Sets and Merges). *Let $d : D$ and $s, t : Set$. For generalised merges for a process $P(d : D)$ we have:*

1. $M_P(s_0) = \delta$,
2. $M_P(in(d, s)) = P(d) \lhd empty(rem(d, s)) \rhd P(d) \parallel M_P(rem(d, s))$,
3. $empty(s) = \mathsf{f} \land empty(t) = \mathsf{f} \land empty(s \cap t) = \mathsf{t} \rightarrow$
   $M_P(s \cup t) = M_P(s) \parallel M_P(t)$.

Proof.

1. Immediately from the definition of M_P.
2. Analogously to the proof of Proposition 17.2. In this proof the axioms for SC are needed instead of AC.
3. By induction on s. We use the scheme in Definition 4. Moreover we use the following intermediate results. If $empty(t) = \mathsf{f} \land empty(in(d, s) \cap t) = \mathsf{t}$ then

   $$\begin{array}{ll} (i) & empty(rem(d, s \cup t)) = \mathsf{f}, \\ (ii) & empty(rem(d, s) \cap t) = \mathsf{t}, \\ (iii) & rem(d, t) = t. \end{array}$$

   If $s := s_0$ the result follows trivially. The i.h. is $empty(rem(d, s)) = \mathsf{f} \land empty(t) = \mathsf{f} \land empty(rem(d, s) \cap t) = \mathsf{t} \rightarrow M_P(rem(d, s) \cup t) = M_P(rem(d, s)) \parallel M_P(t)$. We assume that $empty(t) = \mathsf{f} \land empty(in(d, s) \cap t) = \mathsf{t}$, so

   $$\begin{array}{rcl} M_P(in(d, s) \cup t) & \stackrel{\text{Sets6}}{=} & M_P(in(d, s \cup t)) \\ & \stackrel{20.2, i}{=} & P(d) \parallel M_P(rem(d, s \cup t)) \\ & \stackrel{3.8, iii}{=} & P(d) \parallel M_P(rem(d, s) \cup t). \end{array}$$

   If $empty(rem(d, s)) = \mathsf{t}$ then $rem(d, s) = s_0$, and according to Proposition 20.2 $P(d) = M_P(in(d, s))$, so in this case we are ready.

   If $empty(rem(d, s)) = \mathsf{f}$ then we can combine (ii), together with our assumption $empty(t) = \mathsf{f}$, with the i.h.:

   $$\begin{array}{rcl} P(d) \parallel M_P(rem(d, s) \cup t) & \stackrel{\text{i.h.}}{=} & P(d) \parallel (M_P(rem(d, s)) \parallel M_P(t)) \\ & \stackrel{\text{SC}}{=} & (P(d) \parallel M_P(rem(d, s))) \parallel M_P(t) \\ & \stackrel{20.2}{=} & M_P(in(d, s)) \parallel M_P(t). \end{array}$$

□

The first two identities of the above proposition imply that we can argue about any process term M_P without using f. The third property formula implies that if $t_1 \cup t_2 \cup ... \cup t_n$ is a partition of a set s then

$$M_P(s) = M_P(t_1) \parallel (M_P(t_2) \parallel (... \parallel M_P(t_n) ...)).$$

As we argued above, the relation between S_P and Σ is rather straightforward. We now show that this is also the case with M_P and Σ; there are some very interesting relations, and again we can use a variety of induction principles for the proofs.

Proposition 21 (Merges and Σ). *Let $s : Set$. For generalised merges for a process $P(d : D)$ and an atomic action $a(d : D)$ we have:*

1. $size(s) > 0 = \mathsf{t} \rightarrow x \parallel M_P(s) = \Sigma_{d:D}(x \parallel M_P(s) \lhd test(d, s) \rhd \delta)$,
2. $M_P(s) = \Sigma_{d:D}(M_P(s) \lhd test(d, s) \rhd \delta)$,
3. $M_a(s) = \Sigma_{d:D}((a(d) \lhd empty(rem(d, s)) \rhd a(d) \cdot M_a(rem(d, s)))$
$\lhd test(d, s) \rhd \delta)$.

Proof.

1. For a proof of this property formula we use the rule for hybrid set induction, defined in Section 3. Moreover we use the following identity:

   $$(i) \quad (p \lhd empty(s) \rhd q) \lhd test(d, s) \rhd \delta = q \lhd test(d, s) \rhd \delta.$$

   We apply induction on s, and $f(x, y) \stackrel{\text{def}}{=} x \parallel y$. So if $s := in(e, t)$, where $e : D$ and $t : Set$, and if $\phi(s, x)$ is defined as property formula 21.1, the i.h. is $\phi(rem(e, t), x \parallel y)$. The clause $\phi(s_0, x)$ follows trivially. Furthermore we derive:

   $$\begin{aligned}
   &\Sigma_{d:D}(x \parallel M_P(in(e, t)) \lhd test(d, in(e, t)) \rhd \delta) \\
   \stackrel{18.iii}{=} \quad &\Sigma_{d:D}(x \parallel M_P(in(e, t)) \lhd test(d, rem(e, t)) \rhd \delta) + \\
   &x \parallel M_P(in(e, t)) \\
   \stackrel{20.2,CD}{=} \quad &\Sigma_{d:D}((x \parallel P(e) \lhd empty(rem(e, t)) \rhd \\
   &\qquad x \parallel (P(e) \parallel M_P(rem(e, t)))) \lhd \ test(d, rem(e, t)) \rhd \delta) + \\
   &x \parallel M_P(in(e, t)) \\
   \stackrel{i,SC}{=} \quad &\Sigma_{d:D}((x \parallel P(e)) \parallel M_P(rem(e, t)) \lhd test(d, rem(e, t)) \rhd \delta) + \\
   &x \parallel M_P(in(e, t)).
   \end{aligned}$$

   If $empty(rem(e, t)) = \mathsf{t}$ we can apply 18.i, and if $empty(rem(e, t)) = \mathsf{f}$ we can apply the i.h.

2. By standard set induction on s. The previous property formula can be used.

3. We use the first scheme for special set induction. If $s := s_0$ the result follows using Proposition 18.*i*. We derive

$$
\begin{aligned}
&M_a(in(e,t))\\
&\stackrel{20.2}{=} a(e) \lhd empty(rem(e,t)) \rhd a(e) \parallel M_a(rem(e,t))\\
&\stackrel{M1,M2}{=} a(e) \lhd empty(rem(e,t)) \rhd (a(e)\cdot M_a(rem(e,t)) +\\
&\qquad M_a(rem(e,t)) \mathbin{\|\!\!\_} a(e))\\
&\stackrel{i.h.,Sum6}{=} a(e) \lhd empty(rem(e,t)) \rhd (a(e)\cdot M_a(rem(e,t)) +\\
&\qquad \Sigma_{d:D}(((a(d) \lhd empty(rem(d,rem(e,t))) \rhd\\
&\qquad\quad a(d)\cdot M_a(rem(d,rem(e,t))))\\
&\qquad\qquad \lhd test(d,rem(e,t)) \rhd \delta) \mathbin{\|\!\!\_} a(e)))\\
&\stackrel{CD,M2,M3,A7}{=} a(e) \lhd empty(rem(e,t)) \rhd (a(e)\cdot M_a(rem(e,t)) +\\
&\qquad \Sigma_{d:D}((a(d)\cdot a(e) \lhd empty(rem(d,rem(e,t))) \rhd\\
&\qquad\quad a(d)\cdot (a(e) \parallel M_a(rem(d,rem(e,t)))))\\
&\qquad\qquad \lhd test(d,rem(e,t)) \rhd \delta))\\
&\stackrel{CD,3.3,20.2}{=} a(e) \lhd empty(rem(e,t)) \rhd (a(e)\cdot M_a(rem(e,t)) +\\
&\qquad \Sigma_{d:D}(a(d)\cdot M_a(in(e,rem(d,t)))\\
&\qquad\quad \lhd test(d,rem(e,t)) \rhd \delta))\\
&\stackrel{CD,3.10}{=} a(e) \lhd empty(rem(e,t)) \rhd a(e)\cdot M_a(rem(e,t)) +\\
&\qquad \Sigma_{d:D}(a(d)\cdot M_a(rem(d,in(e,t)))\\
&\qquad\quad \lhd test(d,rem(e,t)) \rhd \delta)\\
&\stackrel{CD,Sets4,3.10}{=} a(e) \lhd empty(rem(e,in(e,t))) \rhd\\
&\qquad a(e)\cdot M_a(rem(e,in(e,t))) +\\
&\qquad \Sigma_{d:D}((a(d) \lhd empty(rem(d,in(e,t))) \rhd\\
&\qquad\quad a(d)\cdot M_a(rem(d,in(e,t)))) \lhd test(d,rem(e,t)) \rhd \delta)\\
&\stackrel{18.iii}{=} \Sigma_{d:D}((a(d) \lhd empty(rem(d,in(e,t))) \rhd\\
&\qquad a(d)\cdot M_a(rem(d,in(e,t)))) \lhd test(d,in(e,t)) \rhd \delta).
\end{aligned}
$$

□

4.4 Broadcasting

The process in Definition 22 models a *broadcasting* mechanism in μCRL. It is inspired by [4], and it can be interpreted as follows. There is a set D that contains all possible sources, or senders of messages, in a communication network. As soon as the network is activated, the broadcast process waits for a message. If a message m comes in from some source d, it is sent to all the destinations in some fixed set E, except d. After this is finished, the process waits again for a message.

Definition 22 (Broadcast Process). Let $send(d: D, m: Msg), read(d: D, m: Msg)$ be atomic actions, where Msg is some non-empty message set, and $E: Set$ a fixed set (closed term) with elements from D. A broadcast process for a set s of destinations, and a source m is

$$\begin{aligned}
&READ \stackrel{\text{def}}{=} \Sigma_{d:D,m:Msg}(read(d,m) \cdot BC(rem(d,E),m))\\
&BC(s: Set, m: Msg) \stackrel{\text{def}}{=}\\
&\quad \Sigma_{d:D}(send(d,m) \cdot BC(rem(d,s),m) \triangleleft test(d,s) \triangleright \delta) +\\
&\quad READ \triangleleft empty(s) \triangleright \delta.
\end{aligned}$$

The actual broadcasting must be regarded as the successive sending of messages to all destinations in the set s. The following equation captures the intuition that broadcasting can be modelled as a parallel process.

Proposition 23 (Broadcasting and Merges). *Let M_{send_m} denote the generalised merge for the atomic action $send(d: D, m: Msg)$, then for a broadcast process for a set s of destinations, and a source m we have:*

$$BC(s,m) = READ \triangleleft empty(s) \triangleright M_{send_m}(s) \cdot READ.$$

Proof. By induction on s, according to the induction scheme in Definition 10. For $s := s_0$ the result follows easily. The i.h. is

$$\begin{aligned}
&test(d,t) = \mathsf{t} \wedge\\
&BC(rem(d,t),m) \; = \\
&\quad READ \triangleleft empty(rem(d,t)) \triangleright M_{send_m}(rem(d,t)) \cdot READ,
\end{aligned}$$

where $t: Set$, and we derive

$$\begin{aligned}
&BC(t,m)\\
&\stackrel{\text{i.h.,CD}}{=} \quad \Sigma_{d:D}(send(d,m) \cdot\\
&\qquad (READ \triangleleft empty(rem(d,t)) \triangleright M_{send_m}(rem(d,t)) \cdot READ)\\
&\qquad\quad \triangleleft test(d,t) \triangleright \delta) + READ \triangleleft empty(t) \triangleright \delta\\
&\stackrel{\text{CD}}{=} \quad \Sigma_{d:D}((send(d,m) \triangleleft empty(rem(d,t)) \triangleright\\
&\qquad send(d,m) \cdot M_{send_m}(rem(d,t))) \cdot READ \triangleleft test(d,t) \triangleright \delta) +\\
&\quad READ \triangleleft empty(t) \triangleright \delta\\
&\stackrel{\text{CD,Sum5,A7}}{=} \quad \Sigma_{d:D}((send(d,m) \triangleleft empty(rem(d,t)) \triangleright\\
&\qquad send(d,m) \cdot M_{send_m}(rem(d,t))) \triangleleft test(d,t) \triangleright \delta) \cdot READ +\\
&\quad READ \triangleleft empty(t) \triangleright \delta\\
&\stackrel{21.3}{=} \quad M_{send_m}(t) \cdot READ + READ \triangleleft empty(t) \triangleright \delta.
\end{aligned}$$

The result now follows easily by case distinction on $empty(t)$. □

References

[1] M.A. Bezem and J.F. Groote. A formal verification of the alternating bit protocol in the calculus of constructions. Technical Report Logic Group Preprint Series No. 88, Utrecht University, 1993.

[2] J.A. Bergstra and J.W. Klop. Process algebra for synchronous communication. *Information and Control*, 60(1/3):109–137, 1984.

[3] J.C.M. Baeten and W.P. Weijland. *Process Algebra.* Cambridge Tracts in Theoretical Computer Science 18. Cambridge University Press, 1990.

[4] J.J. Brunekreef, J.P. Katoen, R.L.C. Koymans and S. Mauw. Design and analysis of dynamic leader election protocols in broadcast networks. Technical report P9324, University of Amsterdam, 1993.

[5] J.F. Groote and H. Korver. A correctness proof of the bakery protocol in μCRL. In this volume: A. Ponse, C. Verhoef and S.F.M. van Vlijmen, editors. *Proceedings of ACP94.* Workshops in Computing, Springer-Verlag, 1994.

[6] J.F. Groote and A. Ponse. The syntax and semantics of μCRL. In this volume: A. Ponse, C. Verhoef and S.F.M. van Vlijmen, editors. *Proceedings of ACP94.* Workshops in Computing, Springer-Verlag, 1994.

[7] J.F. Groote and A. Ponse. Proof theory for μCRL. Technical Report CS-R9138, CWI, Amsterdam, 1991.

[8] J.F. Groote and A. Ponse. Proof theory for μCRL: a language for processes with data. In D.J. Andrews, J.F. Groote and C.A. Middelburg, editors, *Proceedings of the International Workshop on Semantics of Specification Languages*, pages 232–251. Workshops in Computing, Springer-Verlag, 1994.

[9] J.F. Groote and J.J. van Wamel. Algebraic data types and induction in μCRL. Report P9409, University of Amsterdam, 1994.

[10] H.P. Korver and J. Springintveld. A computer-checked verification of Milner's scheduler. In H.P. Korver, *Protocol Verification in μCRL*, PhD thesis, University of Amsterdam, 1994.

[11] J. Meseguer and J.A. Goguen. Initiality, induction and computability. In M. Nivat and J. Reynolds, editors, *Algebraic Methods in Semantics*, pages 459–541. Cambridge University Press, 1985.

[12] R. Milner. *Communication and Concurrency.* Prentice-Hall International, Englewood Cliffs, 1989.

[13] A.S. Troelstra and D. van Dalen. *Constructivism in Mathematics, An Introduction (vol I).* North-Holland, 1988.

[14] M. Wirsing. Algebraic specification. In J. van Leeuwen, editor, *Handbook of Theoretical Computer Science (vol II)*, pages 677–788. Elsevier Science Publishers B.V., 1990.

Formal Semantics of Interworkings with Discrete Absolute Time

J. van den Brink
Dep. of Mathematics and Computer Science, University of Amsterdam, Amsterdam, The Netherlands*

W.O.D. Griffioen
Dep. of Mathematics and Computer Science, University of Amsterdam, Amsterdam, The Netherlands†

Abstract

The formal semantics of interworkings was defined in BPA_{iw}. In this paper, it is extended with discrete absolute time features. In the discrete time setting, the continuous time domain is partitioned into slices. First, a set of axioms is presented for discretely timed interworkings. Then the notion of discrete time intervals is defined. From this definition and the axioms for discretely timed interworkings, a set of propositions for interval timed interworkings is derived.

1 Introduction

1.1 Interworkings

Interworkings are a synchronous variant of message sequence charts [Til91]. They are used in the analysis phase of the development process at PKI[1] Nürnberg, Germany. An interworking is a projection of (a part of) the communication behaviour of a system onto a set of entities [MvWW92a, MvWW93]. An entity may be a single process or a set of processes combined into a functional block. Sequences of communications with other blocks are part of the specification of a functional block. Interworkings are event oriented, and unlike the communication mechanism in message sequence charts, messages cannot be delayed. A graphical representation of an interworking is shown in Fig. 1. The vertical lines (called *time lines*) represent the entities in the system, and the arrows represent communications between them. The time progresses downwards in this figure.

The concrete syntax IW [MvWW92b] provides a means to use the concept of interworkings in real life applications. It describes a way to textually represent interworkings. In order to provide a formal semantics for IW an intermediate language, T, is used [MvWW92c]. This language forms the connection between

*email: brink@fwi.uva.nl
†Current adress: CWI, email: griffioe@cwi.nl
[1]Philips Kommunikations Industrie

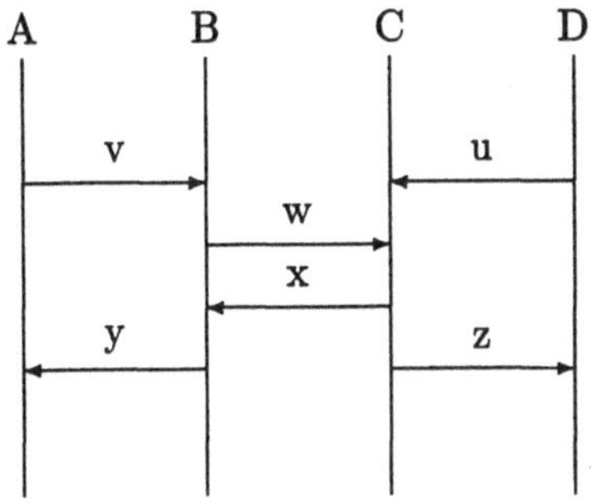

Figure 1: An example interworking

the concrete syntax and the formal semantics by mapping 'programs' written in IW onto process expressions in BPA_{iw}.

Interworkings are useful in describing the message interactions between functional blocks. However, the suitability of interworkings for describing huge parts of a system is limited. Therefore composition operators are needed to combine small parts of a systems description into a larger one. A major issue with respect to interworkings is consistency. In this paper, the word consistency refers to the absence of conflicting actions in and within the interworkings. With the presented formalization it is possible to determine whether a composition of two interworkings is consistent. We will return to this topic later on.

1.2 Notion of time

The formal semantics of interworkings is based on the algebraic concurrency theory BPA. In this theory [BW90] the notion of time is only implicitly present in the form of the sequential composition operator. The introduction of an explicit time notation is useful in describing a system in more detail, and questions regarding correctness, consistency etc. can be answered more specifically. The goal of this paper is to introduce such an explicit time notation in the formal semantics of interworkings. To accomplish this goal we integrated the formal semantics of interworkings BPA_{iw} from Mauw et al. [MvWW93] with the discretely timed theory $ACP_{d\rho}$ from Baeten en Bergstra [BB92]. The latter is an extension of ACP (Algebra of Communicating Processes) to a discrete time process algebra. The intuitive semantics of discrete timing is that the continuous time domain is partitioned into slices of equal length indexed by natural numbers. Actions of ACP are stamped with a time mark from $\mathbf{N}$, and the implicit order of time is present within a time slice only[2].

Going from discretely timed actions, one can imagine actions timed over several time slices (a similar concept was presented in the real time process

[2]In this paper we assume 0 to be an element of $\mathbf{N}$.

algebra $BPA_{\rho\delta I}$ [BB90], by means of the integration operator). In a discretely timed theory, the mapping from a discrete time interval to a choice from a sequence of adjacent time slices appears very natural. In fact, having defined an axiomatisation for discretely timed processes, this is the only definition needed to derive a set of propositions for interworkings timed over discrete intervals, analogous to the axioms for discretely timed processes. The specification of a protocol typically requires an event to happen within some time interval. With the proposed notation it is possible to write down such an event directly.

In the next section we will discuss the fundamentals of BPA and the specialisation of BPA to BPA_{iw} (for as much as needed in BPA_{diw}). In Section 3 we will present a set of axioms that defines a discrete time framework for interworkings. In Section 4 we will extend this set of axioms with the definition of discrete time intervals, and present a number of useful propositions. The closing section contains the conclusions.

2 BPA and BPA_{iw}

In this section an overview of the fundamentals of BPA is presented. The point of view we take is that BPA defines the building blocks from which the specific theory BPA_{iw} is designed. Our aim is not to be complete on the underlying theory, but to provide the reader with the necessary knowledge and understanding of the material. For a thorough discussion of BPA (and extensions) we refer to [BW90]. The formal semantics of interworkings BPA_{iw} is presented in [MvWW93, MvWW92d].

2.1 An overview of BPA

BPA (Basic Process Algebra) is an algebraic theory for the description of process behaviour. In BPA, events are modeled by atomic actions. Atomic actions are indivisible units of behaviour that take no time. All atomic actions in BPA are collected in the action alphabet A. Atomic actions will be notated as a, b, c etc.

The behaviour of a system is described by a process. Processes are defined recursively in terms of atomic actions and processes. Every atomic action is a process. Two processes can be combined into a new process by using the sequential composition ($X \cdot Y$ or XY for short) or the alternative composition ($X + Y$). The sequential composition $X \cdot Y$ is the process that first executes X and then Y. The alternative composition of two processes is the process that chooses and executes one of its options, but not both. The notation for processes is X, Y, Z etc. The sequential and alternative composition are defined in Table 1.

Several extensions to BPA exist, which add features like deadlock, encapsulation, silent step, abstraction etc. to the basic theory. The special action δ (deadlock) is added to the action alphabet of BPA and denotes the unsuccessful termination of a process. The atomic action δ is the action that does nothing but terminate unsuccessfully. The characteristics of deadlock are listed in Table 2. The second axiom of Table 2 states that after a deadlock no action can follow. For this reason δ is sometimes referred to as time stop. The presence of δ in a process expression denotes an inconsistency in the expression.

$$X + Y = Y + X$$
$$(X + Y) + Z = X + (Y + Z)$$
$$X + X = X$$
$$(X + Y) \cdot Z = X \cdot Z + Y \cdot Z$$
$$(X \cdot Y) \cdot Z = X \cdot (Y \cdot Z)$$

Table 1: BPA

$$X + \delta = X$$
$$\delta \cdot X = \delta$$

Table 2: Deadlock

We need the encapsulation operator $\partial_H()$ from [BW90] to define the interworking sequence and merge in Section 3. This operator reduces an atomic action to δ if the action is an element of its encapsulating set. Encapsulation is defined in Table 3.

$$a \notin H \Rightarrow \partial_H(a) = a$$
$$a \in H \Rightarrow \partial_H(a) = \delta$$
$$\partial_H(X + Y) = \partial_H(X) + \partial_H(Y)$$
$$\partial_H(X \cdot Y) = \partial_H(X) \cdot \partial_H(Y)$$

Table 3: Encapsulation

2.2 Building blocks from BPA$_{iw}$

From here, we will restrict attention to the specific case of interworkings. In the following section we will present parts of the work of [MvWW92d]. We also follow their notation.

The previously defined alphabet of atomic actions is restricted to the actions that are defined within the interworkings framework. The most important of these actions is the communication. The other (primitive) actions are executed inside a functional block of the system, and are not visible outside the functional block. They affect the interaction between the functional blocks only implicitly.

Let EID and MID be finite sets, containing entity identifiers and message identifiers. The collection of communication actions is defined by:

$$A_c = \{c(p, q, m) \mid p, q \in EID, m \in MID\}$$

The action $c(p, q, m)$ denotes a communication of message m from entity p to entity q.

Let AN, TID and DID be finite sets of primitive action names, timer identifiers and duration identifiers respectively. The collection of primitive actions consists of four types:

$$\begin{aligned} A_p \quad = \quad & \{act(p, an) \mid p \in EID, an \in AN\} \cup \\ & \{lost(p, q, m) \mid p, q \in EID, m \in MID\} \cup \\ & \{timerset(p, t, d) \mid p \in EID, t \in TID, d \in DID\} \cup \\ & \{timeout(p, t) \mid p \in EID, t \in TID\} \end{aligned}$$

The first, $act(p, an)$ is interpreted as a noncommunication action an performed by entity p. A failed communication from entity p to entity q is denoted by $lost(p, q, m)$, where m is the message identifier. The actions $timerset(p, t, d)$ and $timeout(p, t)$ denote respectively the initialisation of a timer t with duration d by entity p and a timeout signal generated by timer t to entity p.

The specific interworking action alphabet $A_{iw,\delta}$ is the union of A_c and A_p, completed by δ:

$$A_{iw,\delta} = A_c \cup A_p \cup \{\delta\}$$

An auxiliary set generating function is needed to build the theory BPA_{iw}. The entity function $E()$ takes a process as its argument and returns the set of all entities that are involved in the process. This function is also taken from [MvWW92d]. Axioms for the entity function are listed in Table 4.

$$\begin{aligned} & E(\delta) = \emptyset \\ & E(c(p, q, m)) = \{p, q\} \\ & E(act(p, an)) = \{p\} \\ & E(lost(p, q, m)) = \{p\} \\ & E(timerset(p, t, d)) = \{p\} \\ & E(timeout(p, t)) = \{p\} \\ & a \not\equiv \delta \;\Rightarrow\; E(a \cdot X) = E(a) \cup E(X) \\ & E(X + Y) = E(X) \cup E(Y) \end{aligned}$$

Table 4: Entity function

Note that only the initiating entity is involved in a failed communication (Axiom 4 of Table 4).

3 Discrete timing

Now we will introduce discrete timing in the actions of the interworking alphabet. Three operators from the discrete process algebra $\text{BPA}_{d\rho}$ [BB92] are added to the building blocks discussed in the previous section. To specialize the untimed semantics BPA_{iw} to BPA_{diw}, the independent ultimate delay is proposed.

3.1 BPA extended

From the interworking action alphabet $A_{iw,\delta}$ we generate the alphabet for discretely timed actions $A_{diw,\delta}$:

$$A_{diw,\delta} = \{\underline{a}(n) \mid a \in A_{iw,\delta}, n \in \mathbf{N}\}$$

The underscoring of the actions stresses the fact that they are discretely timed. The relation between a slice in discrete time and a point in continuous time (which can be seen as an element of $\mathbf{R} > 0$) is straightforward. The n-th slice starts at real time point $n - 1$ and ends at real time point n.

In Table 5 BPA is extended to $BPA_{d\rho}$ [BB92]. The action $\underline{\underline{\delta}}$ is the discrete time version of the symbolic[3] deadlock. Its intuitive meaning is that of the symbolic deadlock, timed at the end of the current time slice. The characteristics of $\underline{\underline{\delta}}$ are expressed in the first axiom. Then the initialisation operator $\gg_d$ is defined. The process $t \gg_d X$ is the process X starting at time t. All actions contained in X that have to be performed at or before time t are turned into deadlocks because their execution was delayed too long.

The ultimate delay operator $U()$ and the discrete timeout operator $\gg_d$ are also defined in Table 5. The ultimate delay operator takes a process expression as its argument and returns a natural number. X can never reach the time $U(X)$ by just idling. $X \gg_d t$ denotes the process X with the restriction that its first action must be performed at a time before t. With respect to the discrete timeout operator, ω is the neutral element, since $X \gg_d \omega = X$.

To enable the symbolic operator for encapsulation to work within the framework of discrete process algebra, it suffices to define a timed encapsulation operator in terms of its symbolic version. Thus an axiom was designed that only splits an atomic action from its time stamp, and uses the axioms of the symbolic operator. The same applies to the entity function $E()$. These additional axioms are defined in Table 6.

A special variant of the ultimate delay operator was needed because the ultimate delay operator of [BB92] does not take entities into account. If there is a violation of the Global Time Constraint (to be defined shortly), when using the regular ultimate delay operator, some deadlock options may disappear. The independent ultimate delay $U_a()$ is defined in Table 7.

3.2 Formal semantics of interworkings with discrete time

Now we get to the crux of this paper, being the definition of the discrete time interworking sequence and merge.

[3] We will use the term *symbolic* if an untimed version of the theory is denoted, following the terminology in [BB92].

$$\underline{a}(0) = \underline{\delta}(0) = \underline{\underline{\delta}}$$
$$\underline{\delta}(n) \cdot X = \underline{\delta}(n)$$
$$n < m \Rightarrow \underline{\delta}(n) + \underline{\delta}(m) = \underline{\delta}(m)$$
$$\underline{a}(n) + \underline{\delta}(n) = \underline{a}(n)$$
$$\underline{a}(n+1) \cdot X = \underline{a}(n+1) \cdot (n \gg_d X)$$

$$n < m \Rightarrow n \gg_d \underline{a}(m) = \underline{a}(m)$$
$$n \geq m \Rightarrow n \gg_d \underline{a}(m) = \underline{\delta}(n)$$
$$n \gg_d (X + Y) = (n \gg_d X) + (n \gg_d Y)$$
$$n \gg_d (X \cdot Y) = (n \gg_d X) \cdot Y$$

$$U(\underline{a}(t)) = t$$
$$U(X + Y) = \max\{U(X), U(Y)\}$$
$$U(X \cdot Y) = U(X)$$

$$r > t \Rightarrow \underline{a}(r) \gg_d t = \underline{\delta}(t)$$
$$r \leq t \Rightarrow \underline{a}(r) \gg_d t = \underline{a}(r)$$
$$(X + Y) \gg_d t = (X \gg_d t + Y \gg_d t)$$
$$(X \cdot Y) \gg_d t = (X \gg_d t) \cdot Y$$

Table 5: $\mathrm{BPA}_{d\rho}$

$$\partial_H(\underline{a}(n)) = \underline{\partial_H(a)}(n)$$
$$E(\underline{a}(n)) = E(a)$$
$$a \not\equiv \delta \Rightarrow E(\underline{a}(n) \cdot X) = E(\underline{a}(n)) \cup E(X)$$

Table 6: Additional axioms for $\partial_H()$ and $E()$

$E(a) \cap E(b) = \emptyset \Rightarrow U_a(b(t)) = t$
$E(a) \cap E(b) \neq \emptyset \Rightarrow U_a(b(t)) = \omega$
$E(a) \cap E(b) = \emptyset \Rightarrow U_a(b(t) \cdot X) = t$
$E(a) \cap E(b) \neq \emptyset \Rightarrow U_a(b(t) \cdot X) = U_a(X)$
$U_a(X + Y) = \max\{U_a(X), U_a(Y)\}$

Table 7: Independent ultimate delay

3.2.1 Discrete interworking sequence

Informally, the result of sequencing two interworkings is their arbitrary interleaved composition restricted by:

Common Entity Constraint Actions involving common entities are performed first by the left process and then by the right process.

Global Time Constraint Time is not allowed to run backwards, i.e. causality must be preserved.

The formal semantics of the discrete interworking sequence $\circ_{diw}$ is expressed in terms of the operators left-sequence $\mathbf{L}\circ_{diw}$ and right-sequence $\mathbf{R}\circ_{diw}$. The left sequence forces the left process to do the first step. The right sequence works similarly, but it also implements the aforementioned Common Entity Constraint by means of the encapsulation operator and the function $\alpha_E()$. The discrete interworking sequence is defined in Table 8. Table 8 strongly reflects the symbolic interworking sequence from [MvWW93].

$X \circ_{diw} Y = X \,\mathbf{L}\circ_{diw}\, Y + X \,\mathbf{R}\circ_{diw}\, Y$
$\underline{a}(n) \,\mathbf{L}\circ_{diw}\, X = (\underline{a}(n) \gg_d U_a(X)) \cdot X$
$\underline{a}(n) \cdot X \,\mathbf{L}\circ_{diw}\, Y = (\underline{a}(n) \gg_d U_a(Y)) \cdot (X \circ_{diw} Y)$
$(X + Y) \,\mathbf{L}\circ_{diw}\, Z = X \,\mathbf{L}\circ_{diw}\, Z + Y \,\mathbf{L}\circ_{diw}\, Z$
$X \,\mathbf{R}\circ_{diw}\, \underline{a}(n) = \partial_{\alpha_E(X)}(\underline{a}(n) \gg_d U(X)) \cdot X$
$X \,\mathbf{R}\circ_{diw}\, \underline{a}(n) \cdot Y = \partial_{\alpha_E(X)}(\underline{a}(n) \gg_d U(X)) \cdot (X \circ_{diw} Y)$
$X \,\mathbf{R}\circ_{diw}\, (Y + Z) = X \,\mathbf{R}\circ_{diw}\, Y + X \,\mathbf{R}\circ_{diw}\, Z$

Table 8: Discrete Timed Interworking Sequence

The function $\alpha_E(X)$ (in axioms 5 and 6) generates the set of actions that are performed by the entities that are involved in a process X:

$$\alpha_E(X) = \{a \in A_{iw,\delta} | E(a) \cap E(X) \neq \emptyset\}$$

3.2.2 *Discrete interworking merge*

The intuitive idea of the interworking merge is again the interleaved composition of its operands restricted by:

Synchronisation Constraint Actions involving a common pair of entities are forced to synchronise.

Global Time Constraint As defined before.

The formal semantics of the interworking merge is implemented similarly to that of the sequence. The merge splits up into two interleaving options and one synchronisation (or communication) option. The discrete interworking merge is defined in Table 9. Table 9 strongly reflects the symbolic interworking merge from [MvWW93].

$$
\begin{aligned}
&X \|_{diw}^{S} Y = X \mathbin{\|\!\!\lfloor}_{diw}^{S} Y + Y \mathbin{\|\!\!\lfloor}_{diw}^{S} X + X |_{diw}^{S} Y\\
&\underline{a}(n) \mathbin{\|\!\!\lfloor}_{diw}^{S} X = \partial_S((\underline{a}(n) \gg_d U(X)) \cdot X)\\
&\underline{a}(n) \cdot X \mathbin{\|\!\!\lfloor}_{diw}^{S} Y = \partial_S(\underline{a}(n) \gg_d U(Y)) \cdot (X \|_{diw}^{S} Y)\\
&(X+Y) \mathbin{\|\!\!\lfloor}_{diw}^{S} Z = X \mathbin{\|\!\!\lfloor}_{diw}^{S} Z + Y \mathbin{\|\!\!\lfloor}_{diw}^{S} Z\\
&\underline{a}(n) |_{diw}^{S} \underline{b}(m) = \gamma_S(\underline{a}(n), \underline{b}(m))\\
&\underline{a}(n) \cdot X |_{diw}^{S} \underline{b}(m) = (\underline{a}(n) |_{diw}^{S} \underline{b}(m)) \cdot \partial_S(X)\\
&\underline{a}(n) |_{diw}^{S} \underline{b}(m) \cdot X = (\underline{a}(n) |_{diw}^{S} \underline{b}(m)) \cdot \partial_S(X)\\
&\underline{a}(n) \cdot X |_{diw}^{S} \underline{b}(m) \cdot Y = (\underline{a}(n) |_{diw}^{S} \underline{b}(m)) \cdot (X \|_{diw}^{S} Y)\\
&(X+Y) |_{diw}^{S} Z = X |_{diw}^{S} Z + Y |_{diw}^{S} Z\\
&X |_{diw}^{S} (Y+Z) = X |_{diw}^{S} Y + X |_{diw}^{S} Z\\
&\gamma_S(\underline{a}(n), \underline{b}(m)) = \begin{cases} \underline{\gamma_S(a,b)(n)} & \text{if } n = m \\ \underline{\delta}(\min(n,m)) & \text{otherwise} \end{cases}\\
&\gamma_S(a,b) = \begin{cases} a & \text{if } a = b \wedge a \in S \\ \delta & \text{otherwise} \end{cases}
\end{aligned}
$$

Table 9: Discrete Timed Interworking Merge

Note that the ultimate delay from [BB92] is used in the axioms for the right sequence and all axioms of Table 9. The independent operator is not needed here because the global time constraint is already guarded by the encapsulation operator. To complete the definition of the discretely timed interworking merge we define a function $\alpha_{CE}(X,Y)$:

$$\alpha_{CE}(X,Y) = \{c(p,q,m) \in A_c \mid p,q \in E(X) \cap E(Y)\}$$

This function generates the set of communication actions on which processes X and Y are forced to synchronize. The interworking merge becomes:

$$X \|_{diw} Y = X \|_{diw}^{\alpha_{CE}(X,Y)} Y$$

4 Extension to interval timing

In this section we'll introduce interval timing into the interworking concept. We start by defining discrete intervals. Then a number of propositions is presented that extends the timed interworking sequence and merge to an analogous framework for discrete intervals. In the last subsection a number of auxiliary propositions is given.

4.1 Definition of discrete intervals

Intuitively, a discrete time interval is a sequence of adjacent time slices. An action timed over an interval can execute in any of the slices contained in the interval, so an interval can be defined as a choice from the slices in it. However, to prevent the time to run backwards, intervals must start at a slice with an index less than or equal to the index of the slice where they end. If an action is timed over an interval running from n to m and $n > m$, we define it to be a deadlock at m.

$$n \leq m \Rightarrow \underline{a}(n,m) = \textstyle\sum_{t=n}^{m} \underline{a}(t)$$
$$n > m \Rightarrow \underline{a}(n,m) = \underline{\delta}(m)$$

Table 10: Definition of discrete intervals

It should be noted that this definition is just a notational issue. The interval notation is just a shorthand for a repeated alternative composition of the same action over several adjacent time slices. In the case that $n = m$, an interval timed action reduces to a discretely timed action. A process X in which all actions are timed from 0 to ω reduces to a symbolic process. The (interval) time stamps then have become redundant.

4.2 Interworking sequence extended

Now we can extend the axioms presented in Table 8 and 9 to intervals. The extension is one of practical nature. It introduces no essential new features to the theory in the sense that deductions are possible with the extension that weren't without it. However, it provides a set of rules that reduces writing to a minimum. To accomplish this we derive a number of propositions that leave the intervals intact as much as possible. For the proofs of the propositions we refer to appendix A.

Propositions 1 and 2 are useful to prevent redundant calculations, i.e. calculations (of subexpressions) leading to deadlocks which disappear from the process expression. The first proposition states that the application of the timeout operator on an interval timed action can change the duration of the interval; the end can be chopped off, or it can disappear completely, reducing the action to deadlock.

Proposition 1 $\underline{a}(p,q) \gg_d r = \underline{a}(p, \min(q,r))$

Table 11 lists axioms 2, 3, 5 and 6 of Table 8 that are adapted using proposition 1. The other axioms of Table 8 remain valid within the interval framework.

$$
\begin{aligned}
&\underline{a}(n,m)\ \mathbf{L}\!\circ_{diw} X = (\underline{a}(n,\min(m,U_a(X))) \cdot X\\
&\underline{a}(n,m)\cdot X\ \mathbf{L}\!\circ_{diw} Y = (\underline{a}(n,\min(m,U_a(Y)))) \cdot (X \circ_{diw} Y)\\
&X\ \mathbf{R}\!\circ_{diw}\ \underline{a}(n,m) = \partial_{\alpha_E(X)}(\underline{a}(n,\min(m,U(X)))) \cdot X\\
&X\ \mathbf{R}\!\circ_{diw}\ \underline{a}(n,m)\cdot Y = \partial_{\alpha_E(X)}(\underline{a}(n,\min(m,U(X)))) \cdot (X \circ_{diw} Y)
\end{aligned}
$$

Table 11: Additional propositions for the Interworking Sequence

4.3 Interworking merge extended

Proposition 2 states that communication between two interval timed actions can only take place at the intersection of the two intervals.

Proposition 2 $\underline{a}(n,m)|^S_{diw}\underline{b}(p,q) = \underline{\gamma_S(a,b)}(\max(n,p),\min(m,q))$

Table 12 lists axioms 2, 3 and 5 through 8 of Table 9 that are adapted using Proposition 1 and 2. The other axioms of Table 9 remain valid within the interval framework. The last three axioms of Table 12 follow directly from the definition of intervals and the distributivity of $|^S_{diw}$.

$$
\begin{aligned}
&\underline{a}(n,m) \mathbin{\lfloor\!\lfloor}^S_{diw} X = \partial_S((\underline{a}(n,\min(m,U(X))) \cdot X)\\
&\underline{a}(n,m)\cdot X \mathbin{\lfloor\!\lfloor}^S_{diw} Y = \partial_S(\underline{a}(n,\min(m,U(Y)) \cdot (X\|^S_{diw}Y)\\
&\underline{a}(n,m)|^S_{diw}\underline{b}(p,q) = \underline{\gamma_S(a,b)}(\max(n,p),\min(m,q))\\
&\underline{a}(n,m)\cdot X|^S_{diw}\underline{b}(p,q) = (\underline{a}(n,m)|^S_{diw}\underline{b}(p,q)) \cdot \partial_S(X)\\
&\underline{a}(n,m)|^S_{diw}\underline{b}(p,q)\cdot X = (\underline{a}(n,m)|^S_{diw}\underline{b}(p,q)) \cdot \partial_S(X)\\
&\underline{a}(n,m)\cdot X|^S_{diw}\underline{b}(p,q)\cdot Y = (\underline{a}(n,m)|^S_{diw}\underline{b}(p,q)) \cdot (X\|^S_{diw}Y)
\end{aligned}
$$

Table 12: Additional propositions for the Interworking Merge

4.4 Auxiliary propositions

Propositions 3, 4 and 5 can be used to recombine different alternatives into a compact interval expression. For example, one often encounters expressions of the form $\underline{a}(5)\underline{b}(5,7) + \underline{a}(6)\underline{b}(6,7) + \underline{a}(7)\underline{b}(7)$. Using these propositions, this expression reduces to $\underline{a}(5,7)\underline{b}(5,7)$.

The following proposition states that under certain conditions options of execution in an earlier time slice can be added to a discretely timed action, thus transforming it into an interval timed action. This transformation should only be possible if it doesn't create new possibilities for execution. The aforementioned conditions ensure that this is the case.

Proposition 3 $(0 \leq k \leq l \leq n) \Rightarrow \underline{a}(n) \cdot \underline{b}(l) = \underline{a}(n) \cdot \underline{b}(k, l)$

Proposition 4 is very similar to Proposition 3, but now the interval of an already interval timed action is extended 'into the past'. Again, the conditions of the proposition ensure that this proposition remains consistent within BPA_{diw}.

Proposition 4 $(0 \leq k \leq l \leq n) \Rightarrow \underline{a}(n) \cdot \underline{b}(l, p) = \underline{a}(n) \cdot \underline{b}(k, p)$

Proposition 5 expresses the fact that in a sequential composition of two interval timed actions the beginning of the interval of the trailing action can be extended or chopped off if the beginning of the interval of the leading action falls in the interval of the trailing action (i.e. if the interval of the trailing action spans over the first slice of the interval of the leading action).

Proposition 5 $(0 \leq k \leq l \leq n) \Rightarrow \underline{a}(n, m) \cdot \underline{b}(l, p) = \underline{a}(n, m) \cdot \underline{b}(k, p)$

In Proposition 6 is expressed that the beginning of an interval can be chopped off or even completely vanish due to application of the initialisation operator. This proposition can be used to prevent redundant calculations. Note that Proposition 6 is the counterpart of Proposition 1, which states a similar fact about the discrete time out operator.

Proposition 6
$p < r \Rightarrow p \gg_d \underline{a}(q, r) = \underline{a}(\max(p+1, q), r)$
$p \geq r \Rightarrow p \gg_d \underline{a}(q, r) = \underline{\delta}(p)$

In Proposition 7, the interval of a leading action is split into two options, according to the ultimate delay of the trailing process expression.

Proposition 7
$U(X) \geq p \Rightarrow \underline{a}(p, q) \cdot X = \underline{a}(p, \min(q, U(X))) \cdot X + \underline{a}(\min(q, U(X)) + 1, q) \cdot \underline{\underline{\delta}}$
$U(X) < p \Rightarrow \underline{a}(p, q) \cdot X = \underline{a}(p, q) \cdot \underline{\underline{\delta}}$

5 Conclusions

By presenting BPA_{diw} we provided the formal semantics of a timed version of T and thereby of discrete absolute timed interworkings. We constructed the action alphabet of BPA_{diw} by labeling the actions of BPA_{iw} [MvWW92d] with a time stamp. We completed BPA_{diw} by adapting the axioms of BPA_{iw} for the Interworking Sequence and Interworking Merge. It turned out that we needed the Independent Ultimate Delay to do this. Next we presented a notation for discrete time intervals. We proved a number of propositions which help to reduce calculations to a minimum.

The definition of a timed version of IW and T [MvWW92c, MvWW92b] was beyond the scope of this paper. We think time can easily be added to IW and T, and we assumed that a mapping from a timed version of T to BPA_{diw} exists. Furthermore, a proof of the completeness and consistency of the theory BPA_{diw} has been beyond the scope of our project.

We conclude that time can easily be added to BPA_{iw}. This also holds for the case of intervals.

Acknowledgements We want to thank Thijs Winter for introducing us into the world of interworkings. Special thanks are for Jan Bergstra for guiding us during the project.

References

[BB90] J.C.M. Baeten and J.A. Bergstra. Real time process algebra. Technical Report CS-R9053, Computer Science/Department of Software Technology, Centre for Mathematics and Computer Science, Amsterdam, October 1990.

[BB92] J.C.M. Baeten and J.A. Bergstra. Discrete time process algebra. Technical Report P9208b, Programming Research Group, Faculty of Mathematics and Computer Science, University of Amsterdam, December 1992. An extended abstract appeared in Proc. CONCUR'92, Stony Brook (W.R. Cleaveland, ed.), LNCS 630, Springer 1992, pp. 401-420.

[BW90] J.C.M. Baeten and W.P. Weijland. *Process algebra.* Cambridge tracts in theoretical computer science 18. Cambridge University Press, 1990. ISBN 0–521–40043–0.

[MvWW92a] S. Mauw, M. van Wijk, and T. Winter. Syntax and semantics of synchronous interworkings: Informal semantics. Philips Research Laboratorium Eindhoven, Technical University Eindhoven, 1992.

[MvWW92b] S. Mauw, M. van Wijk, and T. Winter. Syntax and semantics of synchronous interworkings: The concrete syntax iw. Philips Research Laboratorium Eindhoven, Technical University Eindhoven, 1992.

[MvWW92c] S. Mauw, M. van Wijk, and T. Winter. Syntax and semantics of synchronous interworkings: The syntax of t and denotational semantics of iw. Philips Research Laboratorium Eindhoven, Technical University Eindhoven, 1992.

[MvWW92d] S. Mauw, M. van Wijk, and T. Winter. Syntax and semantics of synchronous interworkings: A formal semantics. Philips Research Laboratorium Eindhoven, Technical University Eindhoven, 1992.

[MvWW93] S. Mauw, M. van Wijk, and T. Winter. A formal semantics of synchronous interworkings. In *SDL'93: Using objects*, pages 167–178. Elsevier Science Publishers B.V., North-Holland, 1993. ISBN 0–444–81486–8.

[Til91] P.A.J. Tilanus. A formalisation of message sequence charts. In *SDL'91: Evolving methods*, pages 273–288. Elsevier Science Publishers B.V., North-Holland, 1991. ISBN 0–444–88976–0.

A Proofs

Proof of proposition 1
$\underline{a}(p,q) \gg_d r = \underline{a}(p,\min(q,r))$

Case 1 Assume $(p > q)$
$\underline{a}(p,q) \gg_d r = \underline{\delta}(q) \gg_d r = \cdots$

1.a Assume $(q \leq r) \cdots = \underline{\delta}(q) = \underline{a}(p,q) = \underline{a}(p,\min(q,r))$

1.b Assume $(q > r) \cdots = \underline{\delta}(r) = \underline{a}(p,r) = \underline{a}(p,\min(q,r))$

Case 2 Assume $(p \leq q)$
$\underline{a}(p,q) \gg_d r = \sum_{i=p}^{q} \underline{a}(i) \gg_d r = \underline{a}(p) \gg_d r + \cdots + \underline{a}(q) \gg_d r = \cdots$

2.a Assume $(q \leq r) \cdots = \underline{a}(p) + \cdots + \underline{a}(q) = \sum_{i=p}^{q} \underline{a}(i) = \underline{a}(p,q) = \underline{a}(p,\min(q,r))$

2.b Assume $(p \leq r < q) \cdots =$
$\underline{a}(p) \gg_d r + \cdots + \underline{a}(r) \gg_d r + \underline{a}(r+1) \gg_d r + \cdots + \underline{a}(q) \gg_d r =$
$\underline{a}(p) + \cdots + \underline{a}(r) + \underline{\delta}(r) + \cdots + \underline{\delta}(r) = \underline{a}(p) + \cdots + \underline{a}(r) = \sum_{i=p}^{r} \underline{a}(i) =$
$\underline{a}(p,r) = \underline{a}(p,\min(q,r))$

2.c Assume $(r < p) \cdots = \underline{\delta}(r) + \cdots + \underline{\delta}(r) = \underline{\delta}(r) = \underline{a}(p,r) = \underline{a}(p,\min(q,r))$ □

Proof of proposition 2
$\underline{a}(n,m)|_{diw}^{S}\underline{b}(p,q) = \underline{\gamma_S(a,b)}(\max(n,p),\min(m,q))$

The communication function $\gamma_S()$ is commutative, so the following cases cover all possible combinations of two intervals.

Case 1 Assume $(n > m)$
$\underline{a}(n,m)|_{diw}^{S}\underline{b}(p,q) = \underline{\delta}(m)|_{diw}^{S}\underline{b}(p,q) = \underline{\delta}(m)|_{diw}^{S}(\sum_{i=p}^{q} \underline{b}(i)) =$
$\underline{\delta}(m)|_{diw}^{S}\underline{b}(p) + \underline{\delta}(m)|_{diw}^{S}\underline{b}(p+1) + \cdots + \underline{\delta}(m)|_{diw}^{S}\underline{b}(q) = \cdots$

1.a Assume $(m \leq q) \cdots = \underline{\delta}(p) + \underline{\delta}(p+1) + \cdots + \underline{\delta}(m) + \cdots + \underline{\delta}(m) = \underline{\delta}(m) =$
$\underline{\gamma_S(a,b)}(n,m) = \underline{\gamma_S(a,b)}(\max(n,p),\min(m,q))$

1.b Assume $(m > q) \cdots = \underline{\delta}(p) + \cdots + \underline{\delta}(q) = \underline{\delta}(q) = \underline{\gamma_S(a,b)}(n,q) =$
$\underline{\gamma_S(a,b)}(\max(n,p),\min(m,q))$

Case 2 Assume $(n \leq m)$
$\underline{a}(n,m)|_{diw}^{S}\underline{b}(p,q) = (\sum_{i=n}^{m} \underline{a}(i))|_{diw}^{S}(\sum_{j=p}^{q} \underline{b}(j)) =$

$$\begin{array}{lll} \underline{a}(n)|_{diw}^{S}\underline{b}(p) & + \cdots + \underline{a}(n)|_{diw}^{S}\underline{b}(q) & + \cdots \\ \underline{a}(m)|_{diw}^{S}\underline{b}(p) & + \cdots + \underline{a}(m)|_{diw}^{S}\underline{b}(q) & = \end{array}$$

$$\gamma_S(\underline{a}(n), \underline{b}(p)) \quad + \cdots + \gamma_S(\underline{a}(n), \underline{b}(q)) \quad + \cdots$$
$$\gamma_S(\underline{a}(m), \underline{b}(p)) \quad + \cdots + \gamma_S(\underline{a}(m), \underline{b}(q)) \quad = \cdots$$

2.a Assume $(n \le m < p \le q) \cdots =$

$$\underline{\delta}(n) \quad + \cdots + \underline{\delta}(n) \quad + \cdots$$
$$\underline{\delta}(m) \quad + \cdots + \underline{\delta}(m) \quad =$$
$$\underline{\delta}(m) = \underline{\gamma_S(a,b)}(p,m) = \underline{\gamma_S(a,b)}(\max(n,p), \min(m,q))$$

2.b Assume $(n \le p \le q \le m) \cdots =$

$$\underline{\delta}(n) \qquad\qquad + \cdots \qquad\qquad +\underline{\delta}(n) \qquad + \cdots$$
$$\underline{\gamma_S(a,b)}(p) \quad +\underline{\delta}(p) \qquad\qquad + \cdots \qquad +\underline{\delta}(p) \qquad +$$
$$\underline{\delta}(p+1) \qquad +\underline{\gamma_S(a,b)}(p+1) \quad + \cdots \qquad +\underline{\delta}(p+1) \qquad + \cdots$$
$$\underline{\delta}(q) \qquad\qquad + \cdots \qquad\qquad\qquad +\underline{\gamma_S(a,b)}(q) \quad =$$
$$\underline{\gamma_S(a,b)}(p) + \underline{\gamma_S(a,b)}(p+1) + \cdots + \underline{\gamma_S(a,b)}(q) = \sum_{i=p}^{q} \underline{\gamma_S(a,b)}(i) =$$
$$\underline{\gamma_S(a,b)}(p,q) = \underline{\gamma_S(a,b)}(\max(n,p), \min(m,q))$$

2.c Assume $(n \le p \le m \le q) \cdots =$

$$\underline{\delta}(n) \qquad\qquad + \cdots \qquad\qquad\qquad\qquad\qquad +\underline{\delta}(n) \qquad + \cdots$$
$$\underline{\gamma_S(a,b)}(p) \quad +\underline{\delta}(p) \qquad\qquad + \cdots \qquad\qquad\qquad +\underline{\delta}(p) \qquad +$$
$$\underline{\delta}(p+1) \qquad +\underline{\gamma_S(a,b)}(p+1) \quad + \cdots \qquad\qquad\qquad +\underline{\delta}(p+1) \quad + \cdots$$
$$\underline{\delta}(m) \qquad\qquad + \cdots \qquad\qquad +\underline{\gamma_S(a,b)}(m) \quad + \cdots \quad +\underline{\delta}(m) \qquad =$$
$$\underline{\gamma_S(a,b)}(p) + \underline{\gamma_S(a,b)}(p+1) + \cdots + \underline{\gamma_S(a,b)}(m) = \sum_{i=p}^{m} \underline{\gamma_S(a,b)}(i) =$$
$$\underline{\gamma_S(a,b)}(p,m) = \underline{\gamma_S(a,b)}(\max(n,p), \min(m,q)) \qquad\qquad \square$$

Proof of proposition 3
$$(0 \le k \le l \le n) \Rightarrow \underline{a}(n) \cdot \underline{b}(l) = \underline{a}(n) \cdot \underline{b}(k,l)$$

Case 1 Assume $(n = 0)$
$$n = 0 \Rightarrow k = l = 0 \Rightarrow \underline{a}(0)\underline{b}(0) = \underline{a}(0)\underline{b}(0,0)$$

Case 2 Assume $(l < n)$
$$\underline{a}(n)\underline{b}(l) = \underline{a}(n)(n - 1 \gg_d \underline{b}(l)) = \underline{a}(n)\underline{\delta}(n-1) = \cdots$$
$$\underline{a}(n)(\underline{\delta}(n-1) + \cdots + \underline{\delta}(n-1)) = \underline{a}(n)(n - 1 \gg_d \underline{b}(k) + \cdots + n - 1 \gg_d \underline{b}(l)) =$$
$$\underline{a}(n)(n - 1 \gg_d (\underline{b}(k) + \cdots + \underline{b}(l))) = \underline{a}(n)(n - 1 \gg_d \sum_{i=k}^{l} \underline{b}(i)) = \underline{a}(n)\underline{b}(k,l)$$

Case 3 Assume $(l = n)$
$$\underline{a}(n)\underline{b}(l) = \underline{a}(n)\underline{b}(n) = \underline{a}(n)(\underline{\delta}(n) + \underline{b}(n)) = \underline{a}(n)(\underline{\delta}(n-1) + \underline{\delta}(n) + \underline{b}(n)) =$$
$$\underline{a}(n)(\underline{\delta}(n-1) + \underline{b}(n)) = \cdots = \underline{a}(n)(\underline{\delta}(n-1) + \cdots + \underline{\delta}(n-1) + \underline{b}(n)) =$$
$$\underline{a}(n)(n - 1 \gg_d \underline{b}(k) + \cdots + n - 1 \gg_d \underline{b}(n-1) + n - 1 \gg_d \underline{b}(n)) =$$
$$\underline{a}(n)(n - 1 \gg_d (\underline{b}(k) + \cdots + \underline{b}(n))) = \underline{a}(n)(n - 1 \gg_d \sum_{i=k}^{n} \underline{b}(i)) =$$
$$\underline{a}(n)\underline{b}(k,n) = \underline{a}(n)\underline{b}(k,l) \qquad\qquad \square$$

Proof of proposition 4
$(0 \leq k \leq l \leq n) \Rightarrow \underline{a}(n) \cdot \underline{b}(l,p) = \underline{a}(n) \cdot \underline{b}(k,p)$

Case 1 Assume $(n = 0)$
$n = 0 \Rightarrow k = l = 0 \Rightarrow \underline{a}(0)\underline{b}(0,0) = \underline{a}(0)\underline{b}(0,0)$

Case 2 Assume $(l > p)$
$l > p \Rightarrow p < n \Rightarrow \underline{a}(n)\underline{b}(l,p) = \underline{a}(n)\underline{\delta}(p) = \underline{a}(n)(n{-}1 \gg_d \underline{\delta}(p)) = \underline{a}(n)\underline{\delta}(n{-}1) =$
$\underline{a}(n)(\underline{\delta}(n-1) + \underline{\delta}(n-1)) = \cdots = \underline{a}(n)(\underline{\delta}(n-1) + \cdots + \underline{\delta}(n-1)) =$
$\underline{a}(n)(n-1 \gg_d \underline{b}(k) + \cdots + n-1 \gg_d \underline{b}(p)) = \underline{a}(n)(n-1 \gg_d \sum_{i=k}^{p} \underline{b}(i)) =$
$\underline{a}(n)\underline{b}(k,p)$

Case 3 Assume $(l \leq p \leq n)$
$\underline{a}(n)\underline{b}(l,p) \overset{\text{Prop. 3}}{=} \underline{a}(n)\underline{b}(p) \overset{\text{Prop. 3}}{=} \underline{a}(n)\underline{b}(k,p)$

Case 4 Assume $(l < n \leq p)$
$\underline{a}(n)\underline{b}(l,p) = \underline{a}(n)(n-1 \gg_d \sum_{i=l}^{p} \underline{b}(i)) =$
$\underline{a}(n)(n{-}1 \gg_d \underline{b}(l) + \cdots + n{-}1 \gg_d \underline{b}(n{-}1) + n{-}1 \gg_d \underline{b}(n) + \cdots + n{-}1 \gg_d \underline{b}(p)) =$
$\underline{a}(n)(\underline{\delta}(n-1) + \cdots + \underline{\delta}(n-1) + \underline{b}(n) + \cdots + \underline{b}(p)) = \cdots =$
$\underline{a}(n)(\underline{\delta}(n-1) + \cdots + \underline{\delta}(n-1) + \underline{b}(n) + \cdots + \underline{b}(p)) =$
$\underline{a}(n)(n{-}1 \gg_d \underline{b}(k) + \cdots + n{-}1 \gg_d \underline{b}(n{-}1) + n{-}1 \gg_d \underline{b}(n) + \cdots + n{-}1 \gg_d \underline{b}(p)) =$
$\underline{a}(n)(n-1 \gg_d \sum_{i=k}^{p} \underline{b}(i)) = \underline{a}(n)\underline{b}(k,p)$

Case 5 Assume $(l = n \leq p)$
$\underline{a}(n)\underline{b}(l,p) = \underline{a}(n)\underline{b}(n,p) = \underline{a}(n) \sum_{i=n}^{p} \underline{b}(i) = \underline{a}(n)(\underline{b}(n) + \cdots + \underline{b}(p)) =$
$\underline{a}(n)(\underline{\delta}(n) + \underline{b}(n) + \cdots + \underline{b}(p)) = \underline{a}(n)(\underline{\delta}(n-1) + \underline{\delta}(n) + \underline{b}(n) + \cdots + \underline{b}(p)) =$
$\underline{a}(n)(\underline{\delta}(n-1) + \underline{b}(n) + \cdots + \underline{b}(p)) = \cdots$
$\underline{a}(n)(\underline{\delta}(n-1) + \cdots + \underline{\delta}(n-1) + \underline{b}(n) + \cdots + \underline{b}(p)) =$
$\underline{a}(n)(n{-}1 \gg_d \underline{b}(k) + \cdots + n{-}1 \gg_d \underline{b}(n{-}1) + n{-}1 \gg_d \underline{b}(n) + \cdots + n{-}1 \gg_d \underline{b}(p)) =$
$\underline{a}(n)(n-1 \gg_d \sum_{i=k}^{p} \underline{b}(i)) = \underline{a}(n)\underline{b}(k,p)$ □

Proof of proposition 5
$(0 \leq k \leq l \leq n) \Rightarrow \underline{a}(n,m) \cdot \underline{b}(l,p) = \underline{a}(n,m) \cdot \underline{b}(k,p)$

Case 1 Assume $(n > m)$
$\underline{a}(n,m)\underline{b}(l,p) = \underline{\delta}(m)\underline{b}(l,p) = \underline{\delta}(m) = \underline{\delta}(m)\underline{b}(k,p) = \underline{a}(n,m)\underline{b}(k,p)$

Case 2 Assume $(n \leq m)$
$\underline{a}(n,m)\underline{b}(l,p) = \sum_{i=n}^{m} \underline{a}(i)\underline{b}(l,p) = (\underline{a}(n) + \cdots + \underline{a}(m))\underline{b}(l,p) =$
$\underline{a}(n)\underline{b}(l,p) + \cdots + \underline{a}(m)\underline{b}(l,p) \overset{\text{Prop. 4}}{=} \underline{a}(n)\underline{b}(k,p) + \cdots + \underline{a}(m)\underline{b}(k,p) =$
$(\underline{a}(n) + \cdots + \underline{a}(m))\underline{b}(k,p) = \sum_{i=n}^{m} \underline{a}(i)\underline{b}(k,p) = \underline{a}(n,m)\underline{b}(k,p)$ □

Proof of proposition 6
$p < r \Rightarrow p \gg_d \underline{a}(q,r) = \underline{a}(\max(p+1,q),r)$
$p \geq r \Rightarrow p \gg_d \underline{a}(q,r) = \underline{\delta}(p)$

Case 1 Assume $(q > r)$
$p \gg_d \underline{a}(q,r) = p \gg_d \underline{\delta}(r) = \cdots$

1.a Assume $(p < r) \cdots = \underline{\delta}(r) = \underline{a}(q,r) = \underline{a}(\max(p+1,q),r)$

1.b Assume $(p \geq r) \cdots = \underline{\delta}(p)$

Case 2 Assume $(q \leq r)$
$p \gg_d \underline{a}(q,r) = p \gg_d \sum_{i=q}^{r} \underline{a}(i) = \sum_{i=q}^{r}(p \gg_d \underline{a}(i)) = \cdots$

2.a Assume $(p < q) \cdots = \sum_{i=q}^{r} \underline{a}(i) = \underline{a}(q,r) = \underline{a}(\max(p+1,q),r)$

2.b Assume $(q \leq p < r) \cdots =$
$p \gg_d \underline{a}(q) + \cdots + p \gg_d \underline{a}(p) + p \gg_d \underline{a}(p+1) + \cdots + p \gg_d \underline{a}(r) =$
$\underline{\delta}(p) + \cdots + \underline{\delta}(p) + \underline{a}(p+1) + \cdots + \underline{a}(r) = \underline{a}(p+1) + \cdots + \underline{a}(r) =$
$\sum_{i=p+1}^{r} \underline{a}(i) = \underline{a}(p+1,r) = \underline{a}(\max(p+1,q),r)$

2.c Assume $(p \geq r) \cdots = \sum_{i=q}^{r} \underline{\delta}(p) = \underline{\delta}(p)$ □

Proof of proposition 7
$U(X) \geq p \Rightarrow \underline{a}(p,q) \cdot X = \underline{a}(p, \min(q, U(X))) \cdot X + \underline{a}(\min(q,U(X))+1,q) \cdot \underline{\underline{\delta}}$
$U(X) < p \Rightarrow \underline{a}(p,q) \cdot X = \underline{a}(p,q) \cdot \underline{\underline{\delta}}$

Case 1 Assume $(p > q)$
$\underline{a}(p,q)X = \underline{\delta}(q)X = \underline{\delta}(q) = \cdots$

1.a Assume $(U(X) \geq p) \cdots = \underline{\delta}(q) + \underline{\delta}(q) = \underline{\delta}(q)X + \underline{\delta}(q)\underline{\underline{\delta}} =$
$\underline{a}(p,q)X + \underline{a}(q+1,q)\underline{\underline{\delta}} = \underline{a}(p,\min(q,U(X)))X + \underline{a}(\min(q,U(X))+1,q)\underline{\underline{\delta}}$

1.b Assume $(U(X) < p) \cdots = \underline{\delta}(q)\underline{\underline{\delta}} = \underline{a}(p,q)\underline{\underline{\delta}}$

Case 2 Assume $(p \leq q)$
$\underline{a}(p,q)X = \sum_{i=p}^{q} \underline{a}(i)X = (\underline{a}(p) + \underline{a}(p+1) + \cdots + \underline{a}(q))X =$
$\underline{a}(p)(p-1 \gg_d X) + \underline{a}(p+1)(p \gg_d X) + \cdots + \underline{a}(q)(q-1 \gg_d X) = \cdots$

2.a Assume $(U(X) < p)$
$\underline{a}(p)\underline{\delta}(p-1) + \underline{a}(p+1)\underline{\delta}(p) + \cdots + \underline{a}(q)\underline{\delta}(q-1) = \underline{a}(p)\underline{\underline{\delta}} + \underline{a}(p+1)\underline{\underline{\delta}} + \cdots + \underline{a}(q)\underline{\underline{\delta}} =$
$(\underline{a}(p) + \underline{a}(p+1) + \cdots + \underline{a}(q))\underline{\underline{\delta}} = \sum_{i=p}^{q} \underline{a}(i)\underline{\underline{\delta}} = \underline{a}(p,q)\underline{\underline{\delta}}$

2.b Assume $(p \leq U(X) \leq q)$
$\underline{a}(p)X + \cdots + \underline{a}(U(X))X + \underline{a}(U(X)+1)\underline{\delta}(U(x)) + \cdots + \underline{a}(q)\underline{\delta}(q-1) =$
$\underline{a}(p)X + \cdots + \underline{a}(U(X))X + \underline{a}(U(X)+1)\underline{\underline{\delta}} + \cdots + \underline{a}(q)\underline{\underline{\delta}} =$
$\underline{a}(p,U(X))X + \underline{a}(U(X)+1,q)\underline{\underline{\delta}} =$
$\underline{a}(p,\min(q,U(X))X + \underline{a}(\min(q,U(X))+1,q)\underline{\underline{\delta}}$

2.c Assume $(q < U(X))$
$\underline{a}(p)X + \cdots + \underline{a}(q)X = (\underline{a}(p) + \cdots + \underline{a}(q))X = \sum_{i=p}^{q} \underline{a}(i)X = \underline{a}(p,q)X =$
$(\underline{a}(p,q) + \underline{\delta}(q))X = \underline{a}(p,q)X + \underline{\delta}(q)X = \underline{a}(p,q)X + \underline{\delta}(q) = \underline{a}(p,q)X + \underline{\delta}(q)\underline{\underline{\delta}} =$
$\underline{a}(p,q)X + \underline{a}(q+1,q)\underline{\underline{\delta}} = \underline{a}(p, \min(q, U(X)))X + \underline{a}(\min(q, U(X)) + 1, q)\underline{\underline{\delta}}$ $\square$

The ABP and the CABP – a Comparison of Performances in Real Time Process Algebra

Joris Hillebrand
Programming Research Group, University of Amsterdam
Amsterdam, The Netherlands

Abstract

The main goal of this paper is to give a (small) case study in real time ACP. The Alternating Bit Protocol and the Concurrent Alternating Bit Protocol are specified and verified algebraically. Also a short introduction into real time ACP is given. We conclude that real time ACP is a fine formalism to work with, although we need a weaker equivalence for the verification of protocols than branching bisimulation, which we use throughout this paper.

1 Introduction

In recent years, much protocol verification has been done in ACP (Algebra of Communicating Processes) as well as in other process algebras. In 1989, Baeten and Bergstra introduced real time ACP. A revised version appeared in [BB91]. The purpose of this paper is to apply their theory to two simple communication protocols. In [vW93] the Alternating Bit Protocol (ABP) and the Concurrent Alternating Bit Protocol (CABP) are specified in ACP. In this paper, time is added to both protocols. After verification of these timed protocols, we are able to make an exact statement about the performances.

In the following section we will give an introduction into real time ACP. The syntax as well as some simple identities will be presented. Also a way to 'unwind' a recursive equation infinitely many times can be found. Section 3 contains the specification of both protocols. The ABP has been specified using relative time, for the CABP absolute time has been used. In section 4 we rewrite the specifications into linear systems of recursive equations. It is shown that in our setting the general specification of the CABP is too complex to proceed with, so abstract time-constants are replaced by concrete values. We then complete the verification in section 5 and compare the results.

Acknowledgements. I would like to thank Steven Klusener for his general support and for bringing the Unwind Principle to my attention. I thank Wan Fokkink and Jos van Wamel who commented on earlier versions of this paper.

2 Real time ACP

Parts of this section have been taken directly from [FK92] and [Klu91a]. See also [Klu93] and [Fok92].

Real time ACP concerns (closed) process terms, constructed from timed actions that consist of a symbolic action taken from an alphabet A and a time stamp taken from $[0, \infty]$. We also have the special constants δ (deadlock) and τ (internal activity). A timed action $a(t)$ denotes the process which executes an action a at time t, after which it terminates successfully. Process terms are constructed using operators, among which the binary operators $+$ for alternative composition, $\cdot$ for sequential composition and $\|$ for parallel composition. $x|y$ denotes the communication between the processes x and y. There is also an operator concerning time, the *initialisation operator*, written as $\gg$. The initialisation operator takes a nonnegative real number and a process term; $t \gg p$ denotes the part of p that starts after t, so the following equality holds:

$$3 \gg (a(2) + b(4)) = b(4).$$

Adding time results in identities that do not hold in ACP without time, *symbolic* ACP, such as

$$a(2) \cdot (b(1) + c(3)) = a(2) \cdot c(3).$$

After doing $a(2)$ we have passed time 2, so in the remaining subterm $b(1) + c(3)$ the first alternative cannot be chosen anymore and may therefore be removed. This intuition is captured by the following identity:

$$a(t) \cdot p = a(t) \cdot (t \gg p).$$

Then there is the advanced notion of integration, by which it can be expressed that an action occurs somewhere within a (dense) interval. For example, the process that executes an action a somewhere within the interval $\langle 0, 1 \rangle$ is denoted by $\int_{v \in \langle 0,1 \rangle} a(v)$. In this paper we use *prefix* integration, as introduced in [Klu91b].

The *ultimate delay* of a process x, written $U(x)$, is the upperbound of points in time until which x can idle. An example:

$$\begin{array}{lll} p = a(1) \cdot c(4) \; + \; b(3) \cdot c(4) & \Rightarrow & U(p) = 3, \\ p = a(1) + \delta(2) & \Rightarrow & U(p) = 2. \end{array}$$

We shall now give some properties of abstraction in real time process algebra. These properties follow from a notion of abstraction that is called *branching bisimulation.* Symbolic process algebra tells us that sometimes appearances of τ can be removed from a process expression, since τ stands for an internal, invisible action. The most important restriction in doing this is that τ cannot be removed if it determines a choice between processes. For example, in $a(\tau b + c)$ doing τ implies choosing for b; c cannot be performed anymore. Without our notice, because we cannot see τ happen, we have now arrived at a state at which it is impossible to do c, although we have not seen b happen yet. This is different from $a(b+c)$, where up to the moment at which we see either b or c happen, the choice between b and c is free. In a timed setting however, choices are not only determined by the presence of τ or other actions, but also by the lapse of time. We tried to visualise this in figure 1. In $a(1) \cdot (\tau(3) \cdot b(5) + c(2))$ a choice is made at time 2. At that point in time we can either see c happen or not. In the latter case apparently there has been chosen for $\tau(3) \cdot b(5)$. So with or without $\tau(3)$ the choice is made at time 2 and therefore $\tau(3)$ may be

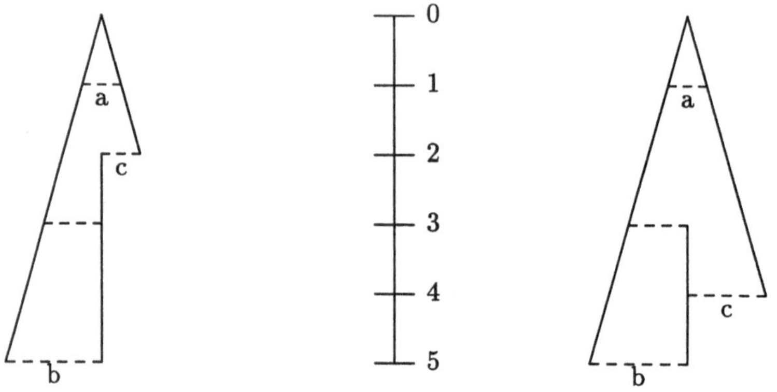

Figure 1: Process diagrams for $a(1)\cdot(\tau(3)\cdot b(5)+c(2))$ and $a(1)\cdot(\tau(3)\cdot b(5)+c(4))$

removed. This is not the case in $a(1)\cdot(\tau(3)\cdot b(5)+c(4))$. Here a choice is made at time 3. Without the τ this choice would be made at time 4. This leads us to the following law:

$$\begin{array}{ll} t < r < U(x) \ \wedge \ U(y) \leq r & \Longrightarrow \\ a(t)\cdot(\tau(r)\cdot x + y) = a(t)\cdot(r \gg x + y) & \tau\text{-law.} \end{array}$$

This identity says that you can remove $\tau(r)$ in $a(t)\cdot(\tau(r)\cdot x + y)$ only if at time r it is too late for actions of y to happen. In that case, not doing y automatically implies choosing for x. This identity axiomatises branching bisimulation.

We can derive two useful identities from this law, being the *τ-swap* and the *τ-removal* property. The τ-swap property is the following rule:

$$\frac{t < r;\ \ x \text{ and } y \text{ can idle until at least } r; \quad x \text{ and } y \text{ cannot execute any actions before or at } r}{a(t)\cdot(\tau(r)\cdot x + y) = a(t)\cdot(x + \tau(r)\cdot y)}\ \tau\text{-swap.}$$

It says that if $\tau(r)$ determines the choice between x and y, because all initial actions of both x and y happen after time r, it does not matter in front of which summand τ is put. r will always be the moment of choice, as not doing τ then automatically implies choosing for the other option.

In symbolic ACP $x\tau = x$ is an axiom. This implies that $a\tau x = ax$ always holds. This is not the case in real time ACP. $a(1)\cdot\tau(3)\cdot(b(2)+c(5)) \neq a(1)\cdot(b(2)+c(5))$ because without the τ, $b(2)$ can still be chosen, whereas with τ it is too late for that. This is captured in the following τ-removal property, saying that we can only remove $\tau(r)$ in $a(t)\cdot\tau(r)\cdot x$ if x has no parts starting at or before r:

$$\frac{t < r;\ \ x \text{ can idle until at least } r; \quad x \text{ cannot execute any actions before or at } r}{a(t) \cdot \tau(r) \cdot x = a(t) \cdot x} \quad \tau\text{-removal.}$$

Relative time. Sometimes one is less interested in the absolute point in time at which an action happens. Often it is sufficient to know how many time steps *after* the previous action the next action is performed. In such cases one can use relative-timed actions.

Relative-timed actions have their time stamp between square brackets. $a[2] \cdot b[1] \cdot c[4]$ denotes: 2 time units after initialisation a is performed, 1 time unit later b happens and then 4 time units later c has been planned. Since we assume that each process starts at time 0, the absolute-timed process $a(2) \cdot b(3) \cdot c(7)$ is equivalent with the relative-timed process above.

Notice that in $a[t] \cdot b[t']$, b always happens after a, so t' need not be greater than t.

Next, we introduce the *time shift*, denoted $(t)x$. It denotes the process x being shifted t steps forward in time. For example:

$$(2)(b[2] \cdot c[1] \ + \ d[1]) \ = \ b[4] \cdot c[1] \ + \ d[3].$$

Note that none of the actions that come after the initial action(s) are affected by the time shift.

Finally, we present the relative-time equivalents of the τ-law, the τ-swap property, and the τ-removal property. The τ-law in relative time looks as follows:

$$U(y) \leq r \quad \Longrightarrow \quad a[t] \cdot (\tau[r] \cdot x + y) = a[t] \cdot ((r)x + y) \quad \tau\text{-law.}$$

Working in relative time, we do not need to demand any more that actions have time stamps with increasing values. For the same reason, actions of x will always happen after $\tau[r]$. Therefore, the only condition is that at time r it is too late for actions of y to happen. Yet, the fact that we work with relative time has another consequence. If we now remove actions from a process expression we must correct the time stamps of later actions. In $a[4] \cdot \tau[1] \cdot b[3]$ b happens 4 time steps after a, whereas in $a[4] \cdot b[3]$ there are 3 time steps in between. Thus in the relative-timed τ-law above x is shifted r time steps forward in time, to make up for the removal of $\tau[r]$.

The τ-swap property looks as follows in relative time:

$$a[t] \cdot (\tau[r] \cdot x + (r)y) = a[t] \cdot ((r)x + \tau[r] \cdot y) \qquad \tau\text{-swap.}$$

Note that we have to use the time shift, because $\tau[r]$ is *added* in front of y. Compare $a[4] \cdot b[3]$ and $a[4] \cdot \tau[1] \cdot b[3]$, where the time between a and b is not the same in both cases.

The τ-removal property looks as follows in relative time:

$$a[t] \cdot \tau[r] \cdot x = a[t] \cdot (r)x \qquad \tau\text{-removal.}$$

The Unwind Principle

In the following we will turn the specification of two communication protocols into a linear system of recursive equations. After abstraction we will have expressions of the form

$$X(t) = \tau(t) \cdot \{Y(t+s) + \tau(t+r_0) \cdot X(t+r_1)\}.$$

If this equation was put in symbolic ACP, i.e., without time, we would have been able to use *Koomen's Fair Abstraction Rule* (KFAR), which says that in abstracting from a closed loop of internal actions, eventually (i.e., after performing a number of τ-steps) an external step will be chosen. So if $X = i \cdot X + Y$ then $\tau \cdot \tau_{\{i\}}(X) = \tau \cdot \tau_{\{i\}}(Y)$. Unfortunately, in a timed setting, doing a τ-step takes time. Therefore the time at which the first external action happens depends on the number of τ-steps that has been done. A way to apply abstraction to a timed, recursively specified process without losing information on time is the *Unwind Principle* (UP) from [Klu91a], [Klu93]. It allows us to 'unwind' a recursive specification infinitely many times. We will now formulate the UP by showing one step of the unwinding. The dots express that the last equality is not provable in a finite derivation.

Let $X(t)$ be defined as above for some $r_0 < r_1$ and $r_0 < s$ and take $Y(t)$ such that it has no parts starting at or before t. Then

$$\begin{aligned}
X(t) &= \tau(t) \cdot \{Y(t+s) + \tau(t+r_0) \cdot X(t+r_1)\} \\
&\overset{\tau\text{-swap}}{=} \tau(t) \cdot \{\tau(t+r_0) \cdot Y(t+s) + X(t+r_1)\} \\
&= \tau(t) \cdot \{\tau(t+r_0) \cdot Y(t+s) + \\
&\qquad \tau(t+r_1) \cdot \{Y(t+s+r_1) + \\
&\qquad\qquad \tau(t+r_0+r_1) \cdot X(t+2 \cdot r_1)\}\} \\
&\overset{\tau\text{-law}}{=} \tau(t) \cdot \{\tau(t+r_0) \cdot Y(t+s) + \\
&\qquad \{Y(t+s+r_1) + \tau(t+r_0+r_1) \cdot X(t+2 \cdot r_1)\}\} \\
&\overset{\tau\text{-swap}}{=} \tau(t) \cdot \{\tau(t+r_0) \cdot Y(t+s) + \\
&\qquad \tau(t+r_0+r_1) \cdot Y(t+s+r_1) + X(t+2 \cdot r_1)\} \\
&\;\vdots \\
&\overset{\text{UP}}{=} \tau(t) \cdot \{\sum_{n=0}^{\infty} \tau(t+r_0+n \cdot r_1) \cdot Y(t+s+n \cdot r_1)\}.
\end{aligned}$$

Here n is the number of times the recursion-loop is being done. The conditions on r_0, r_1, s and $Y(t)$ make it possible to apply the τ-law and the τ-properties. We also state the UP in relative time. In this case, only $r_0 < s$ is necessary.

$$\begin{aligned}
X &= \tau[t] \cdot \{(s)Y + \tau[r_0] \cdot (r_1)X\} \\
&\overset{\tau\text{-swap}}{=} \tau[t] \cdot \{\tau[r_0] \cdot (s-r_0)Y + (r_1+r_0)X\} \\
&= \tau[t] \cdot \{\tau[r_0] \cdot (s-r_0)Y + \\
&\qquad \tau[t+r_1+r_0] \cdot \{(s)Y + \tau[r_0] \cdot (r_1)X\}\} \\
&\overset{\tau\text{-law}}{=} \tau[t] \cdot \{\tau[r_0] \cdot (s-r_0)Y +
\end{aligned}$$

$$\{(s+t+r_1+r_0)Y + \tau[t+r_1+2\cdot r_0]\cdot(r_1)X\}\}$$

$$\overset{\tau\text{-swap}}{=} \quad \tau[t]\cdot\{\tau[r_0]\cdot(s-r_0)Y + \tau[t+r_1+2\cdot r_0]\cdot(s-r_0)Y + (t+2\cdot r_1+2\cdot r_0)X\}$$

$$\vdots$$

$$\overset{\text{UP}}{=} \quad \tau[t]\cdot\{\sum_{n=0}^{\infty}\tau[r_0+n\cdot(t+r_1+r_0)]\cdot(s-r_0)Y\}.$$

3 Specifying the ABP and the CABP

The protocols are provided with time as is done in [BB91]. All *read* actions can happen at any time; time constraints are allowed on *send* actions only. Doing so makes sure that whenever a part of the system is in a state at which it can receive a message, it does not matter at what time that message exactly arrives. Compare this with a telephone: whenever you are able to hear it ringing, it does not matter *when* it starts ringing.

The ABP has been specified with relative time, because we are only interested in the time between actions. The absolute points in time on which the actions happen is less interesting. Moreover, with absolute time, each process would need a time-parameter. This makes the protocol easier to specify in relative time. During the verification, however, we will have to shift process-parts back and forward in time, whenever we add or remove an action (in particular τ's).

We have used absolute time for the specification of the CABP. Because things really happen in parallel in the CABP, we need a time parameter for each component of the system during verification. In this parameter, the time on which the component performed its latest action is stored. This approach would be impossible with relative time, because we would not know the absolute time in that case.

The ABP. See figure 2 for a picture and a specification of the ABP. When a message d from a finite data set D appears at gate 1, the Sender (S) sends it to channel K via gate 3 together with a control bit b from a set $B = \{0,1\}$, and then starts waiting for a positive acknowledgement from the Receiver (R). Channel K can either correctly deliver this *frame*, or give an error to R on gate 4. If R receives such a frame, it computes its *checksum* and checks the control bit. If both are correct, the message is written down at gate 2 and the control bit b is sent to L as a positive acknowledgement. Otherwise the inverse control bit (1-b) is sent to L as a negative acknowledgement. Channel L can deliver the bit correctly or give an error on gate 5. If S receives a positive acknowledgement b, it waits for a new message to appear at gate 1. This message will then be sent to K with the control bit inverted. If S receives a negative acknowledgement (1-b) or an error, the same message is sent again to K.

If R receives a frame of which the computed checksum is incorrect, then channel K has delivered an error. If the checksum is correct but does not have the proper control bit, then it concerns an 'old' frame that has been resent by S because S did not receive the positive acknowledgement correctly. The protocol only behaves correctly if errors can be detected. Therefore, we must presume that when b is sent to L, it either comes out as b, or as an error, but *not* as 1-b. The same holds for the control bit in the frames.

When S writes on gate 3, there is a delay of t_1 time steps. When K writes on gate 4, this delay is t_2. Writing on gate 5 by L has a delay of t_3. Finally, as R sends an acknowledgement on gate 6, this has a delay of t_4 and when R writes a message on gate 2, there is a delay of t_5. $t_1, ..., t_5$ are the so-called *delay-constants* and must be greater than zero.

The CABP. For the CABP, we split the Sender and the Receiver in two parts each. The Sender is split into a part S that just repeatedly sends frames to channel K and a part AR (Acknowledgement Receiver) that does nothing but receiving incoming acknowledgements. Similarly, the Receiver is split into a part R that receives incoming frames and a part AS (Acknowledgement Sender) that repeatedly sends acknowledgements. A picture and a specification can be found in figure 3. The channels do not garble data any more. They either deliver it correctly, or loose it completely. The loss of data is possible now, because both frames and acknowledgements are resent automatically.

S start sending frames with control bit b. AS starts sending negative acknowledgements 1-b. As long as those negative acknowledgements arrive at AR, S keeps sending the same frame. When a frame with the proper control bit arrives at R, this is reported to AS, which then starts sending the control bit b as a positive acknowledgement. When AR receives this positive acknowledgement b, this is reported to S, which then reads a new message from gate 1. This new message is then sent to K in a frame with control bit 1-b.

As in the ABP only one thing at a time actually happens, one can easily see that in the CABP this is not the case. Frames as well as acknowledgements are sent over in parallel, passing different channels in which things also happen. This leads to a situation in which we have to keep track very carefully of what happens on what time in what part of the system. For that purpose we will define a vector of time-parameters in the next section.

We have written an extra encapsulation around S and AR and around R and AS, because together these parts have the function of S and R in the ABP. Therefore we have given the gate between S and K in the CABP the same name as the gate between S and K in the ABP; writing on that gate by S in the CABP also has the same delay-constant as is the case in the ABP. There is a similar correspondence between the gate between AS and L in the CABP and the one between R and L in the ABP. Writing by AS also has the same delay-constant as writing by R in the ABP. Gates 7 and 8 have been added for the internal communication between AR and S and between R and AS. Writing by AR on gate 7 has delay-constant t_6; writing by R on gate 8 has delay-constant t_7. Because each part of the system can handle only one data-element at a time, the following restrictions must hold: $t_1 > t_2, t_5, t_7$ and $t_4 > t_3, t_6$. Now the channels and other parts will not get choked-up.

Taking the encapsulations as described above is different from [vW93]. There S, K and R are taken together as well as AS, L and AR. In our approach a communication on gates between AR and S or between R and AS

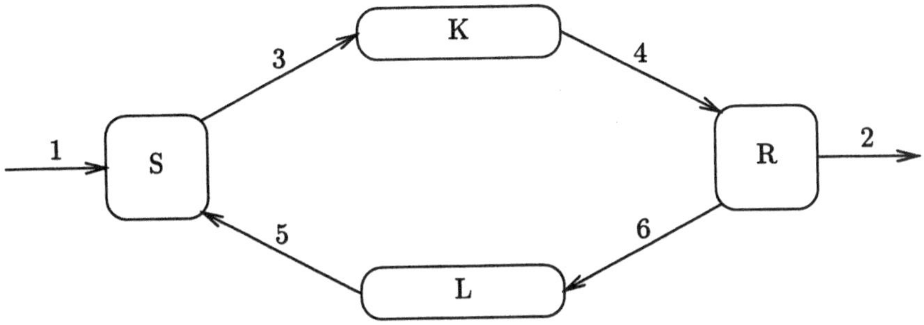

$\text{ABP} = \partial_{H(3,4,5,6)}(S \parallel K \parallel L \parallel R)$, with

Sender

$$\begin{aligned}
S &= RM_0 \\
RM_b &= \textstyle\sum_{d\in D} \int_{v>0} r_1(d)[v] \cdot SF_{db} \\
SF_{db} &= s_3(db)[t_1] \cdot RA_{db} \\
RA_{db} &= (\textstyle\int_{v>0} r_5(1\text{-}b)[v] + \int_{v>0} r_5(e)[v]) \cdot SF_{db} + \\
&\quad \textstyle\int_{v>0} r_5(b)[v] \cdot RM_{1-b}
\end{aligned}$$

Channels

$$\begin{aligned}
K &= \textstyle\sum_{db\in D\times bool} \int_{v>0} r_3(db)[v] \cdot K_{db} \\
K_{db} &= (s_4(e)[t_2] \ + \ s_4(db)[t_2]) \cdot K \\
\\
L &= \textstyle\sum_{b\in bool} \int_{v>0} r_6(b)[v] \cdot L_b \\
L_b &= (s_5(e)[t_3] \ + \ s_5(b)[t_3]) \cdot L
\end{aligned}$$

Receiver

$$\begin{aligned}
R &= RF_0 \\
RF_b &= \textstyle\sum_{d\in D}(\ (\int_{v>0} r_4(d(1\text{-}b))[v] + \int_{v>0} r_4(e)[v]) \cdot SA_{1-b} \\
&\qquad + \textstyle\int_{v>0} r_4(db)[v] \cdot SM_{db}) \\
SA_b &= s_6(b)[t_4] \cdot RF_{1-b} \\
SM_{db} &= s_2(d)[t_5] \cdot SA_b
\end{aligned}$$

$t_1, ..., t_5 > 0$ delay-constants.

Figure 2: The ABP

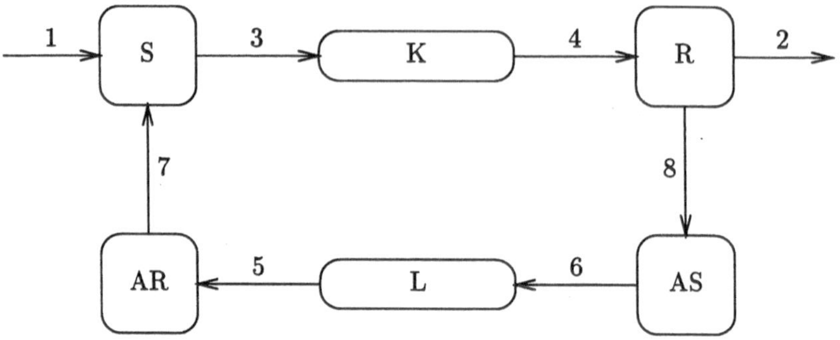

CABP = $\partial_{H(3,4,5,6)}(\partial_{H'(7)}(S \| AR) \| K \| L \| \partial_{H''(8)}(R \| AS))$, with

Sender

$$
\begin{aligned}
S &= RM_0(0)\\
RM_b(t) &= \textstyle\sum_{d \in D} \int_{v>t} r_1(d)(v) \cdot SF_{db}(v)\\
SF_{db}(t) &= s_3(db)(t{+}t_1) \cdot SF_{db}(t{+}t_1) \; + \\
&\quad \textstyle\int_{t<v\le t+t_1} r_7(ac)(v) \cdot RM_{1-b}(v)
\end{aligned}
$$

Ack. Receiver

$$
\begin{aligned}
AR &= AR_0(0)\\
AR_b(t) &= \textstyle\int_{v>t} r_5(1\text{-}b)(v) \cdot AR_b(v) \; + \\
&\quad \textstyle\int_{v>t} r_5(b)(v) \cdot s_7(ac)(v{+}t_6) \cdot AR_{1-b}(v{+}t_6)
\end{aligned}
$$

Channels

$$
\begin{aligned}
K(t) &= \textstyle\sum_{db \in D \times bool} \int_{v>t} r_3(db)(v) \cdot K_{db}(v)\\
K_{db}(t) &= (s_4(db)(t{+}t_2) + errorK(t{+}t_2)) \cdot K(t{+}t_2)
\end{aligned}
$$

$$
\begin{aligned}
L(t) &= \textstyle\sum_{b \in bool} \int_{v>t} r_6(b)(v) \cdot L_b(v)\\
L_b(t) &= (s_5(b)(t{+}t_3) + errorL(t{+}t_3)) \cdot L(t{+}t_3)
\end{aligned}
$$

Receiver

$$
\begin{aligned}
R &= RF_0(0)\\
RF_b(t) &= \textstyle\sum_{d \in D}(\int_{v>t} r_4(d(1\text{-}b))(v) \cdot RF_b(v) \; + \\
&\qquad + \textstyle\int_{v>t} r_4(db)(v) \cdot (s_2(d)(v{+}t_5) \| s_8(ac)(v{+}t_7)) \cdot \\
&\qquad\quad RF_{1-b}(v + max(t_5, t_7)))
\end{aligned}
$$

Ack. Sender

$$
\begin{aligned}
AS &= AS_1(0)\\
AS_b(t) &= s_6(b)(t{+}t_4) \cdot AS_b(t{+}t_4) \; + \\
&\quad \textstyle\int_{t<v\le t+t_4} r8(ac)(v) \cdot AS_{1-b}(v)
\end{aligned}
$$

$t_1, ..., t_7 > 0$ delay-constants.
$t_1 > t_2, t_5, t_7$; $\quad t_4 > t_3, t_6$.

Figure 3: The CABP

is forced to happen, whenever a choice has to be made between such a communication and an other one (for example, when a frame has to be re-sent at the same time that a signal comes in from AR, S has to accept the signal).

4 Linearisation

Now we turn the specifications of the two protocols into linear systems of recursive equations, by applying a head-tail strategy. For each state $\partial_H(C_1\|...\|C_n)$ the set of possible initial actions is computed. This set is the union of the sets of initial actions of C_k for all k. Furthermore, the results of all possible communications between initial actions of two arbitrary different C_k are also initial actions of a state. $\partial_H(P)$ renames all actions of the process expression P that are in the set H into δ. Because in our case almost all read- and send-actions are in the *encapsulation* set H, the initial actions of a state will nearly always be the result of a communication. Now the equations will look like $state = \sum_{ia} ia \cdot next\text{-}state_{ia}$ for each initial action ia of $state$.

4.1 The ABP

Take $H = \{r_j(db), s_j(db), r_k(e), s_k(e), r_l(b), s_l(b)\}$ with $j \in \{3,4\}$, $k \in \{4,5\}$, $l \in \{5,6\}$, $d \in D$ and $b \in B$. Those terms that are being worked out are printed boldface.

$$\begin{aligned} X_0(b) &= \partial_H(\boldsymbol{RM_b} \parallel K \parallel L \parallel RF_b) \\ &= \textstyle\sum_{d\in D} \int_{v>0} r_1(d)[v] \cdot X_1(d,b) \end{aligned}$$

$$\begin{aligned} X_1(d,b) &= \partial_H(\boldsymbol{SF_{db}} \parallel \boldsymbol{K} \parallel L \parallel RF_b) \\ &= c_3(db)[t_1] \cdot X_2(d,b) \end{aligned}$$

$$\begin{aligned} X_2(d,b) &= \partial_H(RA_{db} \parallel \boldsymbol{K_f} \parallel L \parallel \boldsymbol{RF_b}) \\ &= c_4(db)[t_2] \cdot X_3(d,b) \ + \ c_4(e)[t_2] \cdot X_4(d,b) \end{aligned}$$

$$\begin{aligned} X_3(d,b) &= \partial_H(RA_{db} \parallel K \parallel L \parallel \boldsymbol{SM_{db}}) \\ &= s_2(d)[t_5] \cdot X_5(d,b) \end{aligned}$$

$$\begin{aligned} X_4(d,b) &= \partial_H(RA_{db} \parallel K \parallel \boldsymbol{L} \parallel \boldsymbol{SA_{1-b}}) \\ &= c_6(1\text{-}b)[t_4] \cdot X_6(d,b) \end{aligned}$$

$$\begin{aligned} X_5(d,b) &= \partial_H(RA_{db} \parallel K \parallel \boldsymbol{L} \parallel \boldsymbol{SA_b}) \\ &= c_6(b)[t_4] \cdot X_7(d,b) \end{aligned}$$

$$\begin{aligned} X_6(d,b) &= \partial_H(\boldsymbol{RA_{db}} \parallel K \parallel \boldsymbol{L_{1-b}} \parallel RF_b) \\ &= c_5(1\text{-}b)[t_3] \cdot X_1(d,b) \ + \ c_5(e)[t_3] \cdot X_1(d,b) \end{aligned}$$

$$X_7(d,b) = \partial_H(\boldsymbol{RA}_{db} \parallel K \parallel \boldsymbol{L}_b \parallel RF_{1-b})$$
$$= c_5(b)[t_3] \cdot X_0(1\text{-}b) + c_5(e)[t_3] \cdot X_8(d,b)$$

$$X_8(d,b) = \partial_H(\boldsymbol{SF}_{db} \parallel \boldsymbol{K} \parallel L \parallel RF_{1-b})$$
$$= c_3(db)[t_1] \cdot X_9(d,b)$$

$$X_9(d,b) = \partial_H(RA_{db} \parallel \boldsymbol{K}_f \parallel L \parallel \boldsymbol{RF}_{1-b})$$
$$= c_4(db)[t_2] \cdot X_5(d,b) + c_4(e)[t_2] \cdot X_5(d,b)$$

We can see immediately that the ABP is deadlock-free, since in each state $X_i(\_)$ at least one action can be done. Again it becomes clear that in the ABP only one action at a time happens. Only in the channels a choice is made (between correct and incorrect transmission).

4.2 The CABP

As stated above, in the CABP lots of things happen in parallel. Therefore, a state can have different initial actions. In that case it depends on time which initial action will happen first. So we must somehow be able to compute the concrete time at which the actions must happen. The states of the CABP will therefore be labelled with a vector of time-parameters. Each component of that vector is associated with one part of the system and contains the point in time with respect to which the corresponding part of the system has to do its next action. For example, each time a message is sent to channel K, we save the moment at which that happened in the time-parameter for the sender S, t_S. We then know this must be done again at time $t_S + t_1$. Usually, a time-parameter will just contain the moment at which the component performed a last action. Yet sometimes that moment does not matter (for example, in front of a read-integral; the moment of communication is determined by the sending part). In that case "-" is put on the place of the parameter.

Linearising the CABP results in 49 states. Every state has a number of conditions, each followed by a process expression. Because exactly one of the equations can be true, it is now always clear which action happens first. Since this implies that one action will always happen, we can conclude that the CABP is also deadlock-free. We will not give all 49 equations here, but an example of one such state can be found below. So here $\vec{t}$ stands for $(t_S, t_{AR}, t_K, t_L, t_R, t_{AS})$. $H = \{r_j(db), s_j(db), r_k(e), s_k(e), r_l(b), s_l(b)\}$, $H' = \{r_7(ac), s_7(ac)\}$, and $H'' = \{r_8(ac), s_8(ac)\}$ with $j \in \{3,4\}$, $k \in \{4,5\}$, $l \in \{5,6\}$, $d \in D$ and $b \in B$.

$(a_1 \& a_2)(t)$ is called a *multi-action*. It denotes the actions a_1 and a_2 happening at the same time at different places in the system.

$$X_{34}(d,e,b)(\vec{t}) =$$
$$\partial_H(\ \partial_{H'}(SF_{e(1\text{-}b)}(t_S) \| AR_{1\text{-}b}(t_{AR}))$$
$$\| K_{db}(t_K) \| L(t_L)$$
$$\| \partial_{H''}(\boldsymbol{s_2(d)(t_R+t_5)} \cdot \boldsymbol{RF_{1\text{-}b}}(t_R+t_5) \| \boldsymbol{AS}_b(t_{AS})) =$$

$$
\begin{array}{l}
t_{AS}{+}t_4 < t_R{+}t_5 :\to \\
\quad c_6(b)(t_{AS}{+}t_4) \cdot X_{35}(d,e,b)(t_S, -, t_K, t_{AS}{+}t_4, t_R, t_{AS}{+}t_4) \; + \\
t_R{+}t_5 < t_{AS}{+}t_4 :\to \\
\quad s_2(d)(t_R{+}t_5) \cdot X_{36}(d,e,b)(t_S, -, t_K, -, -, t_{AS}) \; + \\
t_{AS}{+}t_4 = t_R{+}t_5 :\to \\
\quad (c_6(b) \;\&\; s_2(d))(t_{AS}{+}t_4) \cdot X_{37}(d,e,b)(t_S, -, t_K, t_{AS}{+}t_4, -, t_{AS}{+}t_4)
\end{array}
$$

with

$$
\begin{array}{rcl}
X_{35}(d,e,b)(\vec{t}) & = & \partial_H(\; \partial_{H'}(SF_{e(1\text{-}b)}(t_S) \| AR_{1\text{-}b}(t_{AR})) \\
& & \| \; K_{db}(t_K) \; \| \; L_b(t_L) \\
& & \| \; \partial_{H''}(s_2(d)(t_R{+}t_5) \cdot RF_{1\text{-}b}(t_R{+}t_5) \| AS_b(t_{AS}))) \;\; ,
\end{array}
$$

$$
\begin{array}{rcl}
X_{36}(d,e,b)(\vec{t}) & = & \partial_H(\; \partial_{H'}(SF_{e(1\text{-}b)}(t_S) \| AR_{1\text{-}b}(t_{AR})) \\
& & \| \; K_{db}(t_K) \; \| \; L(t_L) \\
& & \| \; \partial_{H''}(RF_{1\text{-}b}(t_R) \| AS_b(t_{AS}))) \;\; ,
\end{array}
$$

$$
\begin{array}{rcl}
X_{37}(d,e,b)(\vec{t}) & = & \partial_H(\; \partial_{H'}(SF_{e(1\text{-}b)}(t_S) \| AR_{1\text{-}b}(t_{AR})) \\
& & \| \; K_{db}(t_K) \; \| \; L_b(t_L) \\
& & \| \; \partial_{H''}(RF_{1\text{-}b}(t_R) \| AS_b(t_{AS}))) \;\; .
\end{array}
$$

The complete linearisation of the CABP can be found in [Hil94].

4.3 Assigning values

The system of 49 equations for the CABP is too complex to proceed with. The number of possible execution traces is at most equal to the product of the number of conditions in each state ($> 10^{31}$). Therefore we will now proceed the verification with chosen values for the delay-constants. After we have done this, all conditions can be computed, resulting in a unique execution-trace modulo the choices that are made in the channels. Perhaps with different semantics the general specification of the protocol could have been verified further, but the laws for abstraction (corresponding with branching bisimulation) that we have introduced in section 2 are too distinctive.

When choosing values for $t_1, ..., t_7$ one can imagine a two-layer hierarchy. t_6 and t_7 are relatively small, since these constants belong to send-actions through the internal gates between S and AR and between R and AS. t_2, t_3 and t_5 are then relatively large. t_1 and t_4 can be seen as a retransmission-frequency that is roughly set. These last two must be greater than t_2, t_3 and t_5 (against chocking-up the system), but it would be wise to choose them as small as possible. The higher the retransmission-frequency is set, the sooner a new message or acknowledgement comes in after a mess-up of one of the channels. The chosen values are:

$t_6 = t_7 = 1$
$t_2 = t_3 = t_5 = 7$
$t_1 = t_4 = 8.$

Furthermore, we assume a message to appear from outside at gate 1, 10 time-steps after the Sender is enabled to read from that gate (so $v = 10$). This determines how often a (negative) acknowledgement is sent before the first message is read in. Now all conditions in the system of equations can be computed. These assignments satisfy $t_1 > t_2, t_5, t_7$ and $t_4 > t_3, t_6$. We shall now 'hop' through the system of equations X_0 to X_{48} with instantiated values for t_1 to t_7. For each state it is clear now what will happen, since we can compute the conditions. In the new (much smaller) system, the equations will be called $Y_0, Y_1, \ldots$. It will be stated of which equation X_b each equation Y_a is an instantiation. Two states $Y_a(\vec{t})$ and $Y_b(\vec{u})$ are equal if:

- Y_a and Y_b are instantiations of the same state X_c;
- for all $i, j \in \{S, AR, K, L, R, AS\}: \; t_i - u_i \; = \; t_j - u_j$.

The latter point makes sure that the system, once arrived in an instantiation of the same state X_a for the second time, will follow exactly the same path from that state as it did the first time it was there. On the left hand of '=', the actual values have been substituted for the parameters in $\vec{t}$. Note that, for example, Y_2 and Y_4 are both instantiations of X_3, but the time-parameters have not all been increased by the same value. So Y_2 is not equal to Y_4.

$$\text{CABP} = Y_0(0)(\vec{0})$$

$$Y_0(b)(0,0,0,0,0,0) \stackrel{X_0}{=}$$
$$c_6(1\text{-}b)(t_{AS}+8) \cdot Y_1(b)(t_S, -, -, t_{AS}+8, -, t_{AS}+8)$$

$$Y_1(b)(0,-,-,8,-,8) \stackrel{X_2}{=}$$
$$\textstyle\sum_{d \in D} r_1(d)(t_S+10) \cdot Y_2(d,b)(t_S+10, -, -, t_L, -, t_{AS})$$

$$Y_2(d,b)(10,-,-,8,-,8) \stackrel{X_3}{=}$$
$$c_5(1\text{-}b)(t_L+7) \cdot Y_3(d,b)(t_S, -, -, -, -, t_{AS})$$
$$+ \; errorL(t_L+7) \cdot Y_3(d,b)(t_S, -, -, -, -, t_{AS})$$

$$Y_3(d,b)(10,-,-,-,-,8) \stackrel{X_1}{=}$$
$$c_6(1\text{-}b)(t_{AS}+8) \cdot Y_4(d,b)(t_S, -, -, t_{AS}+8, -, t_{AS}+8)$$

$$Y_4(d,b)(10,-,-,16,-,16) \stackrel{X_3}{=}$$
$$c_3(db)(t_S+8) \cdot Y_5(d,b)(t_S+8, -, t_S+8, t_L, -, t_{AS})$$

$$Y_5(d,b)(18,-,18,16,-,16) \stackrel{X_5}{=}$$
$$c_5(1\text{-}b)(t_L+7) \cdot Y_6(d,b)(t_S, -, t_K, -, -, t_{AS})$$
$$+ \; errorL(t_L+7) \cdot Y_6(d,b)(t_S, -, t_K, -, -, t_{AS})$$

$$Y_6(d,b)(18,-,18,-,-,16) \stackrel{X_4}{=} c_6(1\text{-}b)(t_{AS}+8) \cdot Y_7(d,b)(t_S,-,t_K,t_{AS}+8,-,t_{AS}+8)$$

$$Y_7(d,b)(18,-,18,24,-,24) \stackrel{X_5}{=} c_4(db)(t_K+7) \cdot Y_8(d,b)(t_S,-,-,t_L,t_K+7,t_{AS}) + errorK(t_K+7) \cdot Y_4(d,b)(t_S,-,-,t_L,-,t_{AS})$$

$$Y_8(d,b)(18,-,-,24,25,24) \stackrel{X_7}{=} (c_3(db) \;\&\; c_8(ac))(t_S+8) \cdot Y_9(d,b)(t_S+8,-,t_S+8,t_L,t_R,t_S+8)$$

$$Y_9(d,b)(26,-,26,24,25,26) \stackrel{X_{19}}{=} c_5(1\text{-}b)(t_L+7) \cdot Y_{10}(d,b)(t_S,-,t_K,-,t_R,t_{AS}) + errorL(t_L+7) \cdot Y_{10}(d,b)(t_S,-,t_K,-,t_R,t_{AS})$$

$$Y_{10}(d,b)(26,-,26,-,25,26) \stackrel{X_{13}}{=} s_2(d)(t_R+7) \cdot Y_{11}(d,b)(t_S,-,t_K,-,-,t_{AS})$$

$$Y_{11}(d,b)(26,-,26,-,-,26) \stackrel{X_{17}}{=} c_4(db)(t_K+7) \cdot Y_{12}(d,b)(t_S,-,-,-,-,t_{AS}) + errorK(t_K+7) \cdot Y_{12}(d,b)(t_S,-,-,-,-,t_{AS})$$

$$Y_{12}(d,b)(26,-,-,-,-,26) \stackrel{X_{15}}{=} (c_3(db) \;\&\; c_6(b))(t_S+8) \cdot Y_{13}(d,b)(t_S+8,-,t_S+8,t_S+8,-,t_S+8)$$

$$\begin{aligned} Y_{13}(d,b)(34,-,34,34,-,34) \stackrel{X_{25}}{=} & (c_4(db) \;\&\; c_5(b))(t_K+7) \cdot Y_{14}(d,b)(t_S,t_K+7,-,-,-,t_{AS}) \\ + & (errorK \;\&\; c_5(b))(t_K+7) \cdot Y_{14}(d,b)(t_S,t_K+7,-,-,-,t_{AS}) \\ + & (c_4(db) \;\&\; errorL)(t_K+7) \cdot Y_{12}(d,b)(t_S,-,-,-,-,t_{AS}) \\ + & (errorK \;\&\; errorL)(t_K+7) \cdot Y_{12}(d,b)(t_S,-,-,-,-,t_{AS}) \end{aligned}$$

$$Y_{14}(d,b)(34,41,-,-,-,34) \stackrel{X_{38}}{=} (c_6(b) \;\&\; c_7(ac))(t_{AS}+8) \cdot Y_{15}(1\text{-}b)(t_{AS}+8,-,-,t_{AS}+8,-,t_{AS}+8)$$

$$Y_{15}(d,b)(42,-,-,42,-,42) \stackrel{X_2}{=} c_5(1\text{-}b)(t_L+7) \cdot Y_0(b)(t_S,-,-,-,-t_{AS}) + errorL(t_L+7) \cdot Y_0(b)(t_S,-,-,-,-t_{AS})$$

At this moment, we could replace the delay-constants in the ABP by these actual values as well, but we choose to continue the verification of the general specification. Later on we will of course do the substitution, for a comparison of the performances.

5 Further verification and comparison of performances

The verification of the two protocols will proceed in two steps. First we will simplify our systems of equations by abstracting from internal steps. In this context internal steps are actions that are not interesting for the environment. The next step is to reduce the number of equations by substituting them in each other. Here we will need the Unwind Principle from section 2, because the systems of equations are at some points circular.

5.1 The ABP

Abstraction The environment only deals with two parts of the system: gate 1 and gate 2. When a message comes in at gate 1 we are in fact only interested in when it comes out again at gate 2. So we intend to abstract from all actions other than $r_1(d)$ and $s_2(d)$ for $d \in D$ in the equations for the ABP. $\tau_I(P)$ renames all actions of the process expression P that are in the set I into τ. So I will be the set $\{c_3(db), c_4(db), c_4(e), c_5(b), c_5(e), c_6(b)\}$ with $d \in D$ and $b \in B$.

$$\tau_I(X_0(b)) = \textstyle\sum_{d\in D} \int_{v>0} r_1(d)[v] \cdot \tau_I(X_1(d,b))$$

$$\tau_I(X_1(d,b)) = \tau[t_1] \cdot \tau_I(X_2(d,b))$$

$$\tau_I(X_2(d,b)) = \tau[t_2] \cdot \tau_I(X_3(d,b)) + \tau[t_2] \cdot \tau_I(X_4(d,b))$$

$$\tau_I(X_3(d,b)) = s2(d)[t_5] \cdot \tau_I(X_5(d,b))$$

$$\tau_I(X_4(d,b)) = \tau[t_4] \cdot \tau_I(X_6(d,b))$$

$$\tau_I(X_5(d,b)) = \tau[t_4] \cdot \tau_I(X_7(d,b))$$

$$\tau_I(X_6(d,b)) = \tau[t_3] \cdot \tau_I(X_1(d,b))$$

$$\tau_I(X_7(d,b)) = \tau[t_3] \cdot \tau_I(X_0(1\text{-}b)) + \tau[t_3] \cdot \tau_I(X_8(d,b))$$

$$\tau_I(X_8(d,b)) = \tau[t_1] \cdot \tau_I(X_9(d,b))$$

$$\tau_I(X_9(d,b)) = \tau[t_2] \cdot \tau_I(X_5(d,b))$$

Substitution At this point, we will substitute the equations for states X_k in equations for states X_l with $k > l$. As mentioned above, the system of equations is at some points circular: the system can return to states before the entire protocol has been done. Of course the initial state $X_0(b)$ is also a state at which the system returns, but then the entire protocol has been executed. The other states are, as one can see above, the states $X_1(d,b)$ and $X_5(d,b)$. We will start the substitution with these states. Next the Unwind Principle (UP) will be applied on them, before they are substituted again in equations for other states. During the substitution the identities from section 2 will be applied often. The part(s) of a term that is(are) being worked out are printed boldface.

$$\tau_I(X_5(d,b)) = \tau[t_4]\cdot \boldsymbol{\tau_I(X_7(d,b))}$$

$$= \tau[t_4]\cdot \{\boldsymbol{\tau[t_3]}\cdot\tau_I(X_0(1\text{-}b)) + \tau[t_3]\cdot\tau_I(X_8(d,b))\}$$

$$= \tau[t_4]\cdot \{(t_3)\tau_I(X_0(1\text{-}b)) + \tau[t_3]\cdot\boldsymbol{\tau_I(X_8(d,b))}\}$$

$$= \tau[t_4]\cdot \{(t_3)\tau_I(X_0(1\text{-}b)) + \tau[t_3]\cdot\boldsymbol{\tau[t_1]}\cdot\tau_I(X_9(d,b))\}$$

$$= \tau[t_4]\cdot \{(t_3)\tau_I(X_0(1\text{-}b)) + \tau[t_3]\cdot(t_1)\boldsymbol{\tau_I(X_9(d,b))}\}$$

$$= \tau[t_4]\cdot \{(t_3)\tau_I(X_0(1\text{-}b)) + \tau[t_3]\cdot(\boldsymbol{t_1})(\tau[t_2]\cdot\tau_I(X_5(d,b)))\}$$

$$= \tau[t_4]\cdot \{(t_3)\tau_I(X_0(1\text{-}b)) + \tau[t_3]\cdot\boldsymbol{\tau[t_1+t_2]}\cdot\tau_I(X_5(d,b))\}$$

$$= \tau[t_4]\cdot \{(t_3)\tau_I(X_0(1\text{-}b)) + \tau[t_3]\cdot(t_1+t_2)\tau_I(X_5(d,b))\}$$

$$\overset{UP}{=} \tau[t_4]\cdot \textstyle\sum_{n=0}^{\infty}\tau[t_3+n(t_1+t_2+t_3+t_4)] \cdot \tau_I(X_0(1\text{-}b))$$

$$\tau_I(X_1(d,b)) = \tau[t_1]\cdot \boldsymbol{\tau_I(X_2(d,b))}$$

$$= \tau[t_1]\cdot \{\boldsymbol{\tau[t_2]}\cdot\tau_I(X_3(d,b)) + \tau[t_2]\cdot\tau_I(X_4(d,b))\}$$

$$= \tau[t_1]\cdot \{(t_2)\boldsymbol{\tau_I(X_3(d,b))} + \tau[t_2]\cdot\boldsymbol{\tau_I(X_4(d,b))}\}$$

$$= \tau[t_1]\cdot \{(\boldsymbol{t_2})(s2(d)[t_5]\cdot\tau_I(X_5(d,b))) + \tau[t_2]\cdot\boldsymbol{\tau[t_4]}\cdot\tau_I(X_6(d,b))\}$$

$$= \tau[t_1]\cdot \{\ s2(d)[t_2+t_5]\ \cdot$$
$$\boldsymbol{\tau[t_4]}\cdot\textstyle\sum_{n=0}^{\infty}\tau[t_3+n(t_1+t_2+t_3+t_4)] \cdot \tau_I(X_0(1\text{-}b))$$

$$+ \tau[t_2] \cdot \ (t_4)\boldsymbol{\tau_I}(\boldsymbol{X_6}(\boldsymbol{d},\boldsymbol{b}))\}$$

$$\begin{aligned} = \quad & \tau[t_1] \cdot \{\ s2(d)[t_2{+}t_5] \cdot \\ & \quad \textstyle\sum_{n=0}^{\infty} \tau[t_3{+}t_4 + n(t_1{+}t_2{+}t_3{+}t_4)] \ \cdot \ \tau_I(X_0(1\text{-}b)) \\ & + \tau[t_2] \cdot \ (\boldsymbol{t_4})(\tau[t_3] \cdot \tau_I(X_1(d,b)))\} \end{aligned}$$

$$\begin{aligned} = \quad & \tau[t_1] \cdot \{\ s2(d)[t_2{+}t_5] \cdot \\ & \quad \textstyle\sum_{n=0}^{\infty} \tau[t_3{+}t_4 + n(t_1{+}t_2{+}t_3{+}t_4)] \ \cdot \ \tau_I(X_0(1\text{-}b)) \\ & + \tau[t_2] \cdot \ \boldsymbol{\tau}[\boldsymbol{t_3}{+}\boldsymbol{t_4}] \cdot \tau_I(X_1(d,b))\} \end{aligned}$$

$$\begin{aligned} = \quad & \tau[t_1] \cdot \{\ s2(d)[t_2{+}t_5] \cdot \\ & \quad \textstyle\sum_{n=0}^{\infty} \tau[t_3{+}t_4 + n(t_1{+}t_2{+}t_3{+}t_4)] \ \cdot \ \tau_I(X_0(1\text{-}b)) \\ & + \tau[t_2] \cdot \ (t_3 + t_4)\tau_I(X_1(d,b))\} \end{aligned}$$

$$\begin{aligned} \overset{\text{UP}}{=} \quad & \tau[t_1] \cdot \textstyle\sum_{m=0}^{\infty} \tau[t_2 + m(t_1{+}t_2{+}t_3{+}t_4)] \ \cdot \ (-\boldsymbol{t_2})(s2(d)[t_2{+}t_5] \\ & \cdot \textstyle\sum_{n=0}^{\infty} \tau[t_3{+}t_4 + n(t_1{+}t_2{+}t_3{+}t_4)] \ \cdot \ \tau_I(X_0(1\text{-}b)) \end{aligned}$$

$$\begin{aligned} = \quad & \tau[t_1] \cdot \textstyle\sum_{m=0}^{\infty} \tau[t_2 + m(t_1{+}t_2{+}t_3{+}t_4)] \ \cdot \ s2(d)[t_5] \\ & \cdot \textstyle\sum_{n=0}^{\infty} \tau[t_3{+}t_4 + n(t_1{+}t_2{+}t_3{+}t_4)] \ \cdot \ \tau_I(X_0(1\text{-}b)) \end{aligned}$$

$$\tau_I(X_0(b)) \ = \ \textstyle\sum_{d \in D} \int_{v>0} r_1(d)[v] \cdot \boldsymbol{\tau_I}(\boldsymbol{X_1}(\boldsymbol{d},\boldsymbol{b}))$$

$$\begin{aligned} = \quad & \textstyle\sum_{d \in D} \int_{v>0} r_1(d)[v] \ \cdot \ \boldsymbol{\tau}[\boldsymbol{t_1}] \\ & \cdot \textstyle\sum_{m=0}^{\infty} \tau[t_2 + m(t_1{+}t_2{+}t_3{+}t_4)] \ \cdot \ s2(d)[t_5] \\ & \cdot \textstyle\sum_{n=0}^{\infty} \tau[t_3{+}t_4 + n(t_1{+}t_2{+}t_3{+}t_4)] \ \cdot \ \tau_I(X_0(1\text{-}b)) \end{aligned}$$

$$\begin{aligned} = \quad & \textstyle\sum_{d \in D} \int_{v>0} \boldsymbol{r_1}(\boldsymbol{d})[v] \\ & \cdot \textstyle\sum_{m=0}^{\infty} \tau[t_1{+}t_2 + m(t_1{+}t_2{+}t_3{+}t_4)] \cdot \boldsymbol{s2}(\boldsymbol{d})[t_5] \\ & \cdot \textstyle\sum_{n=0}^{\infty} \tau[t_3{+}t_4 + n(t_1{+}t_2{+}t_3{+}t_4)] \ \cdot \ \tau_I(\boldsymbol{X_0}(\boldsymbol{1\text{-}b})) \end{aligned}$$

In this expression, m stands for the number of mess-ups by channel K before the message is put through correctly; n stands for the number of mess-ups by channel L before the positive acknowledgement is received correctly. We shall now replace the abstract delay-constants $t_1, ..., t_5$ by the actual values, chosen in section 4. This results in the next, final expression for the ABP:

$$\tau_I(\text{ABP}) \ =$$

$$\sum_{d \in D} r_1(d)[10] \cdot \sum_{m=0}^{\infty} \tau[15 + 30m] \cdot s_2(d)[7] \cdot \sum_{n=0}^{\infty} \tau[15 + 30n] \cdot \tau_I(\text{ABP}).$$

It takes, as one can see, 22 time-steps for a message to appear at gate 2 after it has been read in at gate 1, provided that channel K does not make a mistake. If it does, each mistake costs 30 time steps extra. Next, 15 time steps are needed for a positive acknowledgement to be transmitted from Receiver to Sender, provided that channel L makes no mistakes. If it does, each mistakes takes 30 time steps as well.

5.2 The CABP

Abstraction Just as we have done with the ABP, we are going to abstract from CABP-actions that do not deal with the environment. Again, the only actions that are visible to the environment are $r_1(d)$ and $s_2(d)$ for $d \in D$. All other actions are internal actions and will be put into the abstraction-set I, so $I = \{c_3(db), c_4(db), c_5(b), c_6(b), c_7(ac), c_8(ac), errorK, errorL\}$ with $d \in D$ and $b \in B$.

$$\tau_I(Y_0(b)(\vec{t})) = \\ \tau(t_{AS}+8) \cdot \tau_I(Y_1(b)(t_S, -, -, t_{AS}+8, -, t_{AS}+8))$$

$$\tau_I(Y_1(b)(\vec{t})) = \\ \sum_{d \in D} r_1(d)(t_S+10) \cdot \tau_I(Y_2(d,b)(t_S+10, -, -, t_L, -, t_{AS}))$$

$$\tau_I(Y_2(d,b)(\vec{t})) = \\ \tau(t_L+7) \cdot \tau_I(Y_3(d,b)(t_S, -, -, -, -, t_{AS}))$$

$$\tau_I(Y_3(d,b)(\vec{t})) = \\ \tau(t_{AS}+8) \cdot \tau_I(Y_4(d,b)(t_S, -, -, t_{AS}+8, -, t_{AS}+8))$$

$$\tau_I(Y_4(d,b)(\vec{t})) = \\ \tau(t_S+8) \cdot \tau_I(Y_5(d,b)(t_S+8, -, t_S+8, t_L, -, t_{AS}))$$

$$\tau_I(Y_5(d,b)(\vec{t})) = \\ \tau(t_L+7) \cdot \tau_I(Y_6(d,b)(t_S, -, t_K, -, -, t_{AS}))$$

$$\tau_I(Y_6(d,b)(\vec{t})) = \\ \tau(t_{AS}+8) \cdot \tau_I(Y_7(d,b)(t_S, -, t_K, t_{AS}+8, -, t_{AS}+8))$$

$$\tau_I(Y_7(d,b)(\vec{t})) = \\ \tau(t_K+7) \cdot \tau_I(Y_8(d,b)(t_S, -, -, t_L, t_K+7, t_{AS})) \\ + \tau(t_K+7) \cdot \tau_I(Y_4(d,b)(t_S, -, -, t_L, -, t_{AS}))$$

$$\tau_I(Y_8(d,b)(\vec{t})) =$$

$$\tau(t_S{+}8) \;\cdot\; \tau_I(Y_9(d,b)(t_S{+}8,-,t_S{+}8,t_L,t_R,t_S{+}8))$$

$$\begin{aligned}\tau_I(Y_9(d,b)(\vec{t})) \;=\;& \\ &\tau(t_L{+}7) \;\cdot\; \tau_I(Y_{10}(d,b)(t_S,-,t_K,-,t_R,t_{AS}))\end{aligned}$$

$$\begin{aligned}\tau_I(Y_{10}(d,b)(\vec{t})) \;=\;& \\ &s_2(d)(t_R{+}7)\cdot\tau_I(Y_{11}(d,b)(t_S,-,t_K,-,-,t_{AS}))\end{aligned}$$

$$\begin{aligned}\tau_I(Y_{11}(d,b)(\vec{t})) \;=\;& \\ &\tau(t_K{+}7) \;\cdot\; \tau_I(Y_{12}(d,b)(t_S,-,-,-,-,t_{AS}))\end{aligned}$$

$$\begin{aligned}\tau_I(Y_{12}(d,b)(\vec{t})) \;=\;& \\ &\tau(t_S{+}8) \;\cdot\; \tau_I(Y_{13}(d,b)(t_S{+}8,-,t_S{+}8,t_S{+}8,-,t_S{+}8))\end{aligned}$$

$$\begin{aligned}\tau_I(Y_{13}(d,b)(\vec{t})) \;=\;& \\ &\tau(t_K{+}7) \;\cdot\; \tau_I(Y_{14}(d,b)(t_S,t_K{+}7,-,-,-,t_{AS})) \\ +\;&\tau(t_K{+}7) \;\cdot\; \tau_I(Y_{12}(d,b)(t_S,-,-,-,-,t_{AS}))\end{aligned}$$

$$\begin{aligned}\tau_I(Y_{14}(d,b)(\vec{t})) \;=\;& \\ &\tau(t_{AS}{+}8) \;\cdot\; \tau_I(Y_{15}(1\text{-}b)(t_{AS}{+}8,-,-,t_{AS}{+}8,-,t_{AS}{+}8))\end{aligned}$$

$$\begin{aligned}\tau_I(Y_{15}(b)(\vec{t})) \;=\;& \\ &\tau(t_L{+}7) \;\cdot\; \tau_I(Y_0(b)(t_S,-,-,-,-,t_{AS}))\end{aligned}$$

Substitution Finally, also the number of equations for the CABP will be reduced by substitution. Also here there are states at which the system can return before the entire protocol has been done, states $Y_4(d,b)$ and $Y_{12}(d,b)$. We will start the substitution with these states. Again the Unwind Principle will be applied.

Note that if we have

$$\tau_I(Y_4(d,b)(\vec{t})) = \tau(t_S{+}8)\cdot\tau_I(Y_5(d,b)(t_S{+}8,-,t_S{+}8,t_L,-,t_{AS}))$$

and

$$\tau_I(Y_5(d,b)(\vec{t})) = \tau(t_L{+}7)\cdot\tau_I(Y_6(d,b)(t_S,-,t_K,-,-,t_{AS})),$$

we cannot simply replace $Y_5(d,b)$ in the first equation by the right hand side of the second equation. Each time-parameter has to be replaced as well. So t_S and t_K in the right hand side of the second equation have to be replaced by t_S+8 in this example.

$$\tau_I(Y_{12}(d,b)(\vec{t})) \;=\; \tau(t_S{+}8)\cdot\tau_I(\mathbf{Y_{13}(d,b)}(t_S{+}8,-,t_S{+}8,t_S{+}8,-,t_S{+}8))$$

$$\begin{aligned}
= \quad & \tau(t_S{+}8) \cdot \\
& \{\boldsymbol{\tau(t_S{+}15)} \cdot \tau_I(Y_{14}(d,b)(t_S{+}8, t_S{+}15, -, -, -, t_S{+}8)) \ + \\
& \ \tau(t_S{+}15) \cdot \tau_I(Y_{12}(d,b)(t_S{+}8, -, -, -, -, t_S{+}8))\}
\end{aligned}$$

$$\begin{aligned}
= \quad & \tau(t_S{+}8) \cdot \\
& \{\tau_I(\boldsymbol{Y_{14}(d,b)}(t_S{+}8, t_S{+}15, -, -, -, t_S{+}8)) \ + \\
& \ \tau(t_S{+}15) \cdot \tau_I(Y_{12}(d,b)(t_S{+}8, -, -, -, -, t_S{+}8))\}
\end{aligned}$$

$$\begin{aligned}
= \quad & \tau(t_S{+}8) \cdot \\
& \{\boldsymbol{\tau(t_S{+}16)} \cdot \tau_I(Y_{15}(1\text{-}b)(t_S{+}16, -, -, t_S{+}16, -, t_S{+}16)) \ + \\
& \ \tau(t_S{+}15) \cdot \tau_I(Y_{12}(d,b)(t_S{+}8, -, -, -, -, t_S{+}8))\}
\end{aligned}$$

$$\begin{aligned}
= \quad & \tau(t_S{+}8) \cdot \\
& \{\tau_I(\boldsymbol{Y_{15}(1\text{-}b)}(t_S{+}16, -, -, t_S{+}16, -, t_S{+}16)) \ + \\
& \ \tau(t_S{+}15) \cdot \tau_I(Y_{12}(d,b)(t_S{+}8, -, -, -, -, t_S{+}8))\}
\end{aligned}$$

$$\begin{aligned}
= \quad & \tau(t_S{+}8) \cdot \\
& \{\boldsymbol{\tau(t_S{+}23)} \cdot \tau_I(Y_0(1\text{-}b)(t_S{+}16, -, -, -, -, t_S{+}16)) \ + \\
& \ \tau(t_S{+}15) \cdot \tau_I(Y_{12}(d,b)(t_S{+}8, -, -, -, -, t_S{+}8))\}
\end{aligned}$$

$$\begin{aligned}
= \quad & \tau(t_S{+}8) \cdot \\
& \{\tau_I(Y_0(1\text{-}b)(t_S{+}16, -, -, -, -, t_S{+}16)) \ + \\
& \ \tau(t_S{+}15) \cdot \tau_I(Y_{12}(d,b)(t_S{+}8, -, -, -, -, t_S{+}8))\}
\end{aligned}$$

$$\begin{aligned}
\overset{\mathrm{UP}}{=} \quad & \tau(t_S{+}8) \cdot \\
& \textstyle\sum_{n=0}^{\infty} \tau(t_S{+}8n{+}15) \cdot \\
& \tau_I(Y_0(1\text{-}b)(t_S{+}8n{+}16, -, -, -, -, t_S{+}8n{+}16))
\end{aligned}$$

$$\tau_I(Y_4(d,b)(\vec{t})) \ = \ \tau(t_S{+}8) \cdot \tau_I(\boldsymbol{Y_5(d,b)}(t_S{+}8, -, t_S{+}8, t_L, -, t_{AS}))$$

$$\begin{aligned}
= \quad & \tau(t_S{+}8) \cdot \\
& \boldsymbol{\tau(t_L{+}7)} \cdot \tau_I(Y_6(d,b)(t_S{+}8, -, t_S{+}8, -, -, t_{AS}))
\end{aligned}$$

$$\begin{aligned}
= \quad & \tau(t_S{+}8) \cdot \\
& \tau_I(\boldsymbol{Y_6(d,b)}(t_S{+}8, -, t_S{+}8, -, -, t_{AS}))
\end{aligned}$$

$$\begin{aligned}
= \quad & \tau(t_S{+}8) \cdot \\
& \boldsymbol{\tau(t_{AS}{+}8)} \cdot \tau_I(Y_7(d,b)(t_S{+}8, -, t_S{+}8, t_{AS}{+}8, -, t_{AS}{+}8))
\end{aligned}$$

$$
\begin{aligned}
= \quad & \tau(t_S{+}8) \cdot \\
& \tau_I(\boldsymbol{Y_7(d,b)}(t_S{+}8,-,t_S{+}8,t_{AS}{+}8,-,t_{AS}{+}8)) \\
\\
= \quad & \tau(t_S{+}8) \cdot \\
& \{\boldsymbol{\tau(t_S{+}15)} \cdot \tau_I(Y_8(d,b)(t_S{+}8,-,-,t_{AS}{+}8,t_S{+}15,t_{AS}{+}8)) \ + \\
& \ \tau(t_S{+}15) \cdot \tau_I(Y_4(d,b)(t_S{+}8,-,-,t_{AS}{+}8,-,t_{AS}{+}8))\} \\
\\
= \quad & \tau(t_S{+}8) \cdot \\
& \{\tau_I(\boldsymbol{Y_8(d,b)}(t_S{+}8,-,-,t_{AS}{+}8,t_S{+}15,t_{AS}{+}8)) \ + \\
& \ \tau(t_S{+}15) \cdot \tau_I(Y_4(d,b)(t_S{+}8,-,-,t_{AS}{+}8,-,t_{AS}{+}8))\} \\
\\
= \quad & \tau(t_S{+}8) \cdot \\
& \{\boldsymbol{\tau(t_S{+}16)} \cdot \\
& \qquad\qquad \tau_I(Y_9(d,b)(t_S{+}16,-,t_S{+}16,t_{AS}{+}8,t_S{+}15,t_S{+}16)) \ + \\
& \ \tau(t_S{+}15) \cdot \tau_I(Y_4(d,b)(t_S{+}8,-,-,t_{AS}{+}8,-,t_{AS}{+}8))\} \\
\\
= \quad & \tau(t_S{+}8) \cdot \\
& \{\tau_I(\boldsymbol{Y_9(d,b)}(t_S{+}16,-,t_S{+}16,t_{AS}{+}8,t_S{+}15,t_S{+}16)) \ + \\
& \ \tau(t_S{+}15) \cdot \tau_I(Y_4(d,b)(t_S{+}8,-,-,t_{AS}{+}8,-,t_{AS}{+}8))\} \\
\\
= \quad & \tau(t_S{+}8) \cdot \\
& \{\boldsymbol{\tau(t_{AS}{+}15)} \cdot \tau_I(Y_{10}(d,b)(t_S{+}16,-,t_S{+}16,-,t_S{+}15,t_S{+}16)) \ + \\
& \ \tau(t_S{+}15) \cdot \tau_I(Y_4(d,b)(t_S{+}8,-,-,t_{AS}{+}8,-,t_{AS}{+}8))\} \\
\\
= \quad & \tau(t_S{+}8) \cdot \\
& \{\tau_I(\boldsymbol{Y_{10}(d,b)}(t_S{+}16,-,t_S{+}16,-,t_S{+}15,t_S{+}16)) \ + \\
& \ \tau(t_S{+}15) \cdot \tau_I(Y_4(d,b)(t_S{+}8,-,-,t_{AS}{+}8,-,t_{AS}{+}8))\} \\
\\
= \quad & \tau(t_S{+}8) \cdot \\
& \{s_2(d)(t_S{+}22) \cdot \tau_I(\boldsymbol{Y_{11}(d,b)}(t_S{+}16,-,t_S{+}16,-,-,t_S{+}16)) \ + \\
& \ \tau(t_S{+}15) \cdot \tau_I(Y_4(d,b)(t_S{+}8,-,-,t_{AS}{+}8,-,t_{AS}{+}8))\} \\
\\
= \quad & \tau(t_S{+}8) \cdot \\
& \{s_2(d)(t_S{+}22) \cdot \boldsymbol{\tau(t_S{+}23)} \cdot \\
& \qquad\qquad\qquad \tau_I(Y_{12}(d,b)(t_S{+}16,-,-,-,-,t_S{+}16)) \ + \\
& \ \tau(t_S{+}15) \cdot \tau_I(Y_4(d,b)(t_S{+}8,-,-,t_{AS}{+}8,-,t_{AS}{+}8))\} \\
\\
= \quad & \tau(t_S{+}8) \cdot
\end{aligned}
$$

$$\{s_2(d)(t_S{+}22)\cdot\tau_I(\boldsymbol{Y_{12}(d,b)}(t_S{+}16,-,-,-,-,t_S{+}16))\ +$$
$$\tau(t_S{+}15)\cdot\tau_I(Y_4(d,b)(t_S{+}8,-,-,t_{AS}{+}8,-,t_{AS}{+}8))\}$$

$$=\quad \tau(t_S{+}8)\cdot$$
$$\{s_2(d)(t_S{+}22)\cdot\boldsymbol{\tau(t_S{+}24)}\cdot\textstyle\sum_{n=0}^{\infty}\tau(t_S{+}8n{+}31)\ \cdot$$
$$\tau_I(Y_0(1\text{-}b)(t_S{+}8n{+}32,-,-,-,-,t_S{+}8n{+}32))\ +$$
$$\tau(t_S{+}15)\cdot\tau_I(Y_4(d,b)(t_S{+}8,-,-,t_{AS}{+}8,-,t_{AS}{+}8))\}$$

$$=\quad \tau(t_S{+}8)\cdot$$
$$\{s_2(d)(t_S{+}22)\cdot\textstyle\sum_{n=0}^{\infty}\tau(t_S{+}8n{+}31)\ \cdot$$
$$\tau_I(Y_0(1\text{-}b)(t_S{+}8n{+}32,-,-,-,-,t_S{+}8n{+}32))\ +$$
$$\tau(t_S{+}15)\cdot\tau_I(Y_4(d,b)(t_S{+}8,-,-,t_{AS}{+}8,-,t_{AS}{+}8))\}$$

$$\overset{\text{UP}}{=}\quad \tau(t_S{+}8)\cdot$$
$$\textstyle\sum_{m=0}^{\infty}\tau(t_S{+}8m{+}15)\cdot s_2(d)(t_S{+}8m{+}22)\cdot$$
$$\textstyle\sum_{n=0}^{\infty}\tau(t_S{+}8(m{+}n){+}31)\cdot$$
$$\tau_I(Y_0(1\text{-}b)(t_S{+}8(m{+}n){+}32,-,-,-,-,t_S{+}8(m{+}n){+}32))$$

$$\tau_I(Y_0(b)(\vec{t}))\ =\ \tau(t_{AS}{+}8)\cdot\tau_I(\boldsymbol{Y_1(b)}(t_S,-,-,t_{AS}{+}8,-,t_{AS}{+}8))$$

$$=\quad \tau(t_{AS}{+}8)\cdot\textstyle\sum_{d\in D}r_1(d)(t_S{+}10)\cdot$$
$$\tau_I(\boldsymbol{Y_2(d,b)}(t_S{+}10,-,-,t_{AS}{+}8,t_{AS}{+}8))$$

$$=\quad \tau(t_{AS}{+}8)\cdot\textstyle\sum_{d\in D}r_1(d)(t_S{+}10)\cdot$$
$$\boldsymbol{\tau(t_{AS}{+}15)}\cdot\tau_I(Y_3(d,b)(t_S{+}10,-,-,-,-,t_{AS}{+}8))$$

$$=\quad \tau(t_{AS}{+}8)\cdot\textstyle\sum_{d\in D}r_1(d)(t_S{+}10)\cdot$$
$$\tau_I(\boldsymbol{Y_3(d,b)}(t_S{+}10,-,-,-,-,t_{AS}{+}8))$$

$$=\quad \tau(t_{AS}{+}8)\cdot\textstyle\sum_{d\in D}r_1(d)(t_S{+}10)\cdot$$
$$\boldsymbol{\tau(t_{AS}{+}16)}\cdot\tau_I(Y_4(d,b)(t_S{+}10,-,-,t_{AS}{+}16,-,t_{AS}{+}16))$$

$$=\quad \tau(t_{AS}{+}8)\cdot\textstyle\sum_{d\in D}r_1(d)(t_S{+}10)\cdot$$
$$\tau_I(\boldsymbol{Y_4(d,b)}(t_S{+}10,-,-,t_{AS}{+}16,-,t_{AS}{+}16))$$

$$=\quad \tau(t_{AS}{+}8)\cdot\textstyle\sum_{d\in D}r_1(d)(t_S{+}10)\cdot\boldsymbol{\tau(t_S{+}18)}\cdot$$
$$\textstyle\sum_{m=0}^{\infty}\tau(t_S{+}8m{+}25)\cdot s_2(d)(t_S{+}8m{+}32)\cdot$$

$$\textstyle\sum_{n=0}^{\infty}\tau(t_S{+}8(m{+}n){+}41)\cdot$$
$$\tau_I(Y_0(1\text{-}b)(t_S{+}8(m{+}n){+}42,-,-,-,-,t_S{+}8(m{+}n){+}42))$$

$$= \tau(t_{AS}{+}8)\cdot\textstyle\sum_{d\in D}\boldsymbol{r}_1(\boldsymbol{d})(t_S{+}10)\cdot$$
$$\textstyle\sum_{m=0}^{\infty}\tau(t_S{+}8m{+}25)\cdot\boldsymbol{s}_2(\boldsymbol{d})(t_S{+}8m{+}32)\cdot$$
$$\textstyle\sum_{n=0}^{\infty}\tau(t_S{+}8(m{+}n){+}41)\cdot$$
$$\tau_I(\boldsymbol{Y}_0(1\text{-}\boldsymbol{b})(t_S{+}8(m{+}n){+}42,-,-,-,-,t_S{+}8(m{+}n){+}42))$$

We now have the following, final expression for the CABP:

$$\tau_I(\mathrm{CABP}(\vec{t})) =$$
$$\tau(t_{AS}{+}8)\cdot\sum_{d\in D} r_1(d)(t_S{+}10)\cdot$$
$$\sum_{m=0}^{\infty}\tau(t_S{+}8m{+}25)\cdot s_2(d)(t_S{+}8m{+}32)\cdot$$
$$\sum_{n=0}^{\infty}\tau(t_S{+}8(m{+}n){+}41)\cdot\tau_I(\mathrm{CABP}(\vec{t}+8(m{+}n){+}42)).$$

Now one can see that after a message has been read in at gate 1, it takes $(t_S{+}32)-(t_S{+}10) = 22$ time steps for it to appear at gate 2, if channel K does not make any mistakes. Then 10 time steps are needed to transmit a positive acknowledgement if channel L makes no mistakes. A mistake of either one of the channels costs 8 time steps.

5.3 Comparison of Performances

We have specified two simple protocols in real time process algebra. We have also presented laws that made it possible to rewrite our specifications into relatively simple expressions, from which we can read very precisely how much time lies between the execution of external actions (i.e., actions that remain after abstraction).

From these expressions it becomes clear that there is no (significant) difference in performance between the ABP and the CABP as long as the channels do not make mistakes. If the channels do make mistakes, the CABP really turns out to be more efficient. Of course this is due to the fact that mistakes do not have to be reported explicitly in the CABP. In fact, both the Sender and the Acknowledgement Sender presume that the channels loose data and therefore send their announcements repeatedly. The result is that, in our case, for each mistake the CABP is almost 4 times faster than the ABP. Here we have to remark that the retransmission-frequency has been set very efficient, by assigning a value to t_1 and t_4, that is just one time step greater than the value assigned to t_2, t_3 and t_5. t_1 and t_4 must be greater than t_2, t_3 and t_5 (to guard the system against choking up), but the smaller they are, the more efficient the protocol is, since then the system has to wait minimal time to try again after a mistake has occurred.

We can conclude that real time ACP is well-suited for the description of processes in a timed setting. Time constraints can be as precise as one wants (because of the notion of integration), and are denoted in a comprehensible way. Meanwhile, the possibility to read the global behaviour from a process-expression is preserved. Although relative time perhaps makes it easier to describe processes, it has proven to be harder to make calculations in, because process behaviour becomes context-dependent. In absolute time τ's can be removed very easily, though in relative time it forces the use of the time shift. For the verification of protocols, the process equivalence to work with must be weaker than branching bisimulation, which we have used in this paper. A weaker equivalence would perhaps have made it possible to verify the general specification of the CABP. In branching bisimulation we must distinguish between processes that only differ in moments of internal choice, although we are only interested in processes that differ in external behaviour. We did use branching bisimulation, because we wanted to be as precise as possible, but we learnt that this is not always wise.

References

[BB91] J.C.M. Baeten and J.A. Bergstra. Real time process algebra. *Journal of Formal Aspects of Computing Science*, 3(2):142–188, 1991.

[FK92] W.J. Fokkink and A.S. Klusener. Real time process algebra with prefixed integration. Report CS-R9219, CWI, Amsterdam, 1992.

[Fok92] W.J. Fokkink. Regular processes with rational time and silent steps. Report CS-R9231, CWI, Amsterdam, 1992.

[Hil94] J.A. Hillebrand. The ABP and the CABP – a comparison of performances in real time process algebra. report P9211b, Programming Research Group, University of Amsterdam, 1994.

[Klu91a] A.S. Klusener. Abstraction in real time process algebra. report CS-R9144, CWI, amsterdam, 1991. An extended abstract appeared in J.W. de Bakker, C. Huizing, W.P. de Roever and G. Rozenberg , editors, *Proceedings of the REX workshop "Real-Time: Theory in Practice"*, LNCS 600, Springer-Verlag, 1991.

[Klu91b] A.S. Klusener. Completeness in real time process algebra. In J.C.M. Baeten and J.F. Groote, editors, *Proceedings CONCUR 91*, Amsterdam, volume 527 of *Lecture Notes in Computer Science*, pages 376–392. Springer-Verlag, 1991.

[Klu93] A.S. Klusener. *Models and axioms for a fragment of real time process algebra*. PhD thesis, Eindhoven University of Technology, December 1993.

[vW93] J.J. van Wamel. A formal specification of three simple protocols. In S. Mauw and G.J. Veltink, editors, *Algebraic Specification of Communication Protocols*, Cambridge Tracts in Theoretical Computer Science 36, pages 47–70. Cambridge University Press, 1993.

Real Time Process Algebra with Infinitesimals

J.C.M. Baeten
Department of Computer Science, Eindhoven University of Technology, Eindhoven, The Netherlands

J.A. Bergstra
Programming Research Group, University of Amsterdam, Amsterdam, The Netherlands, and
Department of Philosophy, Utrecht University, Utrecht, The Netherlands

We consider a model of the real time process algebra of [1,2,3] based on the nonstandard reals. As a subalgebra, we obtain a theory in which the urgent actions of ATP, TiCCS, TeCCS can be modeled.
Note: This research was supported in part by ESPRIT basic research action 7166, CONCUR2. The second author also is partially supported by ESPRIT basic research action 6454, CONFER.

1. Introduction.

An important motivation for our work on timed process algebra ([1,2,3]) is that we intend to present a uniform framework, in which all constructs that occur in the literature on timed process algebra can be expressed. In [1], we present our real time absolute time process algebra, in [3] we achieve a unified treatment of absolute time and relative time expressions. In [2], we define many discrete time constructs from the literature, both absolute and relative time ones, in our real time process algebra, and define suitable discrete time subalgebras. This paper is concerned with immediate or urgent actions, as they appear in TiCCS [11,12], ATP [13], TeCCS [16], TiCCS [7].

As an example, consider the TiCCS expression (2).a.b.0. Intuitively, this expression denotes the process that, after an initial delay of 2 time units, executes first a and then b instantaneously, and then terminates. Now, in our real time process algebra ACPρ, two actions cannot be executed consecutively at the same point in time, so the process $a(2)\cdot b(2)$ will deadlock after executing the action a at time 2, and b will not be executed. Nevertheless, we can define the process above in an extension of ACPρ. To this end, we extend our time domain to include infinitesimals, and model the TiCCS expression above by a process that first executes a and then b at time points both infinitely close to 2, with the difference of execution times a positive infinitesimal.

This paper describes the extension of ACPρ to a more general time domain, describes the relation of ACPρ over the (standard) positive reals with ACPρ over a time domain including nonstandard reals, and defines suitable subalgebras including urgent actions. Some of these subalgebras correspond to timed process algebras found in the literature.

We give an overview of the systems that we will consider. We start with the system BPAρ√I, real time process algebra with initial abstraction and integration. By adding parallel composition and communication, we obtain ACPρ√I. In this system (over a time domain involving nonstandard reals) we can define urgent actions. We can also directly axiomatise algebras based on urgent actions instead of our usual real time actions. In this way, we obtain the absolute time algebras BPAsρ, PAsρ, ACPsρ, and the relative time algebras BPAst, BPAδst, PAst, ACPst. By adding initial abstraction to the absolute time algebras, obtaining BPAsρ√, PAsρ√, ACPsρ√, we can interpret the relative time ones in them. Figure 1 gives an overview of all algebras we consider, and the embedding relations between them. Boxes with rounded corners denote relative time algebras, square boxes denote absolute time algebras.

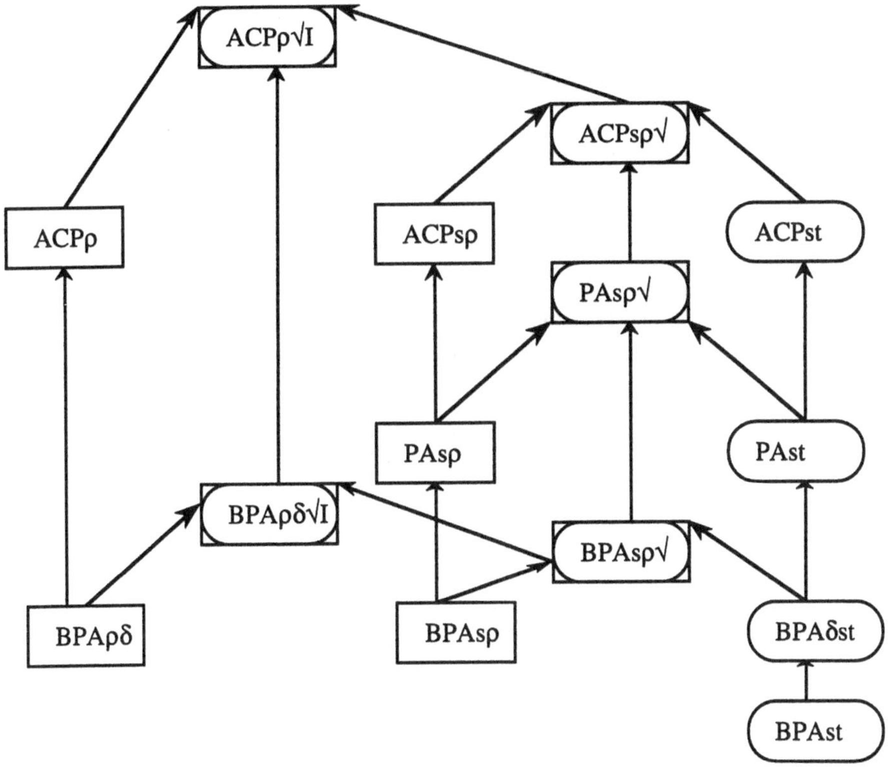

FIGURE 1.

ACKNOWLEDGEMENT. We thank Steven Klusener (CWI Amsterdam) for his useful comments and suggestions, and Peter van Emde Boas (University of Amsterdam) for his suggestion to use nonstandard reals or surreal numbers.

2. Real time process algebra with initial abstraction operator.

We will first consider the axiom system ACPρ√I. It describes the real time process algebra of [1] as presented in [2]. The reader is assumed to be familiar with [2], as we only indicate the differences with that paper. For ACPρ√I, we have a bisimulation model M_A^* as outlined in [2].

Having available M_A^*, the various operators and constants of urgent time process algebra are defined in it. Then, as a subject, nonstandard real time process algebra reduces to an investigation of certain reducts and subalgebras of M_A^* as well as their axiomatic description.

2.1 Time domain.

We assume that we have a time domain $\mathcal{T}$. We have the following requirements on $\mathcal{T}$:

i. $\mathcal{T}$ is linearly ordered by <
ii. 0 is the smallest element of $\mathcal{T}$
iii. $\mathcal{T}$ is closed under addition +.

In [1], we used the supremum of a subset of $\mathcal{T}$ in de definition of the ultimate delay operator, so we needed completeness of $\mathcal{T}$. We reformulate this operator in this paper (as we did in [BAB93]) so that this requirement is not needed anymore. Examples of time domains satisfying these requirements are $\mathbb{N}$, the naturals, $\mathbb{Q}_{\geq 0}$, the non-negative rationals, $\mathbb{R}_{\geq 0}$, the non-negative reals, or $^*\mathbb{R}_{\geq 0}$, the non-standard non-negative reals. Our prefered time domain, the one we used in [1,2,3], is $\mathbb{R}_{\geq 0}$.

2.2 Basic process algebra.

A is the set of (symbolic) atomic actions. The set A is a parameter of the theory. δ denotes inaction. We put $A_\delta = A \cup \{\delta\}$. The set of constants is the set of atomic actions with time parameter, AT, given by $AT = \{a(t) \mid a \in A_\delta, t \in \mathcal{T}\}$.

Real Time Basic Process Algebra with Deadlock (BPAρδ) has two binary operators $+, \cdot : P \times P \to P$; alternative composition (weak choice) and sequential composition, respectively. Moreover, the *initialisation operator* $\gg: \mathcal{T} \times P \to P$. BPAρδ has the axioms given in [1,2].

2.3 Algebra of Communicating Processes.

In order to formulate communication between processes, we assume we have given a *communication function* $|$ on A_δ. An axiomatization of parallel composition with communication uses the left merge operator $\lfloor\!\lfloor$, the communication merge operator $|$, and the encapsulation operator ∂_H of [6]. Moreover, an extra auxiliary operator is needed. In [1,2,3] we used the ultimate delay operator and the bounded initialization operator as auxiliary operators. However, in order to define the ultimate delay of an integral expression, we need the supremum of a subset of $\mathcal{T}$. In the time domain $\mathbb{R}_{\geq 0}$, such a supremum will always exist (or is infinite), but later on, we will consider a time domain where suprema need not exist. For this reason, we will use a different auxiliary operator here. Since the axiomatisation of merge used before only needed expressions of the form $X \gg U(Y)$ (with U the ultimate delay and $\gg$ the bounded initialisation), we will have a new binary operator (left strong choice) for this combination. It turns out that this operator is exactly one half of the strong choice operator $\oplus$ of ATP, and that explains the name.

Thus, we have the operator $\underline{\oplus}$: $P \times P \to P$ (left strong choice) and $X \underline{\oplus} Y$ denotes intuitively that part of X, that can start before an initial action of Y. Notice the similarity of the axioms for the left strong choice operator with the axioms of the unless operator from [4,5]. The latter binary operator also describes a filtering of the first argument by the second. We call the present version of ACPρ ACPρ($\underline{\oplus}$). In table 1, we show the axioms that do not appear in [2], that replace the axioms using ultimate delay or bounded initialisation. The axioms for left merge have a condition because they do not hold for processes involving relative time notation. We have $a,b \in A_\delta$.

$t < s \Rightarrow a(t) \underline{\oplus} b(s) = a(t)$	LSC1
$t \geq s \Rightarrow a(t) \underline{\oplus} b(s) = \delta(s)$	LSC2
$X \underline{\oplus} Y \cdot Z = X \underline{\oplus} Y$	LSC3
$X \underline{\oplus} (Y + Z) = (X \underline{\oplus} Y) + (X \underline{\oplus} Z)$	LSC4
$X \cdot Y \underline{\oplus} Z = (X \underline{\oplus} Z) \cdot Y$	LSC5
$(X + Y) \underline{\oplus} Z = (X \underline{\oplus} Z) + (Y \underline{\oplus} Z)$	LSC6
$X = 0 \gg Z \Rightarrow a(t) \lfloor\!\lfloor X = (a(t) \underline{\oplus} X) \cdot X$	ATCM2⊕
$Y = 0 \gg Z \Rightarrow a(t) \cdot X \lfloor\!\lfloor Y = (a(t) \underline{\oplus} Y) \cdot (X \| Y)$	ATCM3⊕

TABLE 1. New axioms of ACPρ($\underline{\oplus}$).

2.4 Relative time notation.

We follow the approach to relative time of [2,3]. The basis for this approach to relative time notation is the *initial abstraction operator*. Let for $t \in \mathcal{T}$, $F(t)$ be a process in P, then

$\surd t.F(t)$ denotes a process that, when started at time r, proceeds as $F(r)$. In table 2, we show the new axioms for initial abstraction, that give the relation to the left strong choice operator.

$(\surd t.F) \underline{\oplus} X = \surd t.(F \underline{\oplus} t \gg X)$	IA13
$X \underline{\oplus} (\surd t.F) = \surd t.(t \gg X \underline{\oplus} F)$	IA14

TABLE 2. Axioms for initial abstraction and left strong choice.

2.5 Integration.

We extend our real time process algebra by adding the integral operator, denoting a choice over a subset of $\mathcal{T}$. That is, if V is a subset of $\mathcal{T}$, and v is a variable over $\mathcal{T}$, then $\int_{v \in V} P$ denotes the alternative composition of alternatives $P[t/v]$ for $t \in V$ (expression P with $t \in \mathcal{T}$ substituted for variable v). In case the time domain is $\mathbb{R}_{\geq 0}$, this is an uncountable sum, and that explains the use of the notion of an integral. For more information, we refer the reader to [1] and [9].

In table 3, X is a process (that does not contain free variables), and F,G are process expressions (that may contain free variables). We reformulate INT8: this only applies if the set V has a supremum in $\mathcal{T}$. The theory presented thus far is called ACPρ√I, ACPρ plus integration is ACPρI.

$t = \sup V, t \notin V \Rightarrow \int_{v \in V} \delta(v) = \delta(t)$	INT8
$\int_{v \in V} (F \underline{\oplus} X) = (\int_{v \in V} F) \underline{\oplus} X$	INT13
$\int_{v \in V} (X \underline{\oplus} F) = X \underline{\oplus} (\int_{v \in V} F)$	INT14

TABLE 3. Axioms for integration and left strong choice.

2.6 Structured operational semantics.

We obtain a real time transition system $TS(X)$ from a process expression X as indicated in [BAB92]. Here, we just give the SOS rules for the left strong choice operator. We have $t \in \mathcal{T}$, $r \in \mathcal{T}\text{-}\{0\}$, x,x',y are closed process expressions and $a \in A$. On these transition systems, we define bisimulation as usual. We extend with initial abstraction as in [2].

$$\frac{\langle x,t\rangle \rightarrow \langle x,r\rangle,\ \langle y,t\rangle \rightarrow \langle y,r\rangle}{\langle x \underline{\oplus} y,t\rangle \rightarrow \langle x \underline{\oplus} y,r\rangle} \qquad \frac{\langle x,t\rangle \xrightarrow{a(r)} \langle x',r\rangle,\ \langle y,t\rangle \rightarrow \langle y,r\rangle}{\langle x \underline{\oplus} y,t\rangle \xrightarrow{a(r)} \langle x',r\rangle}$$

$$\frac{\langle x,t\rangle \xrightarrow{a(r)} \langle \surd,r\rangle,\ \langle y,t\rangle \rightarrow \langle y,r\rangle}{\langle x \underline{\oplus} y,t\rangle \xrightarrow{a(r)} \langle \surd,r\rangle}$$

TABLE 4. Structured operational semantics for left strong choice.

We present some extra axioms that are useful in the calculations to come, that hold in the model, and that can be derived from the theory for all closed process expressions. The numbering of these axioms corresponds to the numbering in [2].

$s \gg X \mathbin{\|\!\_} r \gg Y = (s \gg X \underline{\oplus} r \gg Y) \mathbin{\|\!\_} s \gg r \gg Y$	SI10
$t \gg (X \underline{\oplus} Y) = t \gg X \underline{\oplus} t \gg Y$	SI14
$X + (X \underline{\oplus} Y) = X$	SI15
$a(t) \underline{\oplus} t \gg X = a(t)$	SI16
$t \gg X \underline{\oplus} a(t) = \delta(t)$	SI17

TABLE 5. Standard initialisation axioms involving left strong choice.

3. Nonstandard Reals.

In this section, we consider two time domains: on the one hand we have the time domain $\mathbb{R}_{\geq 0}$ of non-negative real numbers, on the other hand we consider a subdomain of $^*\mathbb{R}$, the nonstandard model of the real numbers. For more information on nonstandard real numbers, see [10,15]. We use the following notations: $t \approx s$ means that t,s are infinitely close together, $I = \{s \in {^*\mathbb{R}} \mid s \approx 0,\ s > 0\}$ is the set of positive infinitesimals and $t^\circ \in \mathbb{R}$ is the standard part of a finite $t \in {^*\mathbb{R}}$, the standard real infinitely close to t. Note that the supremum over a bounded subset of $^*\mathbb{R}$ need not exist.

As the extended time domain we use $\mathbb{R}_{\geq 0}+I = \mathbb{R}_{\geq 0} \cup \{p + \varepsilon : p \in \mathbb{R}_{\geq 0},\ \varepsilon \in I\}$. Note that this set satisfies the requirements of section 2.1. Thus, this domain is linearly ordered, with 0 as least element, and is closed under addition. Also, it is closed under multiplication, and division by a positive standard real. Note that for $t \in \mathbb{R}_{\geq 0}+I$, always t° exists and $t^\circ \leq t$.

3.1 Definitions.

We have the definitions in table 6 in the algebra over the extended time domain $\mathbb{R}_{\geq 0}+I$. We will always have $t,u \in \mathbb{R}_{\geq 0}+I$, $p,q \in \mathbb{R}_{\geq 0}$, $r \in \mathbb{R}_{>0}$, $\varepsilon, \eta \in I$.

$\tilde{a}[p] = \surd t. \int_{\varepsilon \in I} a(p + t^\circ + \varepsilon)$	$\tilde{\delta}[p] = \surd t. \int_{\varepsilon \in I} \delta(p + t^\circ + \varepsilon)$
$\tilde{\tilde{a}} = \tilde{a}[0]$	$\tilde{\tilde{\delta}} = \tilde{\delta}[0]$
$\tilde{a}(p) = \tilde{\tilde{\delta}} + \int_{\varepsilon \in I} a(p + \varepsilon)$	$\tilde{\delta}(p) = \tilde{\tilde{\delta}} + \int_{\varepsilon \in I} \delta(p + \varepsilon)$
$p \gg_s X = \tilde{\tilde{\delta}} + p \gg X$	$\surd_s p. F = \surd t. F[t^\circ/p]$
$\sigma^r_{\approx}(X) = \surd t. (t+r)^\circ \gg X$	$r \gg_{st} X = \surd t. (t+r)^\circ \gg t^\circ \gg_s X$

TABLE 6. Standard time operators.

The *standard absolute time signature*, $\Sigma(ACP_{s\rho\surd})$, contains the following ingredients: $\tilde{a}(p)$ $(a \in A_\delta)$, $\tilde{\tilde{\delta}}$, $+$, $\cdot$,
$\|$, $\lfloor\!\lfloor$, $|$, $\partial_H (H \subseteq A)$, $\gg_s$, $\oplus$,
$\surd_s$.

The *standard relative time signature*, $\Sigma(ACP_{st})$, contains the following ingredients:
$\tilde{\tilde{a}}$ $(a \in A_\delta)$, $+$, $\cdot$, $\sigma^r_{\approx}$,
$\|$, $\lfloor\!\lfloor$, $|$, $\partial_H (H \subseteq A)$, $r \gg_{st}$, $\oplus$.

Since we have $\tilde{a}[0] = \tilde{\tilde{a}}$ and $\tilde{a}[r] = \sigma^r_{\approx}(\tilde{\tilde{a}})$ for all $r>0$, this signature also contains all constants of the form $\tilde{a}[p]$.

The *standard time signature*, $\Sigma(ST)$, is given by $\Sigma(ST) = \Sigma(ACP_{s\rho\surd}) \cup \Sigma(ACP_{st})$.

4. Standardly initialized processes.

In sections 5 and 6, we will give direct axiomatisations of various subalgebras of $ACP\rho\surd I$, that are generated by signature elements of in section 3. Many of these axioms are not valid in the full algebra M^*_A over $\mathbb{R}_{\geq 0}+I$. In this section we define the notion of a *standardly initialised process*. The set of standardly initialised processes is closed under all operators of the standard time signature, and all axioms to be presented in sections 6 and 7 will hold in this algebra. We prove these facts at length in this section, but omitting some easy proofs. We obtain a subalgebra $si\text{-}M^*_A$ of M^*_A. Similarly, we have a subalgebra $si\text{-}M_A$ of M_A (containing absolute time processes only) and a subalgebra $si\text{-}M^r_A$ of M^r_A(containing relative time processes only).

4.1 Definition.

Let $X \in P^*$. We say X is a *standardly initialised process*, $X \in SIP$, if $X = \surd_s p. p \gg_s X$.

4.2 Theorem. The following statements are equivalent:

i. $X \in SIP$

ii. $X = \surd t.\, t^{\circ} \gg_{S} X$

iii. $X = X + \overset{\approx}{\delta} \wedge X = \surd t.\, t^{\circ} \gg X$.

Proof: i $\Leftrightarrow$ ii: By definition of $\surd_{S}$.

ii $\Leftrightarrow$ iii: If $X = \surd t.\ t^{\circ} \gg_{S} X$, then

$$t \gg X = t \gg \surd u.\, u^{\circ} \gg_{S} X = t \gg (t^{\circ} \gg_{S} X + \overset{\approx}{\delta}) = t \gg (t^{\circ} \gg_{S} X) + t \gg \overset{\approx}{\delta} =$$

$$= t \gg (\surd u.\, u^{\circ} \gg_{S} X) + t \gg \overset{\approx}{\delta} = t \gg X + t \gg \overset{\approx}{\delta} = t \gg (X + \overset{\approx}{\delta}).$$

By extensionality (IA5) $X = X + \overset{\approx}{\delta}$.

Thus $X = \surd t.\ t^{\circ} \gg_{S} X \Rightarrow X = X + \overset{\approx}{\delta}$, and $X = X + \overset{\approx}{\delta}$ holds in both ii and iii. Using this,

$$t \gg \surd u.\, u^{\circ} \gg_{S} X = t \gg (t^{\circ} \gg X + \overset{\approx}{\delta}) = t \gg (t^{\circ} \gg X) + t \gg \overset{\approx}{\delta} =$$

$$= t \gg t^{\circ} \gg X + t \gg t^{\circ} \gg \overset{\approx}{\delta} = t \gg t^{\circ} \gg (X + \overset{\approx}{\delta}) = t \gg t^{\circ} \gg X =$$

$= t \gg (\surd u.\ u^{\circ} \gg X)$. Again by extensionality (IA5) $\surd t.\ t^{\circ} \gg_{S} X = \surd t.\ t^{\circ} \gg X$.

4.3 Note.

A simple example of a process that is not standardly initialised is $\surd t.\ a(t+1)$. This process does satisfy $X = X + \overset{\approx}{\delta}$. On the other hand, the process $\surd t. \int_{r \approx t+1} a(r)$ is standardly initialised. We will nevertheless see that the second process is not in the algebra generated by the signature $\Sigma(ST)$.

We proceed to show that the set of standardly initialised processes is closed under all operators of the standard time signature. Actually, we will show something stronger, viz. that both the condition $X = X + \overset{\approx}{\delta}$ and the condition $X = \surd t.\ t^{\circ} \gg X$ are preserved by all operators of $\Sigma(ST)$. Thus, the algebra $si\text{-}M_A^*$ is actually the intersection of two other interesting subalgebras of M_A^*. We first list some facts about initial abstraction that are proved as in [2].

4.4 Lemma. $\surd t.F + \surd t.G = \surd t.(F + G)$, $\surd t.F = \surd t.\, t \gg F$, $\surd t.F \mid \surd t.G = \surd t.(F \mid G)$, $\surd t.F \mathbin{|\!\lfloor} \surd t.G = \surd t.(F \mathbin{|\!\lfloor} G)$, $\surd t.F \parallel \surd t.G = \surd t.(F \parallel G)$, $\surd t.F \underline{\oplus} \surd t.G = \surd t.(F \underline{\oplus} G)$.

Proof: As in [2].

4.5 Constants.

We start to verify the two conditions of 4.2.iii by looking at the constants of $\Sigma(ST)$.

4.5.1 Proposition: $\tilde{\tilde{\tilde{\delta}}} + \tilde{\tilde{\tilde{\delta}}} = \tilde{\tilde{\tilde{\delta}}}$.

4.5.2 Proposition: $\tilde{\tilde{\tilde{\delta}}} = \surd t.\ t^\circ \gg \tilde{\tilde{\tilde{\delta}}}$.

Proof: This was already used in the proof of 4.2. We give a more complete proof:

$$\tilde{\tilde{\tilde{\delta}}} = \surd t.\ \int_{\varepsilon\in I} \delta(t^\circ+\varepsilon) = \surd t.\ \int_{\varepsilon\in I} t^\circ \gg \delta(t^\circ+\varepsilon) = \surd t.\ t^\circ \gg \int_{\varepsilon\in I} \delta(t^\circ+\varepsilon) = \surd t.\ t^\circ \gg \tilde{\tilde{\tilde{\delta}}}.$$

4.5.3 Proposition: $\tilde{a}(p) + \tilde{\tilde{\tilde{\delta}}} = \tilde{a}(p)$.

4.5.5 Proposition: $\tilde{a}(p) = \surd t.\ t^\circ \gg \tilde{a}(p)$.

Proof: $\tilde{a}(p) = \int_{\varepsilon\in I} a(p+\varepsilon) + \tilde{\tilde{\tilde{\delta}}} = \int_{\varepsilon\in I} a(p+\varepsilon) + \surd t.\ \int_{\varepsilon\in I} \delta(t^\circ+\varepsilon) =$

$$= \surd t.\ (t \gg \int_{\varepsilon\in I} a(p+\varepsilon) + \int_{\varepsilon\in I} \delta(t^\circ+\varepsilon))\ \text{(IA4)}.$$

Now we consider three cases:

i. if $t \le p$, then $t \gg a(p+\varepsilon) = a(p+\varepsilon) = t^\circ \gg a(p+\varepsilon)$ for all $\varepsilon \in I$;

ii. if $t^\circ > p$, then $t \gg a(p+\varepsilon) = \delta(t)$ for all $\varepsilon \in I$, so $t \gg \int_{\varepsilon\in I} a(p+\varepsilon) + \int_{\varepsilon\in I} \delta(t^\circ+\varepsilon)) =$

$$= \int_{\varepsilon\in I} \delta(t^\circ+\varepsilon))\ \text{by ATA3;}$$

iii. if $t = p + \varepsilon$ for some $\varepsilon \in I$, then $t^\circ = p$, so $\surd t.\ (t \gg \int_{\varepsilon\in I} a(p+\varepsilon) + \int_{\varepsilon\in I} \delta(t^\circ+\varepsilon)) =$

$$= \surd t.\ (t \gg \int_{\varepsilon\in I} a(p+\varepsilon) + t \gg \int_{\varepsilon\in I} \delta(t^\circ+\varepsilon)) = \surd t.\ t \gg (\int_{\varepsilon\in I} a(p+\varepsilon) + \int_{\varepsilon\in I} \delta(t^\circ+\varepsilon))\ \text{(ATB3)}$$

$$= \surd t.\ (\int_{\varepsilon\in I} a(p+\varepsilon) + \int_{\varepsilon\in I} \delta(t^\circ+\varepsilon)) = \surd t.\ (t^\circ \gg \int_{\varepsilon\in I} a(p+\varepsilon) + t^\circ \gg \int_{\varepsilon\in I} \delta(t^\circ+\varepsilon))$$

We see that in all cases, we obtain

$$\tilde{a}(p) = \surd t.\ (t^\circ \gg \int_{\varepsilon\in I} a(p+\varepsilon) + t^\circ \gg \int_{\varepsilon\in I} \delta(t^\circ+\varepsilon)) = \surd t.\ t^\circ \gg (\int_{\varepsilon\in I} a(p+\varepsilon) + \int_{\varepsilon\in I} \delta(t^\circ+\varepsilon)) =$$

$$= \surd t.\ t^\circ \gg \tilde{a}(p).$$

4.5.6 Proposition: $\tilde{\tilde{a}} + \tilde{\tilde{\delta}} = \tilde{\tilde{a}}$.

Proof: $\tilde{\tilde{a}} + \tilde{\tilde{\delta}} = (\surd t.\ \int_{\varepsilon \in I} a(t^\circ+\varepsilon)) + (\surd t.\ \int_{\varepsilon \in I} \delta(t^\circ+\varepsilon)) = \surd t.\ (\int_{\varepsilon \in I} a(t^\circ+\varepsilon) + \int_{\varepsilon \in I} \delta(t^\circ+\varepsilon))$

(lemma 4.4) $= \surd t.\ (\int_{\varepsilon \in I} a(t^\circ+\varepsilon) + \delta(t^\circ+\varepsilon))$ (INT5) $= \surd t.\ \int_{\varepsilon \in I} a(t^\circ+\varepsilon)$ (ATA4) $= \tilde{\tilde{a}}$.

4.5.7 Proposition: $\tilde{\tilde{a}} = \surd t.\ t^\circ \gg \tilde{\tilde{a}}$.

Proof: $\tilde{\tilde{a}} = \surd t.\ \int_{\varepsilon \in I} a(t^\circ+\varepsilon) = \surd t.\ t^\circ \gg \int_{\varepsilon \in I} a(t^\circ+\varepsilon) = \surd t.\ t^\circ \gg \surd u.\ \int_{\varepsilon \in I} a(u^\circ+\varepsilon) = \surd t.\ t^\circ \gg \tilde{\tilde{a}}$

4.6 Alternative and sequential composition.

We continue with the operators $+, \cdot$.

4.6.1 Proposition: $(X + \tilde{\tilde{\delta}}) + (Y + \tilde{\tilde{\delta}}) = (X + Y) + \tilde{\tilde{\delta}}$.

4.6.2 Lemma: $\tilde{\tilde{\delta}} \cdot X = \tilde{\tilde{\delta}}$.

Proof: $\tilde{\tilde{\delta}} \cdot X = (\surd t.\ \int_{\varepsilon \in I} \delta(t^\circ+\varepsilon)) \cdot X = \surd t.\ \int_{\varepsilon \in I} \delta(t^\circ+\varepsilon) \cdot X$ (IA7, INT6) $=$

$= \surd t.\ \int_{\varepsilon \in I} \delta(t^\circ+\varepsilon)$ (ATA2) $= \tilde{\tilde{\delta}}$.

4.6.3 Proposition: $(X + \tilde{\tilde{\delta}}) \cdot Y = X \cdot Y + \tilde{\tilde{\delta}}$.

4.6.4 Proposition: $X = \surd t.\ t^\circ \gg X \wedge Y = \surd t.\ t^\circ \gg Y \Rightarrow (X + Y) = \surd t.\ t^\circ \gg (X + Y)$.

4.6.5 Proposition: $X = \surd t.\ t^\circ \gg X \Rightarrow X \cdot Y = \surd t.\ t^\circ \gg X \cdot Y$.

4.7 Standard initialisation and left strong choice.

We continue with the operators $\gg_s, \underline{\oplus}$.

4.7.1 Proposition: $p \gg_s X = p \gg_s X + \tilde{\tilde{\delta}}$.

4.7.2 Proposition: $p \gg_s X = \surd t.\ t^\circ \gg p \gg_s X$.

Proof: $p \gg_s X = p \gg X + \overset{\approx}{\delta} = p \gg X + \surd t.\ \int_{\varepsilon\in I} \delta(t^\circ+\varepsilon) = \surd t.\ (t \gg p \gg X + \int_{\varepsilon\in I} \delta(t^\circ+\varepsilon))$

$$(IA4) = \surd t.\ (t \gg t^\circ \gg p \gg X + t \gg t^\circ \gg \int_{\varepsilon\in I} \delta(t^\circ+\varepsilon))\ \ (SI7) =$$

$$= \surd t.\ t \gg (t^\circ \gg p \gg X + t^\circ \gg \overset{\approx}{\delta})\ \ (ATB3) = \surd t.\ (t^\circ \gg p \gg X + t^\circ \gg \overset{\approx}{\delta})\ \ (\text{lemma } 4.4) =$$

$$= \surd t.\ t^\circ \gg (p \gg X + \overset{\approx}{\delta})\ \ (ATB3) = \surd t.\ t^\circ \gg (p \gg_s X).$$

4.7.3 Lemma: $\overset{\approx}{\delta} \oplus \overset{\approx}{\delta} = \overset{\approx}{\delta}$.

Proof: $\overset{\approx}{\delta} \oplus \overset{\approx}{\delta} = \surd t.\ \int_{\varepsilon\in I} \delta(t^\circ+\varepsilon) \oplus \surd t.\ \int_{\eta\in I} \delta(t^\circ+\eta) = \surd t.\ \int_{\varepsilon\in I} \delta(t^\circ+\varepsilon) \oplus \int_{\eta\in I} \delta(t^\circ+\eta)$

$$(\text{lemma } 4.4) = \surd t.\ \int_{\varepsilon,\eta\in I} \delta(t^\circ+\varepsilon)\ \underline{\oplus}\ \delta(t^\circ+\eta)\ \ (INT13,14) =$$

$$= \surd t.\ \int_{\varepsilon<\eta} \delta(t^\circ+\varepsilon) + \int_{\varepsilon\geq\eta} \delta(t^\circ+\eta)\ \ (LSC1,2) = \surd t.\ \int_{\varepsilon\in I} \delta(t^\circ+\varepsilon) = \overset{\approx}{\delta}.$$

4.7.4 Proposition: $X = X + \overset{\approx}{\delta} \wedge Y = Y + \overset{\approx}{\delta} \Rightarrow X \underline{\oplus}\, Y = X \underline{\oplus}\, Y + \overset{\approx}{\delta}$.

4.7.5 Proposition: $X = \surd t.\ t^\circ \gg X \wedge Y = \surd t.\ t^\circ \gg Y \Rightarrow X \underline{\oplus}\, Y = \surd t.\ t^\circ \gg (X \underline{\oplus}\, Y)$.

4.8 Parallel composition and encapsulation.

We continue with the operators $\parallel$, $\lfloor\!\lfloor$, $|$, ∂_H.

4.8.1 Lemma: $X = X + \overset{\approx}{\delta} \Rightarrow \overset{\approx}{\delta} \lfloor\!\lfloor X = \overset{\approx}{\delta}$.

Proof: Assume X satisfies $X = X + \overset{\approx}{\delta}$. We derive

$$\overset{\approx}{\delta} \lfloor\!\lfloor X = (\surd t.\ \int_{\varepsilon\in I} \delta(t^\circ+\varepsilon)) \lfloor\!\lfloor X = \surd t.\ (\int_{\varepsilon\in I} \delta(t^\circ+\varepsilon) \lfloor\!\lfloor t \gg X)\ \ (IA8, INT9) =$$

$$= \surd t.\ \int_{\varepsilon\in I} (\delta(t^\circ+\varepsilon)\ \underline{\oplus}\ t \gg X)\cdot(t \gg X) \qquad (\text{since } t \gg X = 0 \gg t \gg X \text{ by SI2}) =$$

$$= \surd t.\ \int_{\varepsilon\in I} (\delta(t^\circ+\varepsilon)\ \underline{\oplus}\ t \gg (X + \overset{\approx}{\delta}))\cdot(t \gg X) =$$

$$= \surd t.\ \int_{\varepsilon\in I} ((\delta(t^\circ+\varepsilon)\ \underline{\oplus}\ t \gg X) + (\delta(t^\circ+\varepsilon) \underline{\oplus}\ t \gg \overset{\approx}{\delta}))\cdot(t \gg X) =$$

$= \surd t. \int_{\varepsilon \in I} ((\delta(t^\circ+\varepsilon) \oplus t \gg X) + \delta(t^\circ+\varepsilon))\cdot(t \gg X) =$

$= \surd t. \int_{\varepsilon \in I} \delta(t^\circ+\varepsilon)\cdot(t \gg X)$ (SI15) $= \surd t. \int_{\varepsilon \in I} \delta(t^\circ+\varepsilon)$ (ATA2) $= \tilde{\tilde{\tilde{\delta}}}$.

4.8.2 Proposition: $(X + \tilde{\tilde{\tilde{\delta}}}) \mathbin{\|\!\!\_} (Y + \tilde{\tilde{\tilde{\delta}}}) = X \mathbin{\|\!\!\_} (Y + \tilde{\tilde{\tilde{\delta}}}) + \tilde{\tilde{\tilde{\delta}}}$.

4.8.4 Lemma: $\tilde{\tilde{\tilde{\delta}}} \mid \tilde{\tilde{\tilde{\delta}}} = \tilde{\tilde{\tilde{\delta}}}$.

Proof: $\tilde{\tilde{\tilde{\delta}}} \mid \tilde{\tilde{\tilde{\delta}}} = \surd t. \int_{\varepsilon \in I} \delta(t^\circ+\varepsilon) \mid \surd t. \int_{\eta \in I} \delta(t^\circ+\eta) = \surd t. \int_{\varepsilon \in I} \delta(t^\circ+\varepsilon) \mid \int_{\eta \in I} \delta(t^\circ+\eta)$

(lemma 4.4) $= \surd t. \int_{\varepsilon,\eta \in I} \delta(t^\circ+\varepsilon) \mid \delta(t^\circ+\eta)$ (INT10, INT11) $=$

$= \surd t. \int_{\varepsilon \in I} \delta(t^\circ+\varepsilon)$ (ATC1, ATC2) $= \tilde{\tilde{\tilde{\delta}}}$.

4.8.5 Proposition: $X = X + \tilde{\tilde{\tilde{\delta}}} \wedge Y = Y + \tilde{\tilde{\tilde{\delta}}} \Rightarrow X \mid Y = X \mid Y + \tilde{\tilde{\tilde{\delta}}}$.

4.8.6 Proposition: $X = X + \tilde{\tilde{\tilde{\delta}}} \wedge Y = Y + \tilde{\tilde{\tilde{\delta}}} \Rightarrow X \| Y = X \| Y + \tilde{\tilde{\tilde{\delta}}}$.

4.8.7 Proposition: $X = \surd t.\ t^\circ \gg X \wedge Y = \surd t.\ t^\circ \gg Y \Rightarrow X \mathbin{\|\!\!\_} Y = \surd t.\ t^\circ \gg (X \mathbin{\|\!\!\_} Y)$.

4.8.8 Proposition: $X = \surd t.\ t^\circ \gg X \wedge Y = \surd t.\ t^\circ \gg Y \Rightarrow X \mid Y = \surd t.\ t^\circ \gg (X \mid Y)$.

4.8.9 Proposition: $X = \surd t.\ t^\circ \gg X \wedge Y = \surd t.\ t^\circ \gg Y \Rightarrow X \| Y = \surd t.\ t^\circ \gg (X \| Y)$.

4.8.10 Lemma: $\partial_H(\tilde{\tilde{\tilde{\delta}}}) = \tilde{\tilde{\tilde{\delta}}}$.

Proof: $\partial_H(\tilde{\tilde{\tilde{\delta}}}) = \partial_H(\surd t. \int_{\varepsilon \in I} \delta(t^\circ+\varepsilon)) = \surd t. \int_{\varepsilon \in I} \partial_H(\delta(t^\circ+\varepsilon))$ (IA12, INT12) $=$

$= \surd t. \int_{\varepsilon \in I} \delta(t^\circ+\varepsilon)$ (D1) $= \tilde{\tilde{\tilde{\delta}}}$.

4.8.11 Proposition: $\partial_H(X + \tilde{\tilde{\tilde{\delta}}}) = \partial_H(X) + \tilde{\tilde{\tilde{\delta}}}$.

4.8.12 Proposition: $X = \surd t.\ t^\circ \gg X \Rightarrow \partial_H(X) = \surd t.\ t^\circ \gg \partial_H(X)$.

4.9 Positive time shift, Shifted initialisation.

Next, the operators $\sigma^{r}_{\gtrsim}$, $r \gg_{st}$.

4.9.1 Proposition: $\sigma^{r}_{\gtrsim}(X) = \sigma^{r}_{\gtrsim}(X) + \tilde{\tilde{\delta}}$.

Proof: Note that since $r \in \mathbb{R}_{>0}$, $(t + r)° = t° + r > t° + \varepsilon$ for all $\varepsilon \in I$. Using this,
$\sigma^{r}_{\gtrsim}(X) = \surd t.\ (t+r)° \gg X = \surd t.\ ((t+r)° \gg X + \delta((t+r)°))$ (SI5) $=$

$$= \surd t.\ ((t+r)° \gg X + \delta((t+r)°) + \int_{\varepsilon \in I} \delta(t°+\varepsilon))\ \text{(ATA2)} = \surd t.\ ((t+r)° \gg X + \int_{\varepsilon \in I} \delta(t°+\varepsilon))$$

$$= \surd t.\ ((t+r)° \gg X) + \surd t.\ \int_{\varepsilon \in I} \delta(t°+\varepsilon)\ \text{(lemma 4.4)} = \sigma^{r}_{\gtrsim}(X) + \tilde{\tilde{\delta}}.$$

4.9.2 Proposition: $\sigma^{r}_{\gtrsim}(X) = \surd t.\ t° \gg \sigma^{r}_{\gtrsim}(X)$.

Proof: $\sigma^{r}_{\gtrsim}(X) = \surd t.\ (t+r)° \gg X = \surd t.\ t° \gg (t+r)° \gg X$ (SI2) $= \surd t.\ t° \gg \surd u.\ (u+r)° \gg X =$
$= \surd t.\ t° \gg \sigma^{r}_{\gtrsim}(X)$.

4.9.3 Proposition: $r \gg_{st} X = r \gg_{st} X + \tilde{\tilde{\delta}}$.

4.9.4 Proposition: $r \gg_{st} X = \surd t.\ t° \gg (r \gg_{st} X)$.

Proof: $r \gg_{st} X = \surd t.\ (t+r)° \gg t° \gg_{s} X = \surd t.\ t° \gg (t+r)° \gg t° \gg_{s} X$ (SI7) $=$
$= \surd t.\ t° \gg \surd u.\ (u+r)° \gg u° \gg_{s} X = \surd t.\ t° \gg (r \gg_{st} X)$.

4.10 Standard initial abstraction.

We consider $\surd_{s}$.

4.10.1 Proposition: $F = F + \tilde{\tilde{\delta}} \Rightarrow \surd_{s}p.\ F = \surd_{s}p.\ F + \tilde{\tilde{\delta}}$.

Proof: $\surd_{s}p.\ F = \surd_{s}p.\ (F + \tilde{\tilde{\delta}}) = \surd t.\ (F[t°/p] + t° \gg \tilde{\tilde{\delta}}) = \surd t.\ F[t°/p] + \surd t.\ t° \gg \tilde{\tilde{\delta}} =$
$= \surd_{s}p.\ F + \tilde{\tilde{\delta}}$.

4.10.2 Proposition: $F = \surd t.\ t° \gg F \Rightarrow \surd_{s}p.\ F = \surd t.\ t° \gg \surd_{s}p.\ F$.

Proof: $t \gg \surd_{s}p.\ F = t \gg \surd u.\ F[u°/p] = t \gg F[t°/p] = t \gg t° \gg F[t°/p] =$
$= t \gg (\surd u.\ u° \gg F[u°/p]) = t \gg (\surd u.\ u° \gg \surd_{s}p.\ F)$. Now use IA5.

This finishes the proof that the algebra of standardly initialised processes is closed under all operators of the standard time signature. We proceed to prove additional identities that we will use in the axiomatisations in sections 6 and 7.

4.11 BPA with urgent actions and absolute time.

First, we prove some identities that allow to derive a standard time variant BPAsρδ of BPAρδ as defined in 2.2. The familiar equation $X + \tilde{\tilde{\delta}} = X$ does not hold for all timed processes, as the example $\delta(0) + \tilde{\tilde{\delta}} = \tilde{\tilde{\delta}} \neq \delta(0)$ illustrates. This is one reason to limit ourselves to standardly initialised processes. In each case, we will indicate if we need a condition of standardly initialised processes.

4.11.1 Proposition: $\tilde{\delta}(0) = \tilde{\tilde{\delta}}$.

Proof: $\tilde{\delta}(0) = \tilde{\tilde{\delta}} + \int_{\varepsilon\in I} \delta(\varepsilon) = (\surd t. \int_{\eta\in I} \delta(t^{\circ}+\eta)) + \int_{\varepsilon\in I} \delta(\varepsilon) =$

$$= \surd t. \,(\int_{\eta\in I} \delta(t^{\circ}+\eta) + \int_{\varepsilon\in I} t \gg \delta(\varepsilon)).$$

If $t \gg \delta(\varepsilon) = \delta(t)$, the summand can be dropped by ATA2 since $t < t^{\circ}+\eta$ for some $\eta \in I$; if $t \gg \delta(\varepsilon) = \delta(\varepsilon)$, the summand can also be dropped by ATA2 since $\varepsilon < t^{\circ}+2\varepsilon$. In both cases $\tilde{\delta}(0) = \surd t. \,(\int_{\eta\in I} \delta(t^{\circ}+\eta) + \int_{\varepsilon\in I} t \gg \delta(\varepsilon)) = \surd t. \int_{\eta\in I} \delta(t^{\circ}+\eta) = \tilde{\tilde{\delta}}$.

4.11.2 Proposition: $\tilde{\delta}(p) \cdot X = \tilde{\delta}(p)$.

4.11.3 Proposition: $p < q \Rightarrow \tilde{\delta}(p) + \tilde{\delta}(q) = \tilde{\delta}(q)$.

4.11.4 Proposition: $\tilde{a}(p) + \tilde{\delta}(p) = \tilde{a}(p)$.

4.11.5 Proposition: $X = \surd t.\, t^{\circ} \gg_s X \Rightarrow \tilde{a}(p) \cdot X = \tilde{a}(p) \cdot (p \gg_s X)$.

Comment: This is an example of an identity that only holds for standardly initialised processes.

Proof: $\tilde{a}(p) \cdot X = (\tilde{\tilde{\delta}} + \int_{\varepsilon\in I} a(p+\varepsilon)) \cdot X = \tilde{\tilde{\delta}} \cdot X + \int_{\varepsilon\in I} a(p+\varepsilon) \cdot X$ (INT6) $=$

$$= \tilde{\tilde{\delta}} + \int_{\varepsilon\in I} a(p+\varepsilon) \cdot (p+\varepsilon \gg X) \text{ (ATA5) } = \tilde{\tilde{\delta}} \cdot (p \gg_s X) + \int_{\varepsilon\in I} a(p+\varepsilon) \cdot (p+\varepsilon \gg p \gg_s X)$$

$$(\text{assumption}) = (\tilde{\tilde{\delta}} + \int_{\varepsilon\in I} a(p+\varepsilon))\cdot(p \gg_s X) \;\; (\text{ATA5, INT6}) = \tilde{a}(p)\cdot(p \gg_s X).$$

4.11.6 Proposition: $p \le q \;\Rightarrow\; p \gg_s \tilde{a}(q) = \tilde{a}(q)$.

Proof: Suppose $p \le q$, so $p < q + \varepsilon$ for each $\varepsilon \in I$. Then:

$$p \gg_s \tilde{a}(q) = \tilde{\delta} + p \gg (\tilde{\tilde{\delta}} + \int_{\varepsilon\in I} a(q+\varepsilon)) = \tilde{\delta} + p \gg \tilde{\tilde{\delta}} + \int_{\varepsilon\in I} p \gg a(q+\varepsilon) = \tilde{\tilde{\delta}} +$$

$$+ \int_{\varepsilon\in I} \delta(p+\varepsilon) + \int_{\varepsilon\in I} a(q+\varepsilon) = \tilde{\tilde{\delta}} + \int_{\varepsilon\in I} (\delta(p+\varepsilon) + a(q+\varepsilon)) = \tilde{\tilde{\delta}} + \int_{\varepsilon\in I} a(q+\varepsilon)) = \tilde{a}(q).$$

4.11.7 Proposition: $p > q \;\Rightarrow\; p \gg_s \tilde{a}(q) = \tilde{\delta}(p)$.

Proof: Suppose $p > q$, so $p > q + \varepsilon$ for each $\varepsilon \in I$ as $p,q \in \mathbb{R}$. Then:

$$p \gg_s \tilde{a}(q) = \tilde{\delta} + p \gg (\tilde{\tilde{\delta}} + \int_{\varepsilon\in I} a(q+\varepsilon)) = \tilde{\delta} + p \gg \tilde{\tilde{\delta}} + \int_{\varepsilon\in I} p \gg a(q+\varepsilon) =$$

$$= \tilde{\tilde{\delta}} + \int_{\varepsilon\in I} \delta(p+\varepsilon) + \int_{\varepsilon\in I} \delta(p) = \tilde{\tilde{\delta}} + \int_{\varepsilon\in I} \delta(p+\varepsilon) + \delta(p) = \tilde{\tilde{\delta}} + \int_{\varepsilon\in I} \delta(p+\varepsilon)) = \tilde{\delta}(p).$$

4.11.8 Proposition: $p \gg_s (X + Y) = (p \gg_s X) + (p \gg_s Y)$.

4.11.9 Proposition: $p \gg_s (X \cdot Y) = (p \gg_s X) \cdot Y$.

4.12 PA with urgent actions and absolute time.

We continue with identities that we will use to axiomatise PAspδ, the standard time variant of process algebra with free merge (merge without communication).

4.12.1 Proposition: $p \le q \Rightarrow \tilde{a}(p) \;\underline{\oplus}\; \tilde{b}(q) = \tilde{a}(p)$.

Proof: Straightforward, use INT13, INT14, lemma 4.6.3 and the fact that for each $\varepsilon \in I$, there is a $\eta \in I$ such that $p + \varepsilon < q + \eta$.

4.12.2 Proposition: $p > q \Rightarrow \tilde{a}(p) \;\underline{\oplus}\; \tilde{b}(q) = \tilde{\delta}(q)$.

Proof: In this case, use that for each $\varepsilon \in I$ $p + \varepsilon < q$.

4.12.3 Proposition: $X = 0 \gg_s Z \;\Rightarrow\; \tilde{a}(p) \;⫗\; X = (\tilde{a}(p) \;\underline{\oplus}\; X)\cdot X$.

Proof: We consider 3 cases:

<u>Case 1:</u> Let $\varepsilon \in I$, $r \ge p + \varepsilon$. Then

$r \gg (a(p+\varepsilon) \mathbin{|\!\!\lfloor} X + \tilde{\tilde{\delta}}) = (r \gg a(p+\varepsilon) \mathbin{|\!\!\lfloor} r \gg X) + r \gg \tilde{\tilde{\delta}} = \delta(r) \mathbin{|\!\!\lfloor} r \gg X + r \gg \tilde{\tilde{\delta}} =$

$= (\delta(r) \underline{\oplus}\, r \gg X) \cdot (r \gg X) + r \gg \tilde{\tilde{\delta}}$ (since $r \gg X = 0 \gg r \gg X$ by SI2) $=$

$= \delta(r) \cdot (r \gg X) + r \gg \tilde{\tilde{\delta}}$ (SI16) $= \delta(r) + r \gg \tilde{\tilde{\delta}} = \delta(r) \cdot X + r \gg \tilde{\tilde{\delta}} =$

$= (r \gg a(p+\varepsilon)) \cdot X + r \gg \tilde{\tilde{\delta}} = r \gg (a(p+\varepsilon) + a(p+\varepsilon) \underline{\oplus}\, X) \cdot X + r \gg \tilde{\tilde{\delta}}$ (SI15) $=$

$= r \gg a(p+\varepsilon) \cdot X + r \gg (a(p+\varepsilon) \underline{\oplus}\, X) \cdot X + r \gg \tilde{\tilde{\delta}} =$

$= \delta(r) + r \gg (a(p+\varepsilon) \underline{\oplus}\, X) \cdot X + r \gg \tilde{\tilde{\delta}} = r \gg (a(p+\varepsilon) \underline{\oplus}\, X) \cdot X + r \gg \tilde{\tilde{\delta}} =$

$= r \gg (a(p+\varepsilon) \underline{\oplus}\, X) \cdot X + \tilde{\tilde{\delta}})$.

<u>Case 2:</u> Let $\varepsilon \in I$, $p \le r < p + \varepsilon$. Then

$r \gg (a(p+\varepsilon) \mathbin{|\!\!\lfloor} X + \tilde{\tilde{\delta}}) = (r \gg a(p+\varepsilon) \mathbin{|\!\!\lfloor} r \gg X) + r \gg \tilde{\tilde{\delta}} = a(p+\varepsilon) \mathbin{|\!\!\lfloor} r \gg X + r \gg \tilde{\tilde{\delta}} =$

$= (a(p+\varepsilon) \underline{\oplus}\, r \gg X) \cdot (r \gg X) + r \gg \tilde{\tilde{\delta}}$ (as in case 1) $= a(p+\varepsilon) \cdot (r \gg X) + r \gg \tilde{\tilde{\delta}}$

(since $r \gg X = r \gg (X + \tilde{\tilde{\delta}}) = r \gg X + r \gg \tilde{\tilde{\delta}} = r \gg X + r \gg \tilde{\tilde{\delta}} + \delta(p+2\varepsilon) =$

$= r \gg X + \delta(p+2\varepsilon)) = a(p+\varepsilon) \cdot ((p+\varepsilon) \gg r \gg X) + r \gg \tilde{\tilde{\delta}} =$

$= a(p+\varepsilon) \cdot ((p+\varepsilon) \gg X) + r \gg \tilde{\tilde{\delta}}$ (by SI7, since $X = 0 \gg_s Z$) $= a(p+\varepsilon) \cdot X + r \gg \tilde{\tilde{\delta}} =$

$= (a(p+\varepsilon) \underline{\oplus}\, r \gg X) \cdot X + r \gg \tilde{\tilde{\delta}} = r \gg (a(p+\varepsilon) \underline{\oplus}\, X) \cdot X + r \gg \tilde{\tilde{\delta}}$ (SI14) $=$

$= r \gg (a(p+\varepsilon) \underline{\oplus}\, X) \cdot X + \tilde{\tilde{\delta}})$.

<u>Case 3:</u> Let $\varepsilon \in I$, $r < p$. Then

$r \gg (a(p+\varepsilon) \mathbin{|\!\!\lfloor} X + \tilde{\tilde{\delta}}) = (r \gg a(p+\varepsilon) \mathbin{|\!\!\lfloor} r \gg X) + r \gg \tilde{\tilde{\delta}} = a(p+\varepsilon) \mathbin{|\!\!\lfloor} r \gg X + r \gg \tilde{\tilde{\delta}} =$

$= (a(p+\varepsilon) \underline{\oplus}\, r \gg X) \cdot (r \gg X) + r \gg \tilde{\tilde{\delta}}$. Now if $a(p+\varepsilon) \underline{\oplus}\, r \gg X = a(p+\varepsilon)$, then we proceed as in case 2. Otherwise, by SI15, $a(p+\varepsilon) \underline{\oplus}\, r \gg X$ must be an alternative composition of δ's with timestamps less than $p + \varepsilon$. In this case, we proceed as in case 1. We see that we have for all $\varepsilon \in I$, $r \in \mathcal{T}$

$$r \gg (a(p+\varepsilon) \mathbin{|\!\!\lfloor} X + \tilde{\tilde{\delta}}) = r \gg (a(p+\varepsilon) \underline{\oplus}\, X) \cdot X + \tilde{\tilde{\delta}}).$$

Now let $r \in \mathcal{T}$. Then:

$$r \gg (\tilde{a}(p) \mathbin{|\!\!\lfloor} X) = r \gg (\tilde{\tilde{\delta}} + \int_{\varepsilon \in I} a(p+\varepsilon)) \mathbin{|\!\!\lfloor} X) = \int_{\varepsilon \in I} r \gg (\tilde{\tilde{\delta}} \mathbin{|\!\!\lfloor} X + a(p+\varepsilon) \mathbin{|\!\!\lfloor} X)$$

$$\text{(INT2, INT9, IA16)} = \int_{\varepsilon \in I} r \gg (\tilde{\tilde{\delta}} + a(p+\varepsilon) \mathbin{|\!\!\lfloor} X) \text{ (lemma 4.8.1)} =$$

$$= \int_{\varepsilon \in I} r \gg (a(p+\varepsilon) \underline{\oplus}\, X) \cdot X + \tilde{\tilde{\delta}}) = r \gg \int_{\varepsilon \in I} (a(p+\varepsilon) \underline{\oplus}\, X) \cdot X + \tilde{\tilde{\delta}} \cdot X)$$

$$\text{(lemma 4.6.2)} = r \gg \int_{\varepsilon \in I} (a(p+\varepsilon) \underline{\oplus}\, X) \cdot X + (\tilde{\tilde{\delta}} \underline{\oplus}\, X) \cdot X)$$

$$\text{(use lemma 4.7.3 and SI15)} = r \gg ((\int_{\varepsilon \in I} a(p+\varepsilon) + \tilde{\tilde{\delta}}) \underline{\oplus}\, X) \cdot X =$$

$= r \gg (\tilde{a}(p) \underline{\oplus}\, X) \cdot X$. The proposition now follows by extensionality (IA5).

4.12.4 Proposition: $Y = 0 \gg_s Z \Rightarrow \tilde{a}(p) \cdot X \mathbin{|\!\!\lfloor} Y = (\tilde{a}(p) \underline{\oplus}\, Y) \cdot (X \parallel Y)$.

4.13 Communication and encapsulation.

Next, we add axioms for the communication merge and encapsulation operator, in order to axiomatise ACPsρ, the standard time variant of ACPρ.

4.13.1 Proposition: $p \neq q \Rightarrow \tilde{a}(p) \mid \tilde{b}(q) = \tilde{\delta}(\min(p,q))$.

Proof: Assume without loss of generality $p < q$. Note that since $p,q \in \mathbb{R}$, for all $\varepsilon, \eta \in I$ we have $p+\varepsilon < q+\eta$. Then

$$\tilde{a}(p) \mid \tilde{b}(q) = (\tilde{\tilde{\delta}} + \int_{\varepsilon\in I} a(p+\varepsilon)) \mid (\tilde{\tilde{\delta}} + \int_{\eta\in I} b(q+\eta)) = \tilde{\tilde{\delta}} + \int_{\varepsilon\in I} \int_{\eta\in I} a(p+\varepsilon) \mid b(q+\eta)$$

$$(\text{INT10, 11}) = \tilde{\tilde{\delta}} + \int_{\varepsilon\in I} \int_{\eta\in I} \delta(p+\varepsilon) = \tilde{\tilde{\delta}} + \int_{\varepsilon\in I} \delta(p+\varepsilon)) = \tilde{\delta}(p).$$

4.13.2 Proposition: $\tilde{a}(p) \mid \tilde{b}(p) = (a \mathbin{\top} b)(p)$.

Proof: $$\tilde{a}(p) \mid \tilde{b}(p) = (\tilde{\tilde{\delta}} + \int_{\varepsilon\in I} a(p+\varepsilon)) \mid (\tilde{\tilde{\delta}} + \int_{\eta\in I} b(p+\eta)) = \tilde{\tilde{\delta}} + \int_{\varepsilon\in I} \int_{\eta\in I} a(p+\varepsilon) \mid b(p+\eta)$$

$$= \tilde{\tilde{\delta}} + \int_{\varepsilon\in I} \int_{\eta\in I} a \mid b(p+\varepsilon) = \tilde{\tilde{\delta}} + \int_{\varepsilon\in I} a \mid b(p+\varepsilon)) = (a \mathbin{\top} b)(p).$$

4.13.3 Proposition: $\tilde{a}(p) \cdot X \mid \tilde{b}(q) = (\tilde{a}(p) \mid \tilde{b}(q)) \cdot X$.

4.13.4 Proposition: $\tilde{a}(p) \mid \tilde{b}(q) \cdot X = (\tilde{a}(p) \mid \tilde{b}(q)) \cdot X$.

4.13.5 Proposition: $\tilde{a}(p) \cdot X \mid \tilde{b}(q) \cdot Y = (\tilde{a}(p) \mid \tilde{b}(q)) \cdot (X \parallel Y)$.

4.13.6 Proposition: $\partial_H(\tilde{a}(p)) = (\partial_{\tilde{H}}(a))(p)$.

4.14 Standard initial abstraction.

We present identities used to axiomatise $ACP_{s\rho\surd}$, the standard time variant of ACPρ√.

4.14.1 Proposition: $\tilde{a}[p] = \surd_s q.\ \tilde{a}(p+q)$.

Proof: $\tilde{a}[p] = \surd t. \int_{\varepsilon\in I} a(p+t°+\varepsilon) = \surd_s q. \int_{\varepsilon\in I} a(p+q+\varepsilon) = \surd_s q. \int_{\varepsilon\in I} a(p+q+\varepsilon) + \delta(q+\varepsilon)$

=

$$= \surd_s q. \int_{\varepsilon\in I} a(p+q+\varepsilon) + \int_{\varepsilon\in I} \delta(q+\varepsilon) = \surd_s q. \int_{\varepsilon\in I} a(p+q+\varepsilon) + \tilde{\tilde{\delta}} = \surd_s q. \tilde{a}(p+q).$$

4.14.2 Proposition: $\surd_s p.F = \surd_s q.F[q/p]$ if q is not free in F.
Proof: Suppose q is not free in F. Then
$\surd_s p.F = \surd t. F[t°/p] = \surd_s q. F[t°/p][q/t°] = \surd_s q.F[q/p]$.

4.14.3 Proposition: $p \gg_s \surd_s q.F = p \gg_s F[p/q]$.

Proof: $p \gg_s \surd_s q.F = p \gg \surd t. F[t°/q] + \tilde{\tilde{\delta}} = p \gg F[t°/q][p/t] + \tilde{\tilde{\delta}} = p \gg F[p/q] + \tilde{\tilde{\delta}} =$
$= p \gg_s F[p/q]$.

4.14.4 Proposition: $\surd_s q. \surd_s p. F = \surd_s q. F[q/p]$.
Proof: $\surd_s q. \surd_s p.F = \surd t. \surd u. F[u°/p][t°/q] = \surd t. F[u°/p][t°/q][t/u] = \surd t. F[t°/p][t°/q] =$
$= \surd t. F[q/p][t°/q] = \surd_s q.F[q/p]$.

4.14.5 Proposition: $X = \surd_s p. X$.
Proof: $X = \surd t. X = \surd_s p. X$.

4.14.6 Proposition: $X = \surd t.\ t° \gg_s X \wedge Y = \surd t.\ t° \gg_s Y \wedge \forall p \in \mathbb{R}_{\geq 0}\ p \gg_s X = p \gg_s Y$
$\Rightarrow X = Y$.
Proof: Suppose the conditions hold. Let $t \in \mathbb{R}_{\geq 0}$+I. Then
$t \gg X = t \gg (\surd u. u° \gg_s X) = t \gg (t° \gg_s X) = t \gg (t° \gg_s Y) = t \gg (\surd u. u° \gg_s Y) = t \gg Y$.
Now apply extensionality, IA5.

4.14.7 Proposition: $X = \surd t.\ t° \gg_s X \Rightarrow (\surd_s p.F) + X = \surd_s p.(F + p \gg_s X)$.
Proof: $(\surd_s p.F) + X = \surd t. F[t°/p] + \surd t. t° \gg_s X = \surd t. (F[t°/p] + t° \gg_s X) =$
$= \surd_s p.(F + p \gg_s X)$.

4.14.8 Proposition: $(\surd_s p.F) \cdot X = \surd_s p.(F \cdot X)$, $\partial_H(\surd_s p.F) = \surd_s p.\partial_H(F)$

4.14.9 Proposition: $X = \surd t.\ t° \gg_s X \Rightarrow (\surd_s p.F) \mathbin{⫡} X = \surd_s p.(F \mathbin{⫡} p \gg_s X)$,
$X \mathbin{⫡} (\surd_s p.F) = \surd_s p.(p \gg_s X \mathbin{⫡} F)$, $(\surd_s p.F) \mid X = \surd_s p.(F \mid p \gg_s X)$,
$X \mid (\surd_s p.F) = \surd_s p.(p \gg_s X \mid F)$, $(\surd_s p.F) \underline{\oplus} X = \surd_s p.(F \underline{\oplus}\ p \gg_s X)$,
$X \underline{\oplus} (\surd_s p.F) = \surd_s p.(p \gg_s X \underline{\oplus} F)$.

4.15 BPA with urgent actions and relative time.

Now we present identities that will be used in order to directly axiomatize the relative time algebras in section 7. First, BPA_{st}.

4.15.1 Proposition: $\sigma^{r}_{\approx} \circ \sigma^{p}_{\approx}(X) = \sigma^{r+p}_{\approx}(X)$.

Proof: $\sigma^{r}_{\approx} \circ \sigma^{p}_{\approx}(X) = \surd t.\,(t+r)^{\circ} \gg (\surd u.\,(u+p)^{\circ} \gg X) = \surd t.\,(t+r)^{\circ} \gg ((t+r)^{\circ}+p)^{\circ} \gg X) =$

$= \surd t.\,(t+r)^{\circ} \gg (t+r+p)^{\circ} \gg X = \surd t.\,(t+r+p)^{\circ} \gg X$ (SI2) $= \sigma^{r+p}_{\approx}(X)$.

4.15.2 Proposition: $\sigma^{r}_{\approx}(X) + \sigma^{r}_{\approx}(Y) = \sigma^{r}_{\approx}(X+Y)$, $\sigma^{r}_{\approx}(X) \cdot Y = \sigma^{r}_{\approx}(X \cdot Y)$.

4.16 Left strong choice and shifted initialisation.

4.16.1 Proposition: $\overset{\approx}{a} \,\underline{\oplus}\, \overset{\approx}{b} = \overset{\approx}{a}$.

4.16.2 Proposition: $\overset{\approx}{a} \,\underline{\oplus}\, \sigma^{r}_{\approx}(X) = \overset{\approx}{a}$.

Proof: $\overset{\approx}{a} \,\underline{\oplus}\, \sigma^{r}_{\approx}(X) = \surd t.\, \int_{\varepsilon \in I} a(t^{\circ}+\varepsilon) \,\underline{\oplus}\, \surd t.\,(t+r)^{\circ} \gg X =$

$= \surd t.\, \int_{\varepsilon \in I} a(t^{\circ}+\varepsilon) \,\underline{\oplus}\, (t+r)^{\circ} \gg X$ (lemma 4.4) $=$

$= \surd t.\, \int_{\varepsilon \in I} a(t^{\circ}+\varepsilon) \,\underline{\oplus}\, (t^{\circ}+\varepsilon) \gg (t+r)^{\circ} \gg X$ (SI2) $= \surd t.\, \int_{\varepsilon \in I} a(t^{\circ}+\varepsilon)$ (SI16) $= \overset{\approx}{a}$.

4.16.3 Proposition: $\sigma^{r}_{\approx}(X) \,\underline{\oplus}\, \overset{\approx}{a} = \overset{\approx}{\delta}$.

Proof: $\sigma^{r}_{\approx}(X) \,\underline{\oplus}\, \overset{\approx}{a} = \surd t.\,(t+r)^{\circ} \gg X \,\underline{\oplus}\, \surd t.\, \int_{\varepsilon \in I} a(t^{\circ}+\varepsilon) =$

$= \surd t.\,(t+r)^{\circ} \gg X \,\underline{\oplus}\, \int_{\varepsilon \in I} a(t^{\circ}+\varepsilon)$ (lemma 4.4) $= \surd t.\, \int_{\varepsilon \in I} (t+r)^{\circ} \gg X \,\underline{\oplus}\, a(t^{\circ}+\varepsilon)$

(INT14)$= \surd t.\, \int_{\varepsilon \in I} (t^{\circ}+\varepsilon) \gg (t+r)^{\circ} \gg X \,\underline{\oplus}\, a(t^{\circ}+\varepsilon)$ (SI2)$= \surd t.\, \int_{\varepsilon \in I} \delta(t^{\circ}+\varepsilon)$ (SI17)$= \overset{\approx}{\delta}$

4.16.4 Proposition: $\sigma^{r}_{\approx}(X) \,\underline{\oplus}\, \sigma^{r}_{\approx}(Y) = \sigma^{r}_{\approx}(X \,\underline{\oplus}\, Y)$.

Proof: $\sigma^{r}_{\approx}(X) \,\underline{\oplus}\, \sigma^{r}_{\approx}(Y) = \surd t.\,(t+r)^{\circ} \gg X \,\underline{\oplus}\, \surd t.\,(t+r)^{\circ} \gg Y =$

$= \surd t.\,(t+r)^{\circ} \gg (X \,\underline{\oplus}\, Y)$ (lemma 4.4, SI14) $= \sigma^{r}_{\approx}(X \,\underline{\oplus}\, Y)$.

4.16.5 Proposition: $r \gg_{st} \tilde{\tilde{a}} = \sigma^{r}_{\approx}(\tilde{\tilde{\delta}})$.

Proof: $r \gg_{st} \tilde{\tilde{a}} = \surd t.\ (t+r)^{\circ} \gg t^{\circ} \gg_{s} \tilde{\tilde{a}} = \surd t.\ (t+r)^{\circ} \gg (\tilde{\tilde{\delta}} + \int_{\varepsilon \in I} a(t^{\circ}+\varepsilon)) =$

$$= \surd t.\ (t+r)^{\circ} \gg \tilde{\tilde{\delta}} + \int_{\varepsilon \in I} (t+r)^{\circ} \gg a(t^{\circ}+\varepsilon) =$$

$$= \surd t.\ \int_{\varepsilon \in I} \delta((t+r)^{\circ}+\varepsilon)) + \delta((t+r)^{\circ}) = \surd t.\ \int_{\varepsilon \in I} \delta((t+r)^{\circ}+\varepsilon)) = \surd t.\ (t+r)^{\circ} \gg \tilde{\tilde{\delta}} = \sigma^{r}_{\approx}(\tilde{\tilde{\delta}}).$$

4.16.6 Proposition: $r \gg_{st} (X + Y) = r \gg_{st} X + r \gg_{st} Y$.

4.16.7 Proposition: $r \gg_{st} (X \cdot Y) = (r \gg_{st} X) \cdot Y$.

4.16.8 Proposition: $X = X + \tilde{\tilde{\delta}} \Rightarrow r \gg_{st} \sigma^{r}_{\approx}(X) = \sigma^{r}_{\approx}(X)$.

Proof: $r \gg_{st} \sigma^{r}_{\approx}(X) = \surd t.\ (t+r)^{\circ} \gg t^{\circ} \gg_{s} (t+r)^{\circ} \gg X =$

$= \surd t.\ (t+r)^{\circ} \gg \tilde{\tilde{\delta}} + (t+r)^{\circ} \gg t^{\circ} \gg (t+r)^{\circ} \gg X = \surd t.\ (t+r)^{\circ} \gg \tilde{\tilde{\delta}} + (t+r)^{\circ} \gg X$ (SI7, SI2) $=$

$= \surd t.\ (t+r)^{\circ} \gg (\tilde{\tilde{\delta}} + X) = \surd t.\ (t+r)^{\circ} \gg X = \sigma^{r}_{\approx}(X)$.

4.16.9 Proposition: $p > r \Rightarrow p \gg_{st} \sigma^{r}_{\approx}(X) = \sigma^{r}_{\approx}((p-r) \gg_{st} X)$.

Proof: $p \gg_{st} \sigma^{r}_{\approx}(X) = \surd t.\ (t+p)^{\circ} \gg t^{\circ} \gg_{s} (t+r)^{\circ} \gg X =$

$= \surd t.\ (t+p)^{\circ} \gg \tilde{\tilde{\delta}} + (t+p)^{\circ} \gg t^{\circ} \gg (t+r)^{\circ} \gg X = \surd t.\ (t+p)^{\circ} \gg \tilde{\tilde{\delta}} + (t+p)^{\circ} \gg (t+r)^{\circ} \gg X$
(SI7) $= \surd t.\ (t+p)^{\circ} \gg (t+r)^{\circ} \gg_{s} X = \surd t.\ (t+r)^{\circ} \gg (t+p)^{\circ} \gg (t+r)^{\circ} \gg_{s} X$ (SI7) $=$
$= \surd t.\ (t+r)^{\circ} \gg \surd u.\ (u+p-r)^{\circ} \gg u^{\circ} \gg_{s} X = \sigma^{r}_{\approx}((p-r) \gg_{st} X)$.

4.17 Free merge.

Next we derive some equations for the left merge in relative standard time, used to axiomatise PA_{st}.

4.17.1 Proposition: $X = \surd t.\ t^{\circ} \gg_{s} X \Rightarrow \tilde{\tilde{a}} \underline{\|} X = \tilde{\tilde{a}} \cdot X$.

Proof: Assume (equivalently) that X satisfies $X = X + \tilde{\tilde{\delta}} \wedge X = \surd t.\ t^{\circ} \gg X$.

Note that $\tilde{\tilde{a}} = \surd t.\ \int_{\varepsilon \in I} a(t^{\circ}+\varepsilon) = \surd t.\ \int_{\varepsilon \in I} t \gg a(t^{\circ}+\varepsilon) = \surd t.\ \int_{\varepsilon \in I} a(t+\varepsilon)$. We derive

$$\tilde{\tilde{a}} \mathbin{\|\!\!\!\lfloor} X = (\surd t.\ \int_{\varepsilon\in I} a(t+\varepsilon)) \mathbin{\|\!\!\!\lfloor} X = \surd t.\ \int_{\varepsilon\in I} a(t+\varepsilon) \mathbin{\|\!\!\!\lfloor} t \gg X \text{ (IA8)} =$$

$$= \surd t.\ (\int_{\varepsilon\in I} a(t+\varepsilon) \mathbin{\underline{\oplus}} (t \gg X))\cdot(t \gg X) = \surd t.\ (\int_{\varepsilon\in I} a(t+\varepsilon) \mathbin{\underline{\oplus}} (t \gg X + t \gg \tilde{\tilde{\delta}}))\cdot(t \gg X) =$$

$$= \surd t.\ (\int_{\varepsilon\in I} a(t+\varepsilon) \mathbin{\underline{\oplus}} (t \gg X + t \gg \tilde{\tilde{\delta}} + \delta(t+2\varepsilon)))\cdot(t \gg X) =$$

$$= \surd t.\ (\int_{\varepsilon\in I} a(t+\varepsilon) \mathbin{\underline{\oplus}} (t \gg X) + a(t+\varepsilon) \mathbin{\underline{\oplus}} \delta(t+2\varepsilon))\cdot(t \gg X) =$$

$$= \surd t.\ (\int_{\varepsilon\in I} a(t+\varepsilon) \mathbin{\underline{\oplus}} (t \gg X) + a(t+\varepsilon))\cdot(t \gg X) = \surd t.\ \int_{\varepsilon\in I} a(t+\varepsilon)\cdot(t \gg X) \text{ (SI15)} =$$

$$= \surd t.\ \int_{\varepsilon\in I} a(t+\varepsilon)\cdot(t \gg t^{\circ} \gg X) \text{ (by assumption)} = \surd t.\ \int_{\varepsilon\in I} a(t+\varepsilon)\cdot((t+\varepsilon) \gg t \gg t^{\circ} \gg X)$$

$$\text{(ATA5)} = \surd t.\ \int_{\varepsilon\in I} a(t+\varepsilon)\cdot((t+\varepsilon) \gg t^{\circ} \gg X) \text{ (SI7)} = \surd t.\ \int_{\varepsilon\in I} a(t+\varepsilon)\cdot((t+\varepsilon) \gg X)$$

$$\text{(by assumption, } (t+\varepsilon)^{\circ} = t^{\circ}) = \surd t.\ \int_{\varepsilon\in I} a(t+\varepsilon)\cdot X \text{ (ATA5)} =$$

$$= (\surd t.\ \int_{\varepsilon\in I} a(t+\varepsilon))\cdot X \text{ (INT6, IA7)} = \tilde{\tilde{a}}\cdot X.$$

4.17.2 Proposition: $Y = \surd t.\ t^{\circ} \gg_s Y \Rightarrow (\tilde{\tilde{a}}\cdot X) \mathbin{\|\!\!\!\lfloor} Y = \tilde{\tilde{a}}\cdot(X \| Y)$.

4.17.3 Proposition: $\sigma^{\underline{r}}_{\approx}(X) \mathbin{\|\!\!\!\lfloor} \sigma^{\underline{r}}_{\approx}(Y) = \sigma^{\underline{r}}_{\approx}(X \mathbin{\|\!\!\!\lfloor} Y)$.

Proof: $\sigma^{\underline{r}}_{\approx}(X) \mathbin{\|\!\!\!\lfloor} \sigma^{\underline{r}}_{\approx}(Y) = (\surd t.\ (t+r)^{\circ} \gg X) \mathbin{\|\!\!\!\lfloor} (\surd t.\ (t+r)^{\circ} \gg Y) =$

$$= \surd t.\ ((t+r)^{\circ} \gg X) \mathbin{\|\!\!\!\lfloor} ((t+r)^{\circ} \gg Y)) = \surd t.\ (t+r)^{\circ} \gg (X \mathbin{\|\!\!\!\lfloor} Y) = \sigma^{\underline{r}}_{\approx}(X \mathbin{\|\!\!\!\lfloor} Y).$$

4.17.4 Proposition: $Y = \surd t.\ t^{\circ} \gg_s Y \Rightarrow \sigma^{\underline{r}}_{\approx}(X) \mathbin{\|\!\!\!\lfloor} Y = (\sigma^{\underline{r}}_{\approx}(X) \mathbin{\underline{\oplus}} Y) \mathbin{\|\!\!\!\lfloor} (r \gg_{st} Y)$

Proof: $\sigma^{\underline{r}}_{\approx}(X) \mathbin{\|\!\!\!\lfloor} Y = (\surd t.\ (t+r)^{\circ} \gg X) \mathbin{\|\!\!\!\lfloor} Y = \surd t.\ ((t+r)^{\circ} \gg X \mathbin{\|\!\!\!\lfloor} t \gg Y)$ (IA8) $=$

$$= \surd t.\ ((t+r)^{\circ} \gg X \mathbin{\underline{\oplus}} t \gg Y) \mathbin{\|\!\!\!\lfloor} (t+r)^{\circ} \gg t \gg Y \text{ (SI10)} =$$

$$= \surd t.\ (t \gg (t+r)^{\circ} \gg X \mathbin{\underline{\oplus}} t \gg Y) \mathbin{\|\!\!\!\lfloor} (t+r)^{\circ} \gg t \gg t^{\circ} \gg_s Y \text{ (SI2, assumption)} =$$

$$= \surd t.\ (t \gg \sigma^{\underline{r}}_{\approx}(X) \mathbin{\underline{\oplus}} t \gg Y) \mathbin{\|\!\!\!\lfloor} (t+r)^{\circ} \gg t^{\circ} \gg_s Y \text{ (SI7)} =$$

$$= \surd t.\ t \gg (\sigma^{\underline{r}}_{\approx}(X) \mathbin{\underline{\oplus}} Y) \mathbin{\|\!\!\!\lfloor} t \gg (r \gg_{st} Y) \text{ (SI14)} =$$

$$= \surd t.\ t \gg ((\sigma^{\underline{r}}_{\approx}(X) \mathbin{\underline{\oplus}} Y) \mathbin{\|\!\!\!\lfloor} (r \gg_{st} Y) \text{ (SI8)} =$$

$= (\sigma^{r}_{\approx}(X) \oplus Y) \mathbin{⫫} (r \gg_{st} Y)$ (lemma 4.4, IA4).

4.18 Merge with communication.

Now we consider identities concerning parallel composition with communication in relative time, to be used in order to axiomatise ACP_{st}.

4.18.1 Proposition: $\tilde{\tilde{a}} \mid \tilde{\tilde{b}} = \widetilde{\widetilde{a \mid b}}$.

Proof: $\tilde{\tilde{a}} \mid \tilde{\tilde{b}} = \surd t.\ \int_{\varepsilon \in I} a(t^{\circ}+\varepsilon) \mid \surd t.\ \int_{\eta \in I} b(t^{\circ}+\eta)) =$

$= \surd t.\ \int_{\varepsilon \in I} a(t^{\circ}+\varepsilon) \mid \int_{\eta \in I} b(t^{\circ}+\eta)$ (lemma 4.7.3) =

$= \surd t.\ \int_{\varepsilon \in I} \int_{\eta \in I} a(t^{\circ}+\varepsilon)) \mid b(t^{\circ}+\eta)$ (INT10, INT11) =

$= \surd t.\ \int_{\varepsilon \in I} \int_{\eta \in I} \delta(t^{\circ}+\varepsilon) + (a \mid b)(t^{\circ}+\eta) + \delta(t^{\circ}+\eta)$ (ATC1, ATC2, INT2) =

$= \surd t.\ \int_{\eta \in I} (a \mid b)(t^{\circ}+\eta)$ (ATA3, INT2) $= \widetilde{\widetilde{a \mid b}}$.

4.18.2 Proposition: $(\tilde{\tilde{a}} \cdot X) \mid \tilde{\tilde{b}} = (\tilde{\tilde{a}} \mid \tilde{\tilde{b}}) \cdot X$, $\tilde{\tilde{a}} \mid (\tilde{\tilde{b}} \cdot X) = (\tilde{\tilde{a}} \mid \tilde{\tilde{b}}) \cdot X$, $(\tilde{\tilde{a}} \cdot X) \mid (\tilde{\tilde{b}} \cdot Y) = (\tilde{\tilde{a}} \mid \tilde{\tilde{b}}) \cdot (X \| Y)$.

4.18.3 Proposition: $\tilde{\tilde{a}} \mid \sigma^{r}_{\approx}(X) = \tilde{\tilde{\delta}}$.

Proof: $\tilde{\tilde{a}} \mid \sigma^{r}_{\approx}(X) = \surd t.\ \int_{\varepsilon \in I} a(t^{\circ}+\varepsilon) \mid \surd t.\ (t+r)^{\circ} \gg X = \surd t.\ \int_{\varepsilon \in I} a(t^{\circ}+\varepsilon) \mid (t+r)^{\circ} \gg X$

(lemma 4.4) $= \surd t.\ \int_{\varepsilon \in I} a(t^{\circ}+\varepsilon) \mid (t^{\circ}+\varepsilon) \gg (t+r)^{\circ} \gg X$ (SI2) $= \surd t.\ \int_{\varepsilon \in I} \delta(t^{\circ}+\varepsilon)$ (SI12)

$= \tilde{\tilde{\delta}}$.

4.18.4 Proposition: $\sigma^{r}_{\approx}(X) \mid \tilde{\tilde{a}} = \tilde{\tilde{\delta}}$, $(\tilde{\tilde{a}} \cdot X) \mid \sigma^{r}_{\approx}(Y) = \tilde{\tilde{\delta}}$, $\sigma^{r}_{\approx}(X) \mid (\tilde{\tilde{a}} \cdot Y) = \tilde{\tilde{\delta}}$.

4.18.5 Proposition: $\sigma^{r}_{\approx}(X) \mid \sigma^{r}_{\approx}(Y) = \sigma^{r}_{\approx}(X|Y)$.

Proof: $\sigma^{r}_{\approx}(X) \mid \sigma^{r}_{\approx}(Y) = \surd t.(t+r)^{\circ} \gg X \mid \surd t.\ (t+r)^{\circ} \gg Y = \surd t.\ (t+r)^{\circ} \gg X \mid (t+r)^{\circ} \gg Y$

(lemma 4.4) $= \surd t.\ (t+r)^{\circ} \gg (X \mid Y)$ (SI9) $= \sigma^{r}_{\approx}(X|Y)$.

4.19 Encapsulation.

We continue with equations for the encapsulation operator in relative time, to be used for ACP_{st}.

4.19.1 Proposition: $\partial_H(\tilde{\tilde{a}}) = \tilde{\tilde{a}}$ if $a \notin H$.

Proof: $\partial_H(\tilde{\tilde{a}}) = \partial_H(\surd t.\ \int_{\varepsilon \in I} a(t^{\circ}+\varepsilon)) = \surd t.\ \partial_H(\int_{\varepsilon \in I} a(t^{\circ}+\varepsilon))$ (IA12) =

$= \surd t.\ \int_{\varepsilon \in I} \partial_H(a(t^{\circ}+\varepsilon))$ (INT12) $= \surd t.\ \int_{\varepsilon \in I} a(t^{\circ}+\varepsilon)$ (D1) $= \tilde{\tilde{a}}$.

4.19.2 Proposition: $\partial_H(\tilde{\tilde{a}}) = \tilde{\tilde{\delta}}$ if $a \in H$.

Proof: $\partial_H(\tilde{\tilde{a}}) = \partial_H(\surd t.\ \int_{\varepsilon \in I} a(t^{\circ}+\varepsilon)) = \surd t.\ \partial_H(\int_{\varepsilon \in I} a(t^{\circ}+\varepsilon))$ (IA12) =

$= \surd t.\ \int_{\varepsilon \in I} \partial_H(a(t^{\circ}+\varepsilon))$ (INT12) $= \surd t.\ \int_{\varepsilon \in I} \delta(t^{\circ}+\varepsilon)$ (D2) $= \tilde{\tilde{\delta}}$.

4.19.3 Proposition: $\partial_H(\sigma^{r}_{\approx}(X)) = \sigma^{r}_{\approx}(\partial_H(X))$.

Proof: $\partial_H(\sigma^{r}_{\approx}(X)) = \partial_H(\surd t.\ (t+r)^{\circ} \gg X) = \surd t.\ \partial_H((t+r)^{\circ} \gg X)$ (IA12) =

$= \surd t.\ (t+r)^{\circ} \gg \partial_H(X)$ (SI13) $= \sigma^{r}_{\approx}(\partial_H(X))$.

4.20 Embedding standard relative time into standard absolute time with initial abstraction.

We finish this section with some identities that are used to embed $BPA_{\delta st}$ into $BPA_{sp\surd}$, PA_{st} into $PA_{sp\surd}$ and ACP_{st} into $ACP_{sp\surd}$.

4.20.1 Proposition: $\tilde{\tilde{a}} = \surd_s p.\ \tilde{a}(p)$.

4.20.2 Proposition: $\sigma^{r}_{\approx}(X) = \surd_s p.\ (p+r) \gg_s X$.

Proof: $\sigma^{r}_{\approx}(X) = \surd t.\ (t+r)^{\circ} \gg X = \surd t.\ (\delta((t+r)^{\circ}) + (t+r)^{\circ} \gg X)$ (SI5) =

$$= \surd t.\ (\int_{\varepsilon \in I} \delta(t^\circ+\varepsilon) + \delta((t+r)^\circ) + (t+r)^\circ \gg X) = \surd t.\ (\int_{\varepsilon \in I} \delta(t^\circ+\varepsilon) + (t+r)^\circ \gg X) =$$

$$= \surd t.\ (\tilde{\tilde{\delta}} + (t+r)^\circ \gg X) = \surd_s p.\ (\tilde{\tilde{\delta}} + (p+r \gg X)) = \surd_s p.\ (p+r) \gg_s X.$$

4.20.3 Proposition: $r \gg_{st} X = \surd_s p.\ (p+r) \gg_s p \gg_s X.$

Proof: $r \gg_{st} X = \surd t.\ (t+r)^\circ \gg t^\circ \gg_s X = \surd t.\ \delta((t+r)^\circ) + (t+r)^\circ \gg t^\circ \gg_s X$ (SI5) =

$$= \surd t.\ \int_{\varepsilon \in I} \delta(t^\circ+\varepsilon) + \delta((t+r)^\circ) + (t+r)^\circ \gg t^\circ \gg_s X = \surd t.\tilde{\tilde{\delta}} + (t+r)^\circ \gg t^\circ \gg_s X =$$

$$= \surd t.\ (t+r)^\circ \gg_s t^\circ \gg_s X = \surd_s p.\ (p+r) \gg_s p \gg_s X.$$

5. Standardised processes.

In section 2, we have sketched a theory ACP$\rho\surd$I and a model M^*_A that contain both standard real time and nonstandard real time processes. In section 4, we concentrated on the subalgebra of standardly initialised processes. In this section, we will concentrate on the further subalgebra and the model generated by the standard time processes. For this reason, we introduce the standardisation operator $\mathcal{S}$.

5.1 Definition.

First , we define standardisation as an operator on P*. We have the axioms in table 7.

$\mathcal{S}(a(t)) = \tilde{a}(t^\circ)$	
$\mathcal{S}(X + Y) = \mathcal{S}(X) + \mathcal{S}(Y)$	$\mathcal{S}(X \cdot Y) = \mathcal{S}(X) \cdot \mathcal{S}(Y)$
$\mathcal{S}(\int_{v \in V} F) = \int_{v \in V} \mathcal{S}(F)$	$\mathcal{S}(\surd t.\ F) = \surd t.\ \mathcal{S}(F)$

TABLE 7. Standardisation.

We give some examples:

1. $\mathcal{S}(\delta) = \tilde{\tilde{\delta}}$ 2. $\mathcal{S}(\tilde{a}[p]) = \tilde{a}[p]$ 3. $\mathcal{S}(\tilde{a}(p)) = \tilde{a}(p)$.

In the model, $\mathcal{S}(p)$ is obtained by replacing each transition $\langle s, t\rangle \xrightarrow{a(r)} \langle s', r\rangle$ by transitions $\langle s, t\rangle \xrightarrow{a(v)} \langle s', v\rangle$ for each $v > t$ of the form $v = r^\circ+\varepsilon$ for some $\varepsilon \in$ I, and if $\langle s, t\rangle$ is a state without outgoing transitions and $s \neq \surd$, then we add transitions $\langle s, t\rangle \rightarrow \langle s, r\rangle$ for each $r > t$ of the form $r = t^\circ+\varepsilon$ for some $\varepsilon \in$ I and also transitions $\langle s', t'\rangle \rightarrow \langle s, r\rangle$ (for the same r's) for each transition $\langle s', t'\rangle \rightarrow \langle s, t\rangle$.

5.2 Properties.

1. $\mathcal{S}$ does not distribute over $\|$, $\lfloor\!\lfloor$, $|$.
2. $\mathcal{S}$ is a projection: $\mathcal{S}(\mathcal{S}(X)) = \mathcal{S}(X)$.
3. $\mathcal{S}(\partial_H(X)) = \partial_H(\mathcal{S}(X))$

Proof: We only prove 1. Consider $\|$. Let $\varepsilon \in I$.

$\mathcal{S}(a(1+\varepsilon) \| b(1+2\varepsilon)) = \mathcal{S}(a(1+\varepsilon)\cdot b(1+2\varepsilon)) = \tilde{a}(1)\cdot\tilde{b}(1) = \tilde{a}(1)\cdot\tilde{\tilde{b}}$, but

$\mathcal{S}(a(1+\varepsilon)) \| \mathcal{S}(b(1+2\varepsilon)) = \tilde{a}(1) \| \tilde{b}(1) = \tilde{a}(1)\cdot\tilde{b}(1) + a \urcorner b(1) + \tilde{b}(1)\cdot\tilde{a}(1) =$

$= \tilde{a}(1)\cdot\tilde{\tilde{b}} + a \urcorner b(1) + \tilde{b}(1)\cdot\tilde{\tilde{a}}$.

5.3 Definition.

We say that a process X is *standardised*, $X \in SIS$, if $\mathcal{S}(X) = X$. Also, we define this concept directly on the model M^*_A. Let $f \in M^*_A$. Then f is standardised, $f \in \mathcal{S}(M^*_A)$, if:

i. for all $t \in \mathbb{R}_{\geq 0}+I$, all actions of f(t) have a timestamp not in $\mathbb{R}_{\geq 0}$.

ii. for all $t \in \mathbb{R}_{\geq 0}+I$ we have the following: let $\langle p,v\rangle \xrightarrow{a(r)} \langle q,r\rangle$ be a transition in f(t). Then:

a. for each $r' = r + \varepsilon$ for some $\varepsilon \in I$ there is a q' with $\langle p,v\rangle \xrightarrow{a(r')} \langle q', r'\rangle$ and $\langle q, r'\rangle \underline{\leftrightarrow} \langle q', r'\rangle$;

b. for each r' with $r = r' + \varepsilon$ for some $\varepsilon \in I$ and $r' > v$ there is a q' with $\langle p,v\rangle \xrightarrow{a(r')} \langle q', r'\rangle$ and $\langle q, r\rangle \underline{\leftrightarrow} \langle q', r\rangle$.

iii. similarly for idle transitions: let $\langle p,v\rangle \rightarrow \langle q,r\rangle$ be a transition in f(t). Then:

a. for each $r' = r + \varepsilon$ for some $\varepsilon \in I$ there is a q' with $\langle p,v\rangle \rightarrow \langle q', r'\rangle$and $\langle q, r'\rangle \underline{\leftrightarrow} \langle q', r'\rangle$;

b. for each r' with $r = r' + \varepsilon$ for some $\varepsilon \in I$ and $r' > v$ there is a q' with $\langle p,v\rangle \rightarrow \langle q', r'\rangle$ and $\langle q, r\rangle \underline{\leftrightarrow} \langle q', r\rangle$.

5.4 Proposition.

1. $\tilde{a}(p), \tilde{a}[p] \in SIS, a(t) \notin SIS$.
2. $X \in SIS \Rightarrow \partial_H(X), \sigma^r_{\approx}(X), r \gg_{st} X \in SIS$.
3. $X, Y \in SIS \Rightarrow X+Y, X\cdot Y, X\|Y, X \lfloor\!\lfloor Y, X \mid Y, X \underline{\oplus} Y \in DIS$.
4. for all $v \in V$ $F \in SIS \Rightarrow \int_{v \in V} F \in SIS$.
5. $X \in SIS \Rightarrow X \in SIP$
6. $\surd t.\ a(t^\circ+1) \in SIP - SIS$.

Proof: We only prove 6. $\mathcal{S}(\surd t.\ a(t^\circ+1)) = \surd t.\ \mathcal{S}(a(t^\circ+1) = \surd t.\ \tilde{a}(t^\circ+1) = \tilde{a}[1] \neq$

$\neq \surd t.\ a(t^\circ+1)$. For the positive result, see 4.3.

Note: 5 ensures that all equations in section 4 hold for all standardised processes. Also notice that a deadlock possibility may not be preserved by standardisation: the process $a(1+\varepsilon) + \delta(1+2\varepsilon)$ ($\varepsilon \in I$) has a deadlock possibility, but its standardisation does not.

5.5 A graph model with actions and time separated.

Consider $\mathcal{S}(M^*_A)$, the collection of processes that satisfy $\mathcal{S}(X) = X$. This collection of processes is also closed under all elements of the standard time signature. This indicates the existence of an interesting subalgebra with domain $\mathcal{S}(M^*_A)$ of the reduct of M^*_A to the standard time signature.

We obtain a subalgebra (of a reduct of) M^*_A with domain $\mathcal{S}(M^*_A)$ and signature $\Sigma(ST)$. In the following section 6, we will discuss a direct axiomatisation of this algebra. In section 7, we will present a direct axiomatisation of the elements of this algebra that have only relative time notation (a further restriction to the signature $\Sigma(ACP_{st})$). Here, we present a simplified presentation of the subalgebra using transition systems with actions and time separated. This is the model $\mathcal{G}^*_A$ that we will use in sections 6 and 7.

We define a set of process graphs as in [BAW90] with labels from $A \cup \mathbb{R}_{>0}$ satisfying two extra conditions:

i. every node has *at most one* outgoing s-labeled edge ($s \in \mathbb{R}_{>0}$);
ii. an s-labeled edge ($s \in \mathbb{R}_{>0}$) may not lead to a termination node.

Let $\mathcal{G}^*_A$ be the set of such process graphs with cardinality $\leq 2^{\aleph_0}$. To state this precisely, an element of $\mathcal{G}^*_A$ is a quadruple $\langle N, E, r, T\rangle$ where N is the set of *nodes*, $E \subseteq N \times A \cup \mathbb{R}_{>0} \times N$ is the set of *edges*, $r \in N$ is the *root node*, and $T \subseteq N$ is the set of *termination nodes*. We will always have that a termination node has no outgoing edges. A node without outgoing edges that is not a termination node is called a *deadlock node*.

We define a mapping from $\mathcal{S}(M^*_A)$ to $\mathcal{G}^*_A$ by defining a mapping ϕ from real time transition systems to process graphs. Let R be a real time transition system. For simplicity, we assume that R is actually a tree, so each node has at most one incoming transition. The set of states of $\phi(R)$ consists of those states of R with time coordinate in $\mathbb{R}_{\geq 0}$. The root of $\phi(R)$ is the root of R. Now we consider the transitions. If $\langle s, t\rangle \rightarrow \langle s', t'\rangle$ is a transition in R and $t'^\circ - t^\circ = p > 0$, then we have a transition $\langle s, t^\circ\rangle \xrightarrow{p} \langle s', t'^\circ\rangle$ in $\phi(R)$. If $\langle s, t^\circ\rangle \xrightarrow{a(t)} \langle s', t\rangle$ is a transition in R (note that for standardised processes always $t^\circ < t$ for any $a(t)$-transition), then we have a transition $\langle s, t^\circ\rangle \xrightarrow{a} \langle s', t^\circ\rangle$ in $\phi(R)$.

This describes ϕ. We then obtain a mapping from $\mathcal{S}(M^*_A)$ to $\mathcal{G}^*_A$ by mapping a representative of $f \in \mathcal{S}(M^*_A)$ to $R(f(0))/\underline{\leftrightarrow}$.

The inverse mapping can be defined along the same lines. We leave the verification that this indeed defines an isomorphism to the reader.

5.6 Note.

Thus, we have obtained a series of algebras M^*_A, si-M^*_A, $\mathcal{S}(M^*_A)$, $\mathcal{G}^*_A$. We can also consider the subalgebras obtained if we look at absolute time notation processes only, i.e. processes X satisfying $X = 0 \gg Z$ for some Z. This restriction gives us algebras M_A, si-M_A, $\mathcal{S}(M_A)$, $\mathcal{G}_A$. On the other hand, we can look at processes involving relative time notation only. We do this as in [2].

5.7 Definition.

We have the *positive time shift operator* σ^r_+ introduced in [8] with the equations of [2], e.g. $\sigma^r_+(a(t)) = a(t+r)$. On M^*_A, this operator is obtained by incrementing each time stamp of a state or an action in a transition system by r.

5.8 Definition.

A process X is *translatable* if for all $r \geq 0$ we have $r \gg X = \sigma^r_+(0 \gg X)$. A process X is *standardly translatable* if for all $r \in \mathbb{R}_{>0}$ we have $r \gg_s X = \sigma^r_+(0 \gg_s X)$.

5.9 Remarks.

1. Let X be translatable and let $t = r + s$. Then $t \gg X = \sigma^r_+(\sigma^s_+(X)) = \sigma^r_+(s \gg X)$.
2. $\tilde{a}(p)$ is not standardly translatable, since $2 \gg_s \tilde{a}(5) = \tilde{a}(5)$ but $\sigma^2_+(0 \gg_s \tilde{a}(5)) =$
   $= \sigma^2_+(\tilde{a}(5)) = \tilde{a}(7)$.
3. $\tilde{a}[p]$ is standardly translatable, since $r \gg_s \tilde{a}[p] = \tilde{a}[p+r] = \sigma^r_+(\tilde{a}(p)) = \sigma^r_+(0 \gg_s \tilde{a}[p])$.
4. $X \in$ si-M^*_A, then X is translatable iff X is standardly translatable.

5.10 Algebras.

We obtain four more algebras. M^r_A is the subalgebra of M^*_A consisting of all translatable processes, si-M^r_A is the subalgebra of si-M^*_A consisting of all standardly translatable processes, $\mathcal{S}(M^r_A)$ is the subalgebra of $\mathcal{S}(M^*_A)$ consisting of all its standardly translatable processes, $\mathcal{G}^r_A$ is the subalgebra of $\mathcal{G}^*_A$ consisting of all its standardly translatable processes. $\mathcal{G}^r_A$ is a model for ACP_{st}, the theory of relative time standard time process algebra (to be defined in 7.13), while all laws used to axiomatise ACP_{st} are valid in si-M^r_A. We give an overview of all algebras introduced in fig. 2. Again, rectangular boxes denote absolute time, boxes with rounded corners denote relative time.

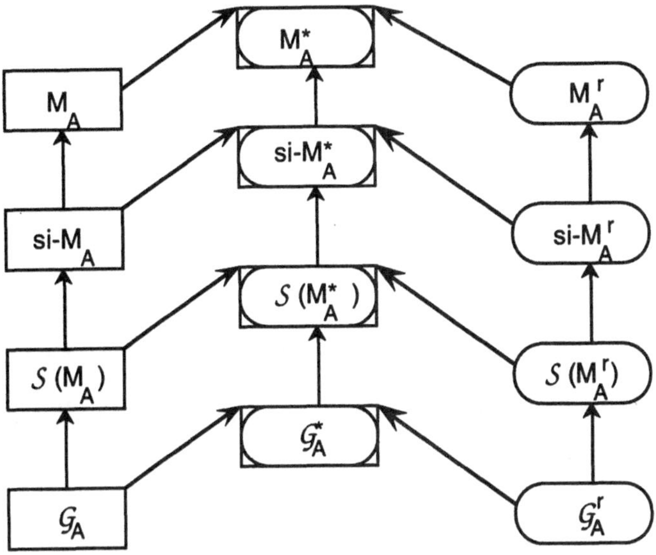

FIGURE 2.

6. Standard real time process algebra: absolute time.

6.1 Basic process algebra.

The process $\tilde{a}(p)$ can let time progress up to and including p, will then execute action a at time p, and then terminate successfully. The process $\tilde{\delta}(p)$ can also let time progress up to and including p, but then nothing more is possible (in particular, time cannot progress anymore).

Standard Real Time Basic Process Algebra with Deadlock (BPAsρδ) is the variant of BPAρδ with urgent actions. BPAsρδ has the axioms from table 8 ($a \in A_\delta$). The time domain is $\mathbb{R}_{\geq 0}$, so we only allow standard real numbers.

•	constants	$\tilde{\tilde{\delta}}$	inaction
		$\tilde{a}(p)$	action ($a \in A \cup \{\delta\}$)
•	functions	$+: P \times P \to P$	alternative composition
		$\cdot: P \times P \to P$	sequential composition
		$\gg_s: \mathbb{R}_{\geq 0} \times P \to P$	standard initialisation

$X+Y=Y+X$	$X=X+\tilde{\tilde{\delta}}$
$(X+Y)+Z=X+(Y+Z)$	$\tilde{\tilde{\delta}}\cdot X=\tilde{\tilde{\delta}}$
$X+X=X$	
$(X+Y)\cdot Z=X\cdot Z+Y\cdot Z$	
$(X\cdot Y)\cdot Z=X\cdot(Y\cdot Z)$	
$\tilde{\delta}(0)=\tilde{\tilde{\delta}}$	$p\leq q \;\Rightarrow\; p\gg_s \tilde{a}(q)=\tilde{a}(q)$
$\tilde{\delta}(p)\cdot X=\tilde{\delta}(p)$	$p>q \;\Rightarrow\; p\gg_s \tilde{a}(q)=\tilde{\delta}(p)$
$p<q \Rightarrow \tilde{\delta}(p)+\tilde{\delta}(q)=\tilde{\delta}(q)$	$p\gg_s(X+Y)=(p\gg_s X)+(p\gg_s Y)$
$\tilde{a}(p)+\tilde{\delta}(p)=\tilde{a}(p)$	$p\gg_s(X\cdot Y)=(p\gg_s X)\cdot Y$
$\tilde{a}(p)\cdot X=\tilde{a}(p)\cdot(p\gg_s X)$	

TABLE 8. BPAsρδ.

6.2 Operational semantics.

The operational semantics is defined on the domain of process graphs defined in 5.5. We have action rules with two types of transitions:

$$\text{idle} \subseteq P\times\mathbb{R}_{\geq 0}\times\mathbb{R}_{>0}\times P\times\mathbb{R}_{\geq 0}, \text{ notation } \langle x,p\rangle \xrightarrow{r} \langle x',p'\rangle$$

$$\text{step} \subseteq P\times\mathbb{R}_{\geq 0}\times A\times P\times\mathbb{R}_{\geq 0}, \text{ notation } \langle x,p\rangle \xrightarrow{a} \langle x',p'\rangle.$$

If $\langle x,p\rangle \xrightarrow{r} \langle x',p'\rangle$, then $r>0$, $x=x'$ and $p'=p+r$; if $\langle x,p\rangle \xrightarrow{a} \langle x',p'\rangle$, then $a\in A$ and $p'=p$.

$$\langle\tilde{a}(p),p\rangle \xrightarrow{a} \langle\surd,p\rangle \qquad q+r\leq p \Rightarrow \langle\tilde{a}(p),q\rangle \xrightarrow{r} \langle\tilde{a}(p),q+r\rangle$$

$$q+r\leq p \;\Rightarrow\; \langle\tilde{\delta}(p),q\rangle \xrightarrow{r} \langle\tilde{\delta}(p),q+r\rangle$$

$$\frac{\langle x,p\rangle \xrightarrow{a} \langle x',p\rangle}{\langle x+y,p\rangle \xrightarrow{a} \langle x',p\rangle,\ \langle y+x,p\rangle \xrightarrow{a} \langle x',p\rangle}$$

$$\frac{\langle x,p\rangle \xrightarrow{a} \langle\surd,p\rangle}{\langle x+y,p\rangle \xrightarrow{a} \langle\surd,p\rangle,\ \langle y+x,p\rangle \xrightarrow{a} \langle\surd,p\rangle}$$

$$\frac{\langle x,p\rangle \xrightarrow{r} \langle x,p+r\rangle}{\langle x+y,p\rangle \xrightarrow{r} \langle x+y,p+r\rangle,\ \langle y+x,p\rangle \xrightarrow{r} \langle y+x,p+r\rangle}$$

$$\frac{\langle x,p\rangle \xrightarrow{a} \langle x',p\rangle}{\langle x\cdot y,p\rangle \xrightarrow{a} \langle x'\cdot y,p\rangle} \qquad \frac{\langle x,p\rangle \stackrel{r}{\mapsto} \langle x,p+r\rangle}{\langle x\cdot y,p\rangle \stackrel{r}{\mapsto} \langle x\cdot y,p+r\rangle}$$

$$\frac{\langle x,p\rangle \xrightarrow{a} \langle \surd,p\rangle}{\langle x\cdot y,p\rangle \xrightarrow{a} \langle y,p\rangle}$$

$$q+r \leq p \Rightarrow \langle p \gg_s x,\ q\rangle \stackrel{r}{\mapsto} \langle p \gg_s x,\ q+r\rangle \qquad \frac{\langle x,q\rangle \xrightarrow{a} \langle x',q\rangle,\ q \geq p}{\langle p \gg_s x,\ q\rangle \xrightarrow{a} \langle x',q\rangle}$$

$$\frac{\langle x,q\rangle \xrightarrow{a} \langle \surd,q\rangle,\ q \geq p}{\langle p \gg_s x,\ q\rangle \xrightarrow{a} \langle \surd,q\rangle} \qquad \frac{\langle x,q\rangle \stackrel{r}{\mapsto} \langle x,q+r\rangle}{\langle p \gg_s x,\ q\rangle \stackrel{r}{\mapsto} \langle p \gg_s x,\ q+r\rangle}$$

TABLE 9. Operational semantics of BPAs$\rho\delta$.

6.3 Parallel composition.

In a setting with urgent actions, it makes sense to define parallel composition without communication, as urgent actions composed in parallel will show interleaving. Thus, $a(2) \| b(2)$ forces synchronisation, but we have $\tilde{a}(2) \| \tilde{b}(2) = \tilde{a}(2)\cdot\tilde{b}(2) + \tilde{b}(2)\cdot\tilde{a}(2)$. In table 10, $H \subseteq A$, $a,b \in A_\delta$. SOS rules are in table 11.

$$p \leq q \Rightarrow \tilde{a}(p) \ \underline{\oplus}\ \tilde{b}(q) = \tilde{a}(p)$$
$$p > q \Rightarrow \tilde{a}(p) \ \underline{\oplus}\ \tilde{b}(q) = \tilde{\delta}(q)$$
$$X \ \underline{\oplus}\ Y\cdot Z = X \ \underline{\oplus}\ Y$$
$$X \ \underline{\oplus}\ (Y + Z) = (X \ \underline{\oplus}\ Y) + (X \ \underline{\oplus}\ Z)$$
$$X\cdot Y \ \underline{\oplus}\ Z = (X \ \underline{\oplus}\ Y)\cdot Z$$
$$(X + Y) \ \underline{\oplus}\ Z = (X \ \underline{\oplus}\ Z) + (Y \ \underline{\oplus}\ Z)$$

$$X \| Y = X \mathbin{|\!\!\lfloor} Y + Y \mathbin{|\!\!\lfloor} X$$
$$X = 0 \gg_s Z \Rightarrow \tilde{a}(p) \mathbin{|\!\!\lfloor} X = (\tilde{a}(p) \ \underline{\oplus}\ X) \cdot X$$
$$Y = 0 \gg_s Z \Rightarrow \tilde{a}(p)\cdot X \mathbin{|\!\!\lfloor} Y = (\tilde{a}(p) \ \underline{\oplus}\ Y) \cdot (X \| Y)$$

TABLE 10. Additional axioms of PAs$\rho\delta$.

$$\frac{\langle x,p\rangle \xrightarrow{a} \langle x',p\rangle,\ p=0 \vee \langle y,0\rangle \xrightarrow{p} \langle y,p\rangle}{\langle x\|y,p\rangle \xrightarrow{a} \langle x'\|y,p\rangle,\ \langle y\|x,p\rangle \xrightarrow{a} \langle y\|x',p\rangle,\ \langle x \mathbin{|\!\!\lfloor} y,p\rangle \xrightarrow{a} \langle x'\|y,p\rangle}$$

$$\frac{\langle x,p\rangle \xrightarrow{a} \langle \surd,p\rangle,\ p=0 \vee \langle y,0\rangle \xrightarrow{p} \langle y,p\rangle}{\langle x\|y,p\rangle \xrightarrow{a} \langle y,p\rangle,\ \langle y\|x,p\rangle \xrightarrow{a} \langle y,p\rangle,\ \langle x \mathbin{|\!\!\lfloor} y,p\rangle \xrightarrow{a} \langle y,p\rangle}$$

$$\frac{\langle x,p\rangle \xrightarrow{r} \langle x,p+r\rangle,\ \langle y,p\rangle \xrightarrow{r} \langle y,p+r\rangle}{\langle x\|y,p\rangle \xrightarrow{r} \langle x\|y,p+r\rangle,\ \langle x \mathbin{|\!\!\lfloor} y,p\rangle \xrightarrow{r} \langle x \mathbin{|\!\!\lfloor} y,p+r\rangle}$$

$$\frac{\langle x,p\rangle \xrightarrow{a} \langle x',p\rangle,\ p=0 \vee \langle y,0\rangle \xrightarrow{p} \langle y,p\rangle}{\langle x \underline{\oplus} y,\ p\rangle \xrightarrow{a} \langle x',p\rangle}$$

$$\frac{\langle x,p\rangle \xrightarrow{a} \langle \surd,p\rangle,\ p=0 \vee \langle y,0\rangle \xrightarrow{p} \langle y,p\rangle}{\langle x \underline{\oplus} y,\ p\rangle \xrightarrow{a} \langle \surd,p\rangle}$$

$$\frac{\langle x,p\rangle \xrightarrow{r} \langle x,p+r\rangle,\ \langle y,p\rangle \xrightarrow{r} \langle y,p+r\rangle}{\langle x \underline{\oplus} y,\ p\rangle \xrightarrow{r} \langle x \underline{\oplus} y,\ p+r\rangle}$$

TABLE 11. Additional operational rules of PAs$\rho\delta$.

6.4 Communication.

Additional axioms of ACPsρ are in table 12. In table 12, $H \subseteq A$, $a,b \in A_\delta$.

$$a \mid b = b \mid a \qquad\qquad a \mid (b \mid c) = (a \mid b) \mid c$$

$$\delta \mid a = \delta$$

$$p \neq q \Rightarrow \tilde{a}(p) \mid \tilde{b}(q) = \tilde{\delta}(\min(p,q))$$

$$\tilde{a}(p) \mid \tilde{b}(p) = (\widetilde{a \mid b})(p)$$

$$X \parallel Y = X \mathbin{\|\!\_} Y + Y \mathbin{\|\!\_} X + X \mid Y$$

$$X = 0 \gg_s Z \Rightarrow \tilde{a}(p) \mathbin{\|\!\_} X = (\tilde{a}(p) \mathbin{\underline{\oplus}} X) \cdot X$$

$$Y = 0 \gg_s Z \Rightarrow \tilde{a}(p) \cdot X \mathbin{\|\!\_} Y = (\tilde{a}(p) \mathbin{\underline{\oplus}} Y) \cdot (X \parallel Y)$$

$$(X + Y) \mathbin{\|\!\_} Z = X \mathbin{\|\!\_} Z + Y \mathbin{\|\!\_} Z$$

$$\tilde{a}(p) \cdot X \mid \tilde{b}(q) = (\tilde{a}(p) \mid \tilde{b}(q)) \cdot X$$

$$\tilde{a}(p) \mid \tilde{b}(q) \cdot X = (\tilde{a}(p) \mid \tilde{b}(q)) \cdot X$$

$$\tilde{a}(p) \cdot X \mid \tilde{b}(q) \cdot Y = (\tilde{a}(p) \mid \tilde{b}(q)) \cdot (X \parallel Y)$$

$$(X + Y) \mid Z = X \mid Z + Y \mid Z \qquad\qquad X \mid (Y + Z) = X \mid Y + X \mid Z$$

$$\partial_H(a) = a \quad \text{if } a \notin H \qquad\qquad \partial_H(a) = \delta \quad \text{if } a \in H$$

$$\partial_H(\tilde{a}(p)) = (\widetilde{\partial_H}(a))(p)$$

$$\partial_H(X + Y) = \partial_H(X) + \partial_H(Y) \qquad\qquad \partial_H(X \cdot Y) = \partial_H(X) \cdot \partial_H(Y)$$

TABLE 12. Additional axioms of ACPsρ.

$$\frac{\langle x,p\rangle \xrightarrow{a} \langle x',p\rangle,\ \langle y,p\rangle \xrightarrow{b} \langle y',p\rangle,\ a \mid b = c \neq \delta}{\langle x \parallel y,p\rangle \xrightarrow{c} \langle x' \parallel y',p\rangle,\ \langle x \mid y,p\rangle \xrightarrow{c} \langle x' \parallel y',p\rangle}$$

$$\frac{\langle x,p\rangle \xrightarrow{a} \langle x',p\rangle,\ \langle y,p\rangle \xrightarrow{b} \langle \surd,p\rangle,\ a \mid b = c \neq \delta}{\langle x \parallel y,p\rangle \xrightarrow{c} \langle x',p\rangle,\ \langle y \parallel x,p\rangle \xrightarrow{c} \langle x',p\rangle,\ \langle x \mid y,p\rangle \xrightarrow{c} \langle x',p\rangle,\ \langle y \mid x,p\rangle \xrightarrow{c} \langle x',p\rangle}$$

$$\frac{\langle x,p\rangle \xrightarrow{a} \langle \surd,p\rangle,\ \langle y,p\rangle \xrightarrow{b} \langle \surd,p\rangle,\ a \mid b = c \neq \delta}{\langle x \parallel y,p\rangle \xrightarrow{c} \langle \surd,p\rangle,\ \langle x \mid y,p\rangle \xrightarrow{c} \langle \surd,p\rangle}$$

$$\frac{\langle x,p\rangle \xrightarrow{r} \langle x,p+r\rangle,\ \langle y,p\rangle \xrightarrow{r} \langle y,p+r\rangle}{\langle x \mid y,p\rangle \xrightarrow{r} \langle x \mid y,p+r\rangle}$$

$$\frac{\langle x,p\rangle \xrightarrow{a} \langle x',p\rangle,\ a\notin H}{\langle \partial_H(x),p\rangle \xrightarrow{a} \langle \partial_H(x'),p\rangle} \qquad \frac{\langle x,p\rangle \xrightarrow{r} \langle x,p+r\rangle}{\langle \partial_H(x),p\rangle \xrightarrow{r} \langle \partial_H(x),p+r\rangle}$$

$$\frac{\langle x,p\rangle \xrightarrow{a} \langle \surd,p\rangle,\ a\notin H}{\langle \partial_H(x),p\rangle \xrightarrow{a} \langle \surd,p\rangle}$$

TABLE 13. Additional operational rules of ACPsρ.

6.5 Initial abstraction.

We extend with initial abstraction as in 2.4, and obtain ACPsρ√. We extend the semantics as in 2.12.

$\tilde{a}[p] = \surd_s q.\ \tilde{a}(p+q)$	
$\surd_s p.F = \surd_s q.F[q/p]$	if q is not free in F
$p \gg_s \surd_s q.F = p \gg_s F[p/q]$	
$\surd_s q.\ \surd_s p.\ F = \surd_s q.\ F[q/p]$	
$X = \surd_s p.\ X$	
$\forall p\in \mathbb{R}_{\geq 0}\ p \gg_s X = p \gg_s Y \ \Rightarrow\ X = Y$	
$(\surd_s p.F) + X = \surd_s p.(F + p \gg_s X)$	$(\surd_s p.F)\cdot X = \surd_s p.(F\cdot X)$
$(\surd_s p.F) \mathbin{\|\!\_} X = \surd_s p.(F \mathbin{\|\!\_} p \gg_s X)$	$X \mathbin{\|\!\_} (\surd_s p.F) = \surd_s p.(p \gg_s X \mathbin{\|\!\_} F)$
$(\surd_s p.F) \mid X = \surd_s p.(F \mid p \gg_s X)$	$X \mid (\surd_s p.F) = \surd_s p.(p \gg_s X \mid F)$
$\partial_H(\surd_s p.F) = \surd_s p.\partial_H(F)$	$(\surd_s p.F) \underline{\oplus}\ X = \surd_s p.(F \underline{\oplus}\ p \gg_s X)$
$X \underline{\oplus}\ (\surd_s p.F) = \surd_s p.(p \gg_s X \underline{\oplus}\ F)$	

TABLE 14. Axioms for initial abstraction.

7. Standard real time process algebra: relative time.

7.1 Basic process algebra.

The signature of BPA_{st} is as follows:

•	constants	$\approx{a}$ ($\overset{\approx}{a}$)	urgent a ($a \in A$).
•	functions	$+: P \times P \to P$	alternative composition
		$\cdot: P \times P \to P$	sequential composition
		$\sigma^r_{\approx}: P \to P$	positive time shift (r>0).

In table 15, r,p > 0.

$X + Y = Y + X$	$\sigma^{r}_{\approx} \circ \sigma^{p}_{\approx}(X) = \sigma^{r+p}_{\approx}(X)$
$(X + Y) + Z = X + (Y + Z)$	$\sigma^{r}_{\approx}(X) + \sigma^{r}_{\approx}(Y) = \sigma^{r}_{\approx}(X + Y)$
$X + X = X$	$\sigma^{r}_{\approx}(X) \cdot Y = \sigma^{r}_{\approx}(X \cdot Y)$
$(X + Y) \cdot Z = X \cdot Z + Y \cdot Z$	
$(X \cdot Y) \cdot Z = X \cdot (Y \cdot Z)$	

TABLE 15. BPA_{st}.

The operational semantics is defined by action rules with two types of transitions:

$\text{idle} \subseteq P \times \mathbb{R}_{>0} \times P$, notation $x \overset{r}{\hookrightarrow} x'$

$\text{step} \subseteq P \times A \times P$, notation $x \overset{a}{\rightarrow} x'$.

Here $r > 0$, $a \in A$.

$\tilde{\tilde{a}} \overset{a}{\rightarrow} \surd$	$\dfrac{x \overset{a}{\rightarrow} x'}{x \cdot y \overset{a}{\rightarrow} x' \cdot y}$	$\dfrac{x \overset{a}{\rightarrow} \surd}{x \cdot y \overset{a}{\rightarrow} y}$
$\dfrac{x \overset{a}{\rightarrow} x'}{x+y \overset{a}{\rightarrow} x',\ y+x \overset{a}{\rightarrow} x'}$		$\dfrac{x \overset{a}{\rightarrow} \surd}{x+y \overset{a}{\rightarrow} \surd,\ y+x \overset{a}{\rightarrow} \surd}$
$\sigma^{r}_{\approx}(x) \overset{r}{\hookrightarrow} x$		$p,r > 0 \Rightarrow \sigma^{p+r}_{\approx}(x) \overset{r}{\hookrightarrow} \sigma^{p}_{\approx}(x)$
$\dfrac{x \overset{r}{\hookrightarrow} x'}{\sigma^{p}_{\approx}(x) \overset{p+r}{\hookrightarrow} x'}$		$\dfrac{x \overset{r}{\hookrightarrow} x'}{x \cdot y \overset{r}{\hookrightarrow} x' \cdot y}$
$\dfrac{x \overset{r}{\hookrightarrow} x',\ y \overset{r}{\hookrightarrow} y'}{x+y \overset{r}{\hookrightarrow} x'+y'}$		$\dfrac{x \overset{r}{\hookrightarrow} x',\ y \overset{r}{\not\hookrightarrow}}{x+y \overset{r}{\hookrightarrow} x',\ y+x \overset{r}{\hookrightarrow} x'}$

TABLE 16. Operational semantics of BPA_{st}.

7.2 Inaction.

$BPA_{\delta st}$ is obtained by introducing $\tilde{\tilde{\delta}}$ as a constant representing inaction. The axioms for $\tilde{\tilde{\delta}}$ are standard (table 17).

The operational meaning of $\tilde{\tilde{\delta}}$ is a process that allows no step whatsoever. In the graph model, $\tilde{\tilde{\delta}}$ will be modeled by the tree with no edges, and one node which is the root but not a termination node.

$X + \overset{\approx}{\delta} = X$	$\overset{\approx}{\delta} \cdot X = \overset{\approx}{\delta}$

TABLE 17. Additional axioms of $BPA_{\delta st}$.

7.3 Parallel composition.

Axioms are in table 18 ($a \in A_\delta$, $x,y,z \in P$, $r,p > 0$).

$\overset{\approx}{a} \underline{\oplus} \overset{\approx}{b} = \overset{\approx}{a}$	$\overset{\approx}{a} \underline{\oplus} \sigma^{r}_{\approx}(X) = \overset{\approx}{a}$
$\sigma^{r}_{\approx}(X) \underline{\oplus} \overset{\approx}{a} = \overset{\approx}{\delta}$	$\sigma^{r}_{\approx}(X) \underline{\oplus} \sigma^{r}_{\approx}(Y) = \sigma^{r}_{\approx}(X \underline{\oplus} Y)$
$X \underline{\oplus} Y \cdot Z = X \underline{\oplus} Y$	$X \underline{\oplus} (Y + Z) = (X \underline{\oplus} Y) + (X \underline{\oplus} Z)$
$X \cdot Y \underline{\oplus} Z = (X \underline{\oplus} Y) \cdot Z$	$(X + Y) \underline{\oplus} Z = (X \underline{\oplus} Z) + (Y \underline{\oplus} Z)$
$r \gg_{st} \overset{\approx}{a} = \sigma^{r}_{\approx}(\overset{\approx}{\delta})$	$r \gg_{st} (X + Y) = r \gg_{st} X + r \gg_{st} Y$
$r \gg_{st} (X \cdot Y) = (r \gg_{st} X) \cdot Y$	$r \gg_{st} \sigma^{r}_{\approx}(X) = \sigma^{r}_{\approx}(X)$
$p > r \Rightarrow p \gg_{st} \sigma^{r}_{\approx}(X) = \sigma^{r}_{\approx}((p-r) \gg_{st} X)$	
$X \| Y = X \mathbin{\|\!\_} Y + Y \mathbin{\|\!\_} X$	$\overset{\approx}{a} \mathbin{\|\!\_} X = \overset{\approx}{a} \cdot X$
$(\overset{\approx}{a} \cdot X) \mathbin{\|\!\_} Y = \overset{\approx}{a} \cdot (X \| Y)$	$(X + Y) \mathbin{\|\!\_} Z = X \mathbin{\|\!\_} Z + Y \mathbin{\|\!\_} Z$
$\sigma^{r}_{\approx}(X) \mathbin{\|\!\_} Y = (\sigma^{r}_{\approx}(X) \underline{\oplus} Y) \mathbin{\|\!\_} (r \gg_{st} Y)$	$\sigma^{r}_{\approx}(X) \mathbin{\|\!\_} \sigma^{r}_{\approx}(Y) = \sigma^{r}_{\approx}(X \mathbin{\|\!\_} Y)$

TABLE 18. Additional axioms of $PA_{\delta st}$.

$$\frac{x \xrightarrow{a} x'}{x \| y \xrightarrow{a} x' \| y,\ y \| x \xrightarrow{a} y \| x',\ x \lfloor\!\lfloor y \xrightarrow{a} x' \| y}$$

$$\frac{x \xrightarrow{a} \surd}{x \| y \xrightarrow{a} y,\ y \| x \xrightarrow{a} y,\ x \lfloor\!\lfloor y \xrightarrow{a} y} \qquad \frac{x \xrightarrow{r} x',\ y \xrightarrow{r} y'}{x \| y \xrightarrow{r} x' \| y',\ x \lfloor\!\lfloor y \xrightarrow{r} x' \lfloor\!\lfloor y'}$$

$$\frac{x \xrightarrow{r} x',\ r \geq p}{p \gg_{st} x \xrightarrow{r} x'} \qquad \frac{x \xrightarrow{r} x',\ r < p}{p \gg_{st} x \xrightarrow{r} (p-r) \gg_{st} x}$$

$$\frac{x \xrightarrow{a} x'}{x \underline{\oplus} y \xrightarrow{a} x'} \qquad \frac{x \xrightarrow{a} \surd}{x \underline{\oplus} y \xrightarrow{a} \surd} \qquad \frac{x \xrightarrow{r} x',\ y \xrightarrow{r} y'}{x \underline{\oplus} y \xrightarrow{r} x' \underline{\oplus} y'}$$

TABLE 19. Additional operational rules of $PA_{\delta st}$.

7.4 Communication.

The axioms of ACP_{st} are in table 20. In table 20, $H \subseteq A$, $r > 0$, $a,b \in A_\delta$.

$$\overset{\approx}{a} \mid \overset{\approx}{b} = \overset{\approx}{a \mid b}$$

$$X \| Y = X \lfloor\!\lfloor Y + Y \lfloor\!\lfloor X + X \mid Y$$

$$\overset{\approx}{a} \lfloor\!\lfloor X = \overset{\approx}{a} \cdot X$$

$$(\overset{\approx}{a} \cdot X) \lfloor\!\lfloor Y = \overset{\approx}{a} \cdot (X \| Y)$$

$$\sigma^{r}_{\approx}(X) \lfloor\!\lfloor Y = (\sigma^{r}_{\approx}(X) \oplus Y) \lfloor\!\lfloor (r \gg_{st} Y)$$

$$\sigma^{r}_{\approx}(X) \lfloor\!\lfloor \sigma^{r}_{\approx}(Y) = \sigma^{r}_{\approx}(X \lfloor\!\lfloor Y)$$

$$(X + Y) \lfloor\!\lfloor Z = X \lfloor\!\lfloor Z + Y \lfloor\!\lfloor Z$$

$$\partial_H(\overset{\approx}{a}) = \overset{\approx}{a} \quad \text{if } a \notin H$$

$$\partial_H(\overset{\approx}{a}) = \overset{\approx}{\delta} \quad \text{if } a \in H$$

$$\partial_H(X + Y) = \partial_H(X) + \partial_H(Y)$$

$$\partial_H(X \cdot Y) = \partial_H(X) \cdot \partial_H(Y)$$

$$\partial_H(\sigma^{r}_{\approx}(X)) = \sigma^{r}_{\approx}(\partial_H(X))$$

$$(\overset{\approx}{a} \cdot X) \mid \overset{\approx}{b} = (\overset{\approx}{a} \mid \overset{\approx}{b}) \cdot X$$

$$\overset{\approx}{a} \mid (\overset{\approx}{b} \cdot X) = (\overset{\approx}{a} \mid \overset{\approx}{b}) \cdot X$$

$$(\overset{\approx}{a} \cdot X) \mid (\overset{\approx}{b} \cdot Y) = (\overset{\approx}{a} \mid \overset{\approx}{b}) \cdot (X \| Y)$$

$$(X + Y) \mid Z = X \mid Z + Y \mid Z$$

$$X \mid (Y + Z) = X \mid Y + X \mid Z$$

$$\overset{\approx}{a} \mid \sigma^{r}_{\approx}(X) = \overset{\approx}{\delta}$$

$$\sigma^{r}_{\approx}(X) \mid \overset{\approx}{a} = \overset{\approx}{\delta}$$

$$(\overset{\approx}{a} \cdot X) \mid \sigma^{r}_{\approx}(Y) = \overset{\approx}{\delta}$$

$$\sigma^{r}_{\approx}(X) \mid (\overset{\approx}{a} \cdot Y) = \overset{\approx}{\delta}$$

$$\sigma^{r}_{\approx}(X) \mid \sigma^{r}_{\approx}(Y) = \sigma^{r}_{\approx}(X \mid Y)$$

TABLE 20. Additional axioms of ACP_{st}.

$$\frac{x \xrightarrow{a} x',\ y \xrightarrow{b} y',\ a \mid b = c \neq \delta}{x \| y \xrightarrow{c} x' \| y',\ x \mid y \xrightarrow{c} x' \| y'}$$

$$\frac{x \xrightarrow{a} \surd,\ y \xrightarrow{b} \surd,\ a \mid b = c \neq \delta}{x \| y \xrightarrow{c} \surd,\ x \mid y \xrightarrow{c} \surd}$$

$$\frac{x \xrightarrow{a} x',\ y \xrightarrow{b} \surd,\ a \mid b = c \neq \delta}{x \| y \xrightarrow{c} x',\ x \mid y \xrightarrow{c} x',\ y \| x \xrightarrow{c} x',\ y \mid x \xrightarrow{c} x'}$$

$$\frac{x \xrightarrow{r} x',\ y \xrightarrow{r} y'}{x \mid y \xrightarrow{r} x' \mid y'}$$

$$\frac{x \xrightarrow{a} x',\ a \notin H}{\partial_H(x) \xrightarrow{a} \partial_H(x')}$$

$$\frac{x \xrightarrow{a} \surd,\ a \notin H}{\partial_H(x) \xrightarrow{a} \surd}$$

$$\frac{x \xrightarrow{r} x'}{\partial_H(x) \xrightarrow{r} \partial_H(x')}$$

TABLE 21. Additional operational rules of ACP_{st}.

7.5 Remark.

Consider the axiom system ACP_{st}, and omit all signature elements refering to timing, i.e. omit the operators $\sigma^{r}_{\approx}$, $r \gg_{st}$, $\gg_s$, $\oplus$. The remaining signature is exactly the signature of the untimed theory ACP of [5,6] (interpreting $\overset{\approx}{a}$ as a, $\overset{\approx}{\delta}$ as δ), and the axioms over this signature are exactly the axioms of ACP. Thus, the untimed theory

ACP is obtained from ACP_{st} by throwing away all timing information, or, put another way, ACP_{st} over the one-point time domain $\mathcal{T} = \{0\}$ gives the untimed theory ACP.

8. Translations.

This section is based on [9]. We provide interpretations of the timing constructs of other process algebras, involving urgent actions, into our real time process algebra. $[\![.]\!]$ denotes the interpretation function. More information, involving the treatment of parallel composition and the modeling of maximal progress, can be found in [9].

8.1 ATP [13].

$$[\![0]\!] = \overset{\approx}{\delta}$$

$$[\![\delta]\!] = \tilde{\delta}[\infty] = \int_{p\geq 0} \tilde{\delta}[p]$$

$$[\![\dot{a}P]\!] = \overset{\approx}{a} \cdot [\![P]\!]$$

$$[\![\tilde{a}P]\!] = \int_{p\geq 0} \tilde{a}[p] \cdot [\![P]\!]$$

$$[\![P \oplus Q]\!] = ([\![P]\!] \underline{\oplus} [\![Q]\!]) + ([\![Q]\!] \underline{\oplus} [\![P]\!]) \qquad \text{(strong choice)}$$

$$[\![P \overset{d}{\lhd} Q]\!] = ([\![P]\!] \underline{\oplus} \tilde{\delta}[d]) + (d \gg_{st} [\![Q]\!]).$$

8.2 TCCS of Wang Yi [16].

$$[\![NIL]\!] = \tilde{\delta}[\infty] = \int_{p\geq 0} \tilde{\delta}[p]$$

$$[\![\tau]\!] = \overset{\approx}{\tau}$$

$$[\![a.P]\!] = \int_{p\geq 0} \tilde{a}[p] \cdot [\![P]\!] \qquad (a \neq \tau)$$

$$[\![\varepsilon(r).P]\!] = \sigma^{r}_{\approx}([\![P]\!]).$$

$$[\![P + Q]\!] = ([\![P]\!] \underline{\oplus} [\![Q]\!]) + ([\![Q]\!] \underline{\oplus} [\![P]\!]) \qquad \text{(strong choice).}$$

Note that [16] uses a maximal progress assumption. In order to model this, we need to use a priority operator, as in [3]. This has been worked out in [9].

8.3 TCCS of Moller & Tofts [11,12].

$$[\![0]\!] = \overset{\approx}{\delta}$$

$[\underline{0}] =$	$\tilde{\delta}[\infty] = \int_{p\geq 0} \tilde{\delta}[p]$	(only in [12])
$[a.P] =$	$\approx{a} \cdot [P]$	
$[\underline{a}.P] =$	$\int_{p\geq 0} \tilde{a}[p] \cdot [P]$	(only in [12])
$[(r).P] =$	$\sigma^{r}_{\approx}([P])$	
$[\![\delta.P]\!] =$	$\int_{r\geq 0} \sigma^{r}_{\approx}([\![P]\!] \underline{\oplus} \tilde{\tilde{\delta}})$	(only in [11])
$[\![P +\!\!+ Q]\!] =$	$([\![P]\!] \underline{\oplus} [\![Q]\!]) + ([\![Q]\!] \underline{\oplus} [\![P]\!])$	(strong choice; denoted + in [11])
$[P + Q] =$	$[P] + [Q]$	(weak choice; denoted ⊕ in [11]).

8.4 Chen [7].

$[NIL] =$	$\tilde{\delta}[\infty] = \int_{p\geq 0} \tilde{\delta}[p]$	
$[a(p)|^{b}_{e}.P] =$	$\int_{p\in[b,e]} \tilde{a}[p]\cdot[P] + \tilde{\delta}[e]$	
$[P + Q] =$	$[P] + [Q]$	(weak choice)
$[(r).P] =$	$\sigma^{r}_{\approx}([P])$.	

9. Conclusion.

We have modeled urgent actions in our real time process algebra, by extending the time domain to include nonstandard real numbers, and interpreting urgent actions as an alternative composition of normal timed actions of ACPρ. In this way, process calculi involving urgent actions can be interpreted in ACPρ. Also, we have indicated subalgebras that allow an operational semantics where actions and time are separated.

We conclude that our real time process algebra provides a general framework, in which many features occurring in the literature can be modeled.

References.

1. J.C.M. Baeten & J.A. Bergstra. Real time process algebra. Formal Aspects of Computing 1991; 3:142-188.

2. J.C.M. Baeten & J.A. Bergstra. Discrete time process algebra. In: W.R. Cleaveland (ed.), Proc. CONCUR'92, Stony Brook. Springer Verlag 1992, pp. 401-420 (Lecture Notes in Computer Science no. 630).

3. J.C.M. Baeten & J.A. Bergstra. Real space process algebra. Formal Aspects of Computing 1993; 5:481-529.

4. J.C.M. Baeten, J.A. Bergstra & J.W. Klop. Syntax and defining equations for an interrupt mechanism in process algebra. Fund. Inf. 1986; IX:127-168.

5. J.C.M. Baeten & W.P. Weijland. Process algebra. Cambridge Tracts in Theor. Comp. Sci. 18, Cambridge University Press 1990.

6. J.A. Bergstra & J.W. Klop. Process algebra for synchronous communication. Information & Control 1984; 60:109-137.

7. L. Chen. Timed processes: models, axioms and decidability. Ph.D. thesis, University of Edinburgh 1993.

8. W.J. Fokkink & A.S. Klusener. Real time process algebra with prefixed integration. Report CS-R9219, CWI, Amsterdam 1992.

9. A.S. Klusener. Models and axioms for a fragment of real time process algebra. Ph.D. Thesis, Eindhoven University of Technology 1993.

10. W.A.J. Luxemburg. Non-standard analysis (lectures on A. Robinson's theory of infinitesimals and infinitely large numbers). California Institute of Technology, Pasadena 1962.

11. F. Moller & C. Tofts. A temporal calculus of communicating systems. In: J.C.M. Baeten & J.W. Klop (eds.), Proc. CONCUR'90, Amsterdam. Springer Verlag 1990, pp. 401-415 (Lecture Notes in Computer Science no. 458).

12. F. Moller & C. Tofts. Behavioural abstraction in TCCS. In: W. Kuich (ed.), Proc. ICALP 92, Vienna. Springer Verlag 1992 (Lecture Notes in Computer Science 623).

13. X. Nicollin & J. Sifakis. The algebra of timed processes ATP: theory and application (revised version). Report RT-C26, IMAG Grenoble 1991.

14. G.D. Plotkin. A structural approach to operational semantics. Report DAIMI FN-19, Comp. Sci. Dept., Aarhus University 1981.

15. A. Robinson. Non-standard analysis. Studies in Logic and the Foundations of Mathematics, North-Holland, Amsterdam 1966.

16. Wang Yi. Real-time behaviour of asynchronous agents. In: J.C.M. Baeten & J.W. Klop (eds.), Proc. CONCUR'90, Amsterdam, Springer Verlag 1990, pp. 502-520 (Lecture Notes in Computer Science 458).

On the Expressiveness of ACP
(extended abstract)

Rob van Glabbeek*
Computer Science Department, Stanford University
Stanford, CA 94305, USA
rvg@cs.stanford.edu

Abstract

De Simone showed that a wide class of languages, including CCS, SCCS, CSP and ACP, are expressible up to strong bisimulation equivalence in Meije. He also showed that every recursively enumerable process graph is representable by a Meije expression. Meije in turn is expressible in aprACP (ACP with action prefixing instead of sequential composition).

Vaandrager established that both results crucially depend on the use of unguarded recursion, and its noncomputable consequences. *Effective* versions of CCS, SCCS, Meije and ACP, not using unguarded recursion, are incapable of expressing all effective De Simone languages. And no effective language can denote all computable process graphs.

In this paper I recreate De Simone's results in aprACP without using unguarded recursion. The price to be payed for this is the use of a partial recursive communication function and—for the second result—a single constant denoting a simple infinitely branching process. Due to the noncomputable communication function, the version of aprACP employed is still not effective.

However, I also define a wide class of De Simone languages that are expressible in an effective version of aprACP. This class includes the effective versions of CCS, SCCS, ACP, Meije and most other languages proposed in the literature, but not CSP. An even wider class, including CSP, turns out to be expressible in an effective version of aprACP to which an effective relational renaming operator has been added.

1 Introduction

In the early 1980's several languages for the description of communicating processes were introduced, most notably CCS [8], SCCS [9], CSP [5] and ACP [3]. Using these languages for purposes of specification and verification showed the necessity, or at least the convenience, of adding many additional constructs tailored to specific applications. In foundational research however, for instance when proving a property by structural induction, it is more convenient to have a fixed and well-defined set of operators. This consideration gives rise to the task of finding a basic language in which all or most useful operators can be expressed. Although the languages CCS and SCCS were designed with this goal in mind, the language Meije, proposed by Austry & Boudol [1], was the first result of a systematic analysis of expressiveness issues. In [13], Robert

*This work was supported by ONR under grant number N00014-92-J-1974.

DE SIMONE observed that all constructs of the languages CCS, SCCS, CSP and ACP, and most constructs that are used in specific applications, can be defined in a particular way—namely by structural operational rules that fit in what is now known as the *De Simone format*—and showed that any operator that can so be defined is expressible in MEIJE, up to strong bisimulation equivalence. I will refer to a language all of whose constructs can so be defined as a *De Simone language.*

The interleaving semantics of CCS-like languages is most conveniently described in terms of *process graphs* or *labelled transition systems.* Depending on how constructive one wants the theory to be, several classes of graphs or transition systems can be considered, as indicated in Figure 1. For any un-

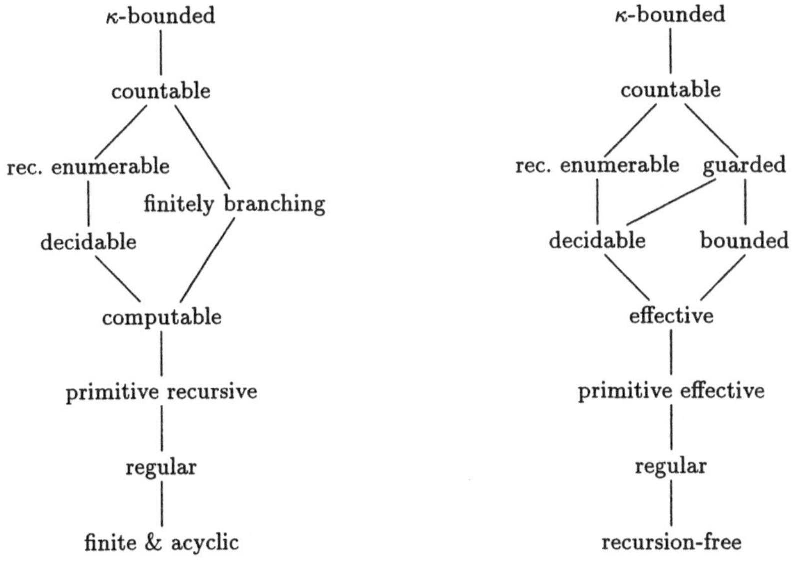

Figure 1: Classes of process graphs ... and process expressions

countable cardinal κ one can define the *κ-bounded* process graphs to be the ones with less than κ (reachable) nodes (states) and edges (transitions), or—clearly equivalent—the ones in which every node has less than κ outgoing edges. For $\kappa = \aleph_0$, however, these definitions would not be equivalent. The *regular* process graphs (having a finite set of states and transitions) form a strictly smaller class than the *finitely branching* ones. Additionally one can consider the *recursively enumerable* graphs, whose nodes form a recursive set (such as the natural numbers), and whose edges form a recursively enumerable set. If, moreover, it is decidable whether a prospective edge is present or not, such a graph is *decidable.* Furthermore, a *computable* process graph is one for which there exists a terminating algorithm that, when given a node, produces as output the finite set of its outgoing edges. This notion was introduced in BAETEN, BERGSTRA & KLOP [2]. Note that it is a stronger requirement than "decidable and finitely branching". Finally, in PONSE [11] a graph is called *primitive recursive* if it is computable by means of a primitive recursive algorithm.

In this paper I propose a similar classification for (expressions in) De Simone languages. For each class of process graphs in Figure 1 I introduce a class of (process) expressions that are guaranteed to denote graphs from that class only. The classes of process expressions are displayed in the second part of Figure 1. The notions of *recursion-free*, *regular*, and *κ-bounded* expressions are just mentioned for the sake of completeness, and are defined for CCS-like languages only. The *countable* process expressions are the ones in which all operators, not counting recursion, have finite arities. Except in the section on κ-bounded expressions, I presume this to be the default. The concepts *bounded* and *effective* are due to VAANDRAGER [14]. These two general but simple restrictions ensure that the effected expressions only denote, respectively, finitely branching and computable process graphs. Applied to CCS-like languages boundedness reduces to guardedness of recursive specifications. The notion of effectivity is also applied to languages as a whole. In that case it presupposes a decidable set of closed terms, and requires the interpretation which maps closed terms to descriptions of computable process graphs to be computable as well.

Whereas for languages like ACP and CCS one can define a κ-bounded version (containing only κ-bounded expressions), a recursively enumerable version, an effective version, etc., the language MEIJE was designed in such a way that all its constructs, except for recursion, are intrinsically effective. Nevertheless, as shown in DE SIMONE [12], many undecidable and infinitely branching processes can be specified by MEIJE-expressions. The expressive power needed to do so can only originate from the use of unguarded recursion. It follows that the restriction to guarded recursion must be a prerequisite in the definitions of decidable and bounded expressions. This is indicated in Figure 1. The language MEIJE with unguarded recursion falls in the class of recursively enumerable languages. In MEIJE only finite recursion is used.

In [13], DE SIMONE shows that MEIJE is universal in two different ways. It is universal among the De Simone languages, and it has a universal power to specify process graphs. Namely, any finitary recursively enumerable De Simone language is expressible in MEIJE, and any recursively enumerable process graph is representable by a MEIJE expression. Here *finitariness* is a syntactic restriction that is met by virtually all De Simone languages encountered in practice. Both expressiveness results hold up to bisimulation equivalence.

VAANDRAGER [14] investigates whether similar expressiveness results can be obtained for effective languages, and comes up with two negative results. First of all he points out that no effective language is capable of denoting all computable process graphs. This had been shown already by BAETEN, BERGSTRA & KLOP [2] up to strong bisimulation equivalence, but Vaandrager sharpens the result to hold for the coarser notion of trace equivalence as well. Secondly he defines a finitary effective De Simone language PC that cannot be translated in any effective version of MEIJE, CCS, SCCS or ACP with only finite guarded recursion. The culprit is a *relational renaming operator* that occurs in this language. By means of this operator a primitive recursive graph can be specified that is not specifiable in the mentioned languages.

The main contributions of the present paper are two effective versions of De Simone's first expressivity result. As unguarded recursion violates effectivity, its use must be eliminated from De Simone's construction. But MEIJE appears to lose most of its power when unguarded recursion is gone. Therefore I will use a variant of ACP [3], called aprACP_F. aprACP_F is a parametrised language, of which an instantiation is obtained by selecting a set A of *actions* and a partial binary *communication function* on A. One particular choice of the communication function yields an extension of MEIJE. Thus all expressiveness results obtained for MEIJE hold for aprACP$_F$ as well. However, thanks to the possibility of other communication functions, aprACP$_F$ is more flexible, and potentially more expressive.

In Section 2 I propose a definition of what it means for one language to be expressible in another. This definition agrees with the notion of a *translation* from BOUDOL [4]. I also introduce (annotated) signatures to determine the syntax of a language and review the method of structural operational semantics for interpreting the closed expressions in a language as (equivalence classes of) process graphs.

In Section 3 I introduce the language aprACP$_F$, and indicate how it relates to ACP, CCS and MEIJE. In short, it is a sublanguage of ACP—thus "apr" (for *action prefix*)—to which renaming operations have been added—thus the subscript F (for *functional renaming*). CCS as well as MEIJE are obtained as sublanguages of aprACP$_F$ under particular instantiations of the parameters. I show that several operators, such as the *left-merge*, the *communication merge* and the *alternative composition*, that play a rôle in ACP, are expressible in terms of the other operators, and hence need not to be introduced as primitives. I also show that one can benefit from all choices of communication functions at the same time. There is namely one canonical instantiation of aprACP$_F$ in which any other instantiation with the same set of actions can be expressed.

In Section 4 I define the classification of aprACP$_F$-expressions of Figure 1. When possible, I make these classes of expressions large enough to denote every graph in the corresponding class. Only in case of the decidable aprACP$_F$-expressions I do not succeed in this. In particular, each computable process graph can be represented by an effective expression. This had been shown earlier by PONSE [11] for the language μCRL. However, it is not possible to combine enough such expressions in one effective language, as follows from the first negative expressiveness result mentioned above. When one tries to do this, the set of closed terms of the language becomes undecidable. For the same reason I couldn't find a canonical candidate for an effective version of aprACP$_F$, in which all other (finitary) effective versions of aprACP$_F$ can be expressed. This leaves not much hope for expressing all finitary effective De Simone languages in aprACP$_F$. Therefore I settle for the *primitive effective* De Simone languages, denoting only primitive recursive graphs. By means of a trivial construction, every such graph can be denoted in a single primitive effective version of aprACP$_F$. As the construction uses infinite—but primitive recursive—guarded recursion, this is still not a primitive effective version of De Simone's second expressiveness result. The same can be said for a more disciplined construction by PONSE [11].

In Section 5 I extend the classification of Figure 1, or actually the part between "countable" and "primitive effective", to arbitrary De Simone languages. Subsequently I state the expressiveness results that I obtained, but in this extended abstract the proofs are not included.

Vaandrager's counterexample-language PC needs only primitive recursion, so not all finitary primitive effective De Simone languages translate in aprACP$_F$ with finite guarded recursion. Therefore I isolate a class that do. These are the *functional* finitary primitive effective De Simone languages. They include the primitive effective versions of CCS, SCCS, ACP and MEIJE. The restriction to primitive recursion is probably not so bad, as I am not aware of any use of effective but not primitive effective De Simone languages. However, the requirement of functionality rules out CSP and Vaandrager's PC. This is solved by adding the relational renaming operator of PC—which to a great extent also occurs as the *(inverse) image operator* in CSP—to aprACP_F, thereby obtaining aprACP_R. All finitary primitive effective De Simone languages are expressible in the primitive recursive version of aprACP_R using only finite guarded recursion. I do not know if these results carry over to MEIJE. They do not carry over to CCS, CSP or PC.

Besides these main contributions I establish similar results for the countable De Simone languages and the bounded ones. Again, I use only aprACP_R with finite guarded recursion. Similarly, I recreate De Simone's first expressiveness result for aprACP_R, without using unguarded recursion. For this, an undecidable communication function is used.

All results announced above are also generalised to non-finitary De Simone languages. This necessitates the use of infinite, but still guarded, recursion.

VAANDRAGER [14] established that any finite effective De Simone language is expressible in a finite version of PC. Here *finite* means specified by means of a finite number of De Simone rules. Such languages only denote graphs with a finite set of actions. It should be noted that any finite De Simone language with guarded recursion is functional, finitary and primitive effective. Finite guarded De Simone languages can also be expressed in aprACP_F and MEIJE.

As far as the power to specify process graphs goes, I recreate De Simone's result while using only finite guarded recursion. For this purpose I use a version of aprACP_F with a partial recursive (but undecidable) communication function, to which a single constant U has been added denoting a simple infinitely branching process. Without adding U no infinitely branching processes can be specified in aprACP_F with guarded recursion. This result did not extend to other classes of process graphs/expressions.

2 Syntax, Semantics and Expressibility

In this section I propose a definition of what it means for one language to be expressible in another. In order for such a notion of expressibility to be meaningful, a language is understood to combine syntax and semantics.

First I will introduce the syntax of a language through the notion of a *annotated signature.* The annotated signature determines what are the valid expressions of the language. I will define a signature as a set of function declarations, thus omitting the possibility of predicates. These do not occur in the languages I want to study. The annotation specifies to what extent recursion is incorporated in the language.

The semantics of a language is given through an interpretation of the (closed) terms in a domain of values. Such an interpretation should be *compositional* and satisfy a few other requirements. My notion of a compositional semantics generalises notions of *denotational semantics*, namely by not insisting that the meaning of recursive expressions is obtained by order-theoretic methods. Thus my concept of expressiveness applies to languages with a denotational semantics as well.

Subsequently I treat the notion of Structural Operational Semantics in PLOTKIN's style [10]. The languages considered in this paper will all be equipped with a structural operational semantics. Such an operational semantics associates in a standard way a *process graph* to every closed process expression.

On process graphs, I divide out *bisimulation equivalence*, which is among the finest congruence relations available. All my expressibility results will be established up to bisimulation equivalence. They will then also hold for most other—less fine—equivalences.

2.1 Syntax

In this paper V is an infinite set of *variables*, ranged over by X, X_i, x, y, x' etc.

Definition 1 (*Signatures*). A *function declaration* is a pair (f, n) of a *function symbol* $f \notin V$ and an *arity* $n \in \mathbb{N}$. A function declaration $(c, 0)$ is also called a *constant declaration.* A *signature* is a set of function declarations. The set $\mathbb{T}^r(\Sigma)$ of *terms with recursion* over a signature Σ is defined inductively by:

- $V \subseteq \mathbb{T}^r(\Sigma)$,
- if $(f, n) \in \Sigma$ and $t_1, \ldots, t_n \in \mathbb{T}^r(\Sigma)$ then $f(t_1, \ldots, t_n) \in \mathbb{T}^r(\Sigma)$,
- If $V_S \subseteq V$, $S : V_S \to \mathbb{T}^r(\Sigma)$ and $X \in V_S$, then $\langle X|S\rangle \in \mathbb{T}^r(\Sigma)$.

A term $c()$ is often abbreviated as c. A function S as appears in the last clause is called a *recursive specification.* A recursive specification S is often displayed as $\{X = S_X \mid X \in V_S\}$. An occurrence of a variable y in a term t is *free* if it does not occur in a subterm of the form $\langle X|S\rangle$ with $y \in V_S$. A term is *closed* if it contains no free occurrences of variables. Let $T^r(\Sigma)$ be the set of closed terms over Σ. The sets $\mathbb{T}(\Sigma)$ and $T(\Sigma)$ of open and closed terms over Σ without recursion are defined likewise, but without the last clause.

The syntax of a language can be given as a signature together with an annotation which places some restrictions on the use of recursion. These can be:

κ-bounded:	the sets S should have cardinality less than κ,
countable:	the sets S should be countable,
enumerable:	the functions S should be partial recursive,
computable:	the sets V_S should be decidable and the functions S recursive,
primitive:	as above, but using only primitive recursion,
finite:	the sets S should be finite,
guarded:	the sets of equations S should satisfy a syntactic criterion that ensures that they have unique solutions under a given interpretation,

or no recursion.

Definition 2 (*Substitutions*). A Σ-*substitution* σ is a partial function from V to $\mathbb{T}^r(\Sigma)$. If σ is a substitution and t a term, then $t[\sigma]$ denotes the term obtained from t by replacing, for x in the domain of σ, every free occurrence of x in t by $\sigma(x)$, while renaming bound variables if necessary to prevent name-clashes. In that case $t[\sigma]$ is called a *substitution instance* of t. A substitution instance $t[\sigma]$ where σ is given by $\sigma(x_i) = s_i$ for $i \in I$ is denoted as $t[s_i/x_i]_{i\in I}$. These notions extend to syntactic objects containing terms, with the understanding that such an object is a substitution instance of another one only if the same substitution has been applied to each of its constituent terms.

2.2 Expressibility

A *language* can be given by an annotated signature, specifying its syntax, and an *interpretation*, assigning to every (closed) term t its meaning $[\![t]\!]$. The meaning of a closed term can simply be a *value* chosen from a set of values $\mathbb{D}$, which is called a *domain*. Usually interpretations are required to satisfy some sanity requirements. One of them is that the meaning of a term $\langle X|S\rangle$ is the X-component of a solution of S. To be precise:

$$[\![\langle X|S\rangle]\!] = [\![S_X[\langle Y|S\rangle/Y]_{Y\in V_S}]\!].$$

Other requirements are *compositionality* and *invariance under α-recursion.* Compositionality demands that the meaning of a term is completely determined by the meaning of its components. This means that for functions $(f,n) \in \Sigma$ and for recursive specifications S and S' with $X \in V_S = V_{S'}$ we have

$$[\![t_i]\!] = [\![t'_i]\!] \ (i = 1,...,n) \quad \Rightarrow \quad [\![f(t_1,...,t_n)]\!] = [\![f(t'_1,...,t'_n)]\!]$$
$$\text{and } [\![S_Y]\!] = [\![S'_Y]\!] \ (Y \in V_S) \quad \Rightarrow \quad [\![\langle X|S\rangle]\!] = [\![\langle X|S'\rangle]\!].$$

Invariance demands that the meaning of a term is independent of the names of its bound variables, i.e. for any injective substitution $\alpha : V_S \to V$

$$[\![\langle\alpha(X)|S[\alpha]\rangle]\!] = [\![\langle X|S\rangle]\!].$$

In order for language L_1 to be expressible in language L_2, I require that for every closed L_1-term t there exists a closed L_2-term $\tilde{t}$ denoting the same value. Usually this can only be the case if the domain $\mathbb{D}_1$ in which L_1-expressions are interpreted is included in the domain $\mathbb{D}_2$ of L_2.

The requirement on closed terms is not sufficient. It says that every value denotable by language L_1 can also be denoted by language L_2. In addition I want that every *operation* of L_1 can be mimicked in L_2. This has to do with the meaning of open terms. For every open term in $\mathbb{T}^r(\Sigma_1)$ I want to find a term in $\mathbb{T}^r(\Sigma_2)$ with the same meaning.

A common approach to open terms is to reduce them to the collection of their closed substitution instances. In this view there is no need to extend the interpretation $[\![\cdot]\!]$ explicitly to open terms. The preconditions $[\![S_Y]\!] = [\![S'_Y]\!]$ of the second compositionality requirement are simply read as "$[\![S_Y[\sigma]]\!] = [\![S'_Y[\sigma]]\!]$ for each closed substitution $\sigma : V \to T^r(\Sigma)$". This approach is often taken when generalising an equivalence relation on closed terms to an equivalence on open terms (over the same signature). Two open terms are then declared equivalent if for every closed substitution the corresponding substitution instances are equivalent.

It is slightly more difficult to employ this approach in defining expressibility. The problem is that open terms over different signatures are compared, so that it is impossible to employ the same substitution at both sides. This is solved as follows:

Definition 3 (*Expressibility*) Let L_i $(i = 1,2)$ be two languages, given as an annotated signature Σ_i and an interpretation $[\![\cdot]\!]_i : T^r(\Sigma_i) \to \mathbb{D}_i$. L_1 is said to be *expressible* in L_2 if there is a translation $\tilde{\cdot} : \mathbb{T}^r(\Sigma_1) \to \mathbb{T}^r(\Sigma_2)$ such that for all $t \in \mathbb{T}^r(\Sigma_1)$ with free variables $x_1,...,x_n$ and for all $t_1,...,t_n \in T^r(\Sigma)$

$$[\![t[t_i/x_i]_{i=1}^n]\!]_1 = [\![\tilde{t}[\tilde{t}_i/x_i]_{i=1}^n]\!]_2.$$

This will be my definition of expressibility for now. In the full version of this paper, however, I plan to be a bit more ambitious. An open term with n free variables is interpreted as an n-ary operator on the domain $\mathbb{D}$, and for L_1 to be expressible in L_2 I require that for every L_1-term t there is an L_2-term $\tilde{t}$, such that t and $\tilde{t}$ denote the same operator, at least when applied to values from the domain of L_1. If L_1 can be expressed into L_2 in that sense, it can certainly be done in the sense of Definition 3.

2.3 Quotient domains

Let $\sim$ be an equivalence relation on a domain $\mathbb{D}$. An interpretation $[\![\cdot]\!]$ in $\mathbb{D}$ is *compositional up to* $\sim$ if it satisfies the requirements for compositionality, but with "=" replaced by "$\sim$". The first requirement for instance reads

$$[\![t_i]\!] \sim [\![t'_i]\!] \ (i = 1, ..., n) \quad \Rightarrow \quad [\![f(t_1, ..., t_n)]\!] \sim [\![f(t'_1, ..., t'_n)]\!].$$

In the same way the other sanity requirements, as well as the notion of expressibility, can be defined *up to* $\sim$. In the case of expressibility, however, it is necessary that $\sim$ is defined on $\mathbb{D}_1 \cup \mathbb{D}_2$.

Given a domain $\mathbb{D}$ for interpreting languages and an equivalence relation $\sim$, the *quotient domain* $\mathbb{D}/_\sim$ consists of the $\sim$-equivalence classes of elements of $\mathbb{D}$. An interpretation $[\![\cdot]\!] : T^R(\Sigma) \rightarrow \mathbb{D}$ of the closed terms of a language in $\mathbb{D}$, is turned into the *quotient interpretation* $[\![\cdot]\!]_\sim : T^R(\Sigma) \rightarrow \mathbb{D}/_\sim$ of these terms by letting $[\![t]\!]_\sim$ be the equivalence class containing $[\![t]\!]$. This quotient interpretation satisfies the sanity requirements of Section 2.2 iff the original interpretation satisfies them up to $\sim$. Likewise, one language is expressible in another under a quotient interpretation obtained by dividing out the same equivalence $\sim$ on their domains of interpretation, iff, under the original interpretation, this language is expressible in the other up to $\sim$.

2.4 Process Graphs

When the expressions in a language are meant to represent processes, they are called *process expressions*, and the language a *process description language*. Suitable domains for interpreting process description languages are the class of *process graphs* and its quotients. In such *graph domains* a process is represented by either a process graph, or an equivalence class of process graphs. Process graphs are also known as *state-transition diagrams* or *automata*. They are *labelled transition systems* equipped with an initial state.

Definition 4 (*Process graphs*) A *process graph*, labelled over a set A of actions, is a triple $G = (S, T, I)$ with

- S a set of *nodes* or *states*,
- $T \subseteq S \times A \times S$ a set of *edges* or *transitions*,
- and $I \in S$ the *root* or *initial* state.

Let $\mathbb{G}(A)$ be the domain of process graphs labelled over A.

Virtually all so-called *interleaving models* for the representation of processes are isomorphic to graph models. The *failure sets* for instance that represent expressions in the process description language CSP [5] can easily be exchanged for equivalence classes of graphs, under a suitable equivalence. In [3] the language ACP is equipped with a process graph semantics, and the semantics of CCS, SCCS and MEIJE given in [8, 9, 1, 13] are operational ones, which, as I will show below, induce a process graph semantics.

Usually the parts of a graph that cannot be reached from the initial state by following a finite path of transitions are considered meaningless for the description of processes. This means that one is only interested in process graphs as a model of system behaviour up to some equivalence, and this equivalence identifies at least graphs with the same reachable parts. Likewise, the particular identity of the states in a process graph is normally not of any importance. Two graphs that only differ in the naming of their states are called *isomorphic* and also isomorphic graphs are semantically identified.

Definition 5 (*Reachability*). Let $G = (S, T, I)$ be a process graph. A *path* is an alternating sequence $s_0, a_1, s_1, a_2, s_2, \ldots, s_{n-1}, a_n, s_n$ of states and actions, such that $s_{i-1} \xrightarrow{a_i} s_i$ for $i = 1, \ldots, n$. Here $s_{i-1} \xrightarrow{a_i} s_i$ is an abbreviation for $(s_{i-1}, a_i, s_i) \in T$. Such a path is said *to go from* s_0 *to* s_n. A state s' is *reachable* from a state s if there is a path from s to s'. The *reachable part* of G is the graph (S^R, T^R, I) with $S^R \subseteq S$ the set of states reachable from the initial state I, and $T^R = T \cap (S^R \times A \times S^R)$.

(*Isomorphism*). Two process graphs $G = (S, T, I)$ and $H = (S', T', I')$ are *isomorphic* if there exists a bijection $f : S \to S'$—called an *isomorphism*—with $f(I) = I'$ and $(s, a, t) \in T \Leftrightarrow (f(s), a, f(t)) \in T'$.

Write $G \cong H$ if the reachable parts of G and H are isomorphic.

Thus $\cong$ may be considered the finest equivalence (the one with the fewest identifications, and the smallest equivalence classes) on $\mathbb{G}$ that makes $\mathbb{G}/_{\cong}$ in a reasonable model of concurrency. However, some languages (interpretations) encountered in this paper fail to be compositional up to $\cong$. Also several expressiveness results will not hold up to $\cong$. Therefore a coarser equivalence (identifying more, and having larger equivalence classes) should be divided out on $\mathbb{G}$. In the literature many equivalences have been proposed. The finest of those is (strong) bisimulation equivalence, due to MILNER [8, 9].

Definition 6 (*Bisimulation equivalence*). Two process graphs G and H are *bisimulation equivalent*—notation $G \leftrightarrows H$—if there exists a binary relation R—called a *bisimulation*—between their states, such that

- the initial states of G and H are related,
- if sRt and $s \xrightarrow{a} s'$ then H has a state t' with $t \xrightarrow{a} t'$ and $s'Rt'$,
- if sRt and $t \xrightarrow{a} t'$ then G has a state s' with $s \xrightarrow{a} s'$ and $s'Rt'$.

All languages mentioned in this paper satisfy the sanity requirements of Section 2.2 up to $\leftrightarrows$. The expressiveness results of this paper will also be established up to bisimulation equivalence. This means that they surely hold for the (coarser) equivalences used elsewhere in the literature, such as *weak bisimulation equivalence*—the standard equivalence of CCS [8]—and (weak) failures equivalence—the standard semantics of CSP [5].

2.5 Operational Semantics

In this section I present PLOTKIN's method of *Structural Operational Semantics* [10] for interpreting expressions as process graphs or labelled transition systems.

Definition 7 (*Transition system specifications;* GROOTE & VAANDRAGER [6]). Let Σ be an annotated signature and A a set (of *actions*). A *(positive)* (Σ, A)-*literal* is an expression $t \xrightarrow{a} t'$ with $t, t' \in \mathbb{T}^r(\Sigma)$ and $a \in A$. An *action rule* over (Σ, A) is an expression of the form $\frac{H}{\alpha}$ with H a set of (Σ, A)-literals (the *premises* of the the rule) and α a (Σ, A)-literal (the *conclusion*). A rule $\frac{H}{\alpha}$ with $H = \emptyset$ is also written α. A *transition system specification (TSS)* is a triple (Σ, A, R) with Σ a signature and R a set of action rules over (Σ, A).

The following definition tells when a transition is provable from a TSS. It generalises the standard definition (see e.g. [6]) by (also) allowing the derivation of rules. The derivation of a transition $t \xrightarrow{a} t'$ corresponds to the derivation of the rule $\frac{H}{t \xrightarrow{a} t'}$ with $H = \emptyset$. The case $H \neq \emptyset$ corresponds to the derivation of $t \xrightarrow{a} t'$ under the assumptions H.

Definition 8 (*Proof*). Let $P = (\Sigma, R)$ be a TSS. A *proof* of an action rule $\frac{H}{\alpha}$ from P is a well-founded, upwardly branching tree of which the nodes are labelled by Σ-literals, such that:

- the root is labelled by α, and
- if β is the label of a node q and K is the set of labels of the nodes directly above q, then
  - either $K = \emptyset$ and $\beta \in H$,
  - or $\frac{K}{\beta}$ is a substitution instance of a rule from R.

If a proof of $\frac{H}{\alpha}$ from P exists, then $\frac{H}{\alpha}$ is *provable* from P, notation $P \vdash \frac{H}{\alpha}$.

Transition system specifications often contain infinitely many rules, yet are presented finitely by giving *rule schemata*, each of which codifies a large set of rules. This practice is formalized in part by the notion of an abstract TSS.

Definition 9 (*Abstract TSSs*). An *abstract Σ-literal* is an expression $t \longrightarrow t'$ with $t, t' \in \mathbb{T}(\Sigma)$. An *abstract action rule* over (Σ, A) is an expression of the form $\frac{H,\ Pr}{\alpha}$ with H a set of abstract (Σ, A)-literals, $Pr \subseteq A^H \times A$, and α an abstract Σ-literal. An *abstract TSS* is a triple (Σ, A, R) with Σ a signature and R a set of abstract action rules over (Σ, A). An abstract TSS (Σ, A, R) *determines* the (concrete) TSS (Σ, A, R') with

$$R' = \left\{ \frac{\{t_i \xrightarrow{a_i} t_i' \mid i \in I\}}{t \xrightarrow{b} t'} \;\middle|\; \frac{\{t_i \longrightarrow t_i' \mid i \in I\},\ Pr}{t \longrightarrow t'} \in R \wedge Pr(\vec{a}, b) \right\}.$$

Finally I will show how the operational semantics of a language, given as a TSS, induces a process graph semantics.

Definition 10 (*Interpreting the closed expressions in a TSS as process graphs*). Let $P = (\Sigma, A, R)$ be a TSS and $t \in T(\Sigma)$. Then $[\![t]\!]$ is defined to be the reachable part of the process graph $(T^r(\Sigma), T, t)$ with T the set of transitions provable from P.

3 Prefix ACP with Relational Renaming

The language that I will use for my expressiveness results is a variant of ACP [3] that could be called *prefix ACP with relational renaming*. Like ACP this language has two parameters: an alphabet A of *actions* and a partial *communication function* $| : A^2 \rightarrow A$, which is commutative and associative, i.e.

- $a|b = b|a$ (commutativity)
- $(a|b)|c = a|(b|c)$ (associativity)

for all $a, b, c \in A$ (and each side of these equations is defined just when the other side is). I will denote this language as $\text{aprACP}_R(A, |)$.

Its signature contains a constant 0 denoting inaction, two binary operators $+$ and $\|$ denoting *alternative* and *parallel composition* respectively, a unary operator a for any action $a \in A$, a unary *encapsulation* operator ∂_H for any $H \subseteq A$ and a *relational renaming* operator ρ_R for any binary relation $R \subseteq A \times A$.

$p\|q$ represents the independent execution of the processes p and q, partly synchronized by the communication function $|$. If $a|b$ is defined, an occurrence of a in p can synchronise with an occurrence of b in q into a communication action $a|b$ between p and q. If $a|b$ is not defined, no such communication is possible. The action a of p can, instead of synchronising with an action of q, (also) appear independent of q, and likewise can b occur independently of p.

The process ap first performs the action $a \in A$ and then behaves like p. $\partial_H(p)$ behaves like p, but without the possibility of performing actions from H. The operator ρ_R is a slight generalisation of the relabelling and (inverse) image operators of CCS and CSP. Process $\rho_R(p)$ behaves just like process p, except that if p has the possibility of doing an a, $\rho_R(p)$ can do any one action b that is related to a via R.

In $\text{aprACP}_R(A, |)$-expressions brackets are omitted under the convention that a binds strongest and $+$ weakest. Besides aprACP with relation renaming I also consider aprACP with functional renaming, denoted $\text{aprACP}_F(A, |)$. This is the same language, but with a renaming operator ρ_f only for every *function* $f : A \rightarrow A$ instead of for every relation.

The action rules for $\text{aprACP}_R(A, |)$ are given in Table 1, thereby completing the formalisation of this language as a TSS. These rules determine an interpretation of the $\text{aprACP}_R(A, |)$-expressions in $\mathbb{G}(A)$. This interpretation agrees, up to bisimulation equivalence, with the more denotational interpretation of $\text{ACP}(A, |)$ in $\mathbb{G}(A)$ given in [Baeten,] Bergstra & Klop [3, 2].

It is common to regard the entries in tables like 1 as schemata, each of which denotes a rule for any proper instantiation of the metavariables a, b, c by real actions from A. Thus in case A is infinite, there are infinitely many rules for every operator. In an attempt at "finitisation" I will in this paper regard each of the first six entries as single rules of an abstract TSS. The rule $\dfrac{x \xrightarrow{a} x'}{x\|y \xrightarrow{a} x'\|y}$ for instance, should be read as $\dfrac{x \longrightarrow x',\ Id}{x\|y \longrightarrow x'\|y}$ where $Id \subseteq A \times A$ is the identity relation on A. In the rules were Pr is not the identity, it is explicitely given. The last three entries in Table 1 remain schemata even when interpreted as abstract action rules. There is namely one rule for every encapsulation operator, one for every relational renaming, and one for every pair $\langle X|S\rangle$ with S a recursive specification and $X \in V_S$. But at least there are now finitely many rules for every operator, even if A is infinite. This will turn out to be a useful property.

$ax \xrightarrow{a} x$	$\dfrac{x \xrightarrow{a} x'}{x+y \xrightarrow{a} x'}$	$\dfrac{y \xrightarrow{a} y'}{x+y \xrightarrow{a} y'}$
$\dfrac{x \xrightarrow{a} x'}{x \parallel y \xrightarrow{a} x' \parallel y}$	$\dfrac{x \xrightarrow{a} x',\ y \xrightarrow{b} y',\ a \mid b = c}{x \parallel y \xrightarrow{c} x' \parallel y'}$	$\dfrac{y \xrightarrow{a} y'}{x \parallel y \xrightarrow{a} x \parallel y'}$
$\dfrac{x \xrightarrow{a} x',\ a \notin H}{\partial_H(x) \xrightarrow{a} \partial_H(x')}$	$\dfrac{S_x[\langle Y \mid S\rangle / Y]_{Y \in V_S} \xrightarrow{a} z}{\langle X \mid S \rangle \xrightarrow{a} z}$	$\dfrac{x \xrightarrow{a} x',\ R(a,b)}{\rho_R(x) \xrightarrow{b} \rho_R(x')}$

Table 1: aprACP_R

3.1 CCS

MILNER's Calulus of Communicating Systems (CCS) [8] can be regarded as (a sublanguage of) an instantiation of aprACP with functional renaming. CCS is parametrised with a set $\mathcal{A}$ of *names*. The set $\bar{\mathcal{A}}$ of *co-names* is given by $\bar{\mathcal{A}} = \{\bar{a} \mid a \in \mathcal{A}\}$, and $\mathcal{L} = \mathcal{A} \cup \bar{\mathcal{A}}$ is the set of *labels*. The function $\bar{\cdot}$ is extended to $\mathcal{L}$ by declaring $\bar{\bar{a}} = a$. Finally $Act = L \cup \{\tau\}$ is the set of *actions*. CCS can now be presented as aprACP$(Act, |)$, where $|$ is the partial function on Act given by $a|\bar{a} = \tau$.

In CCS there is a renaming operator only for every function $f : Act \to Act$ that satisfies $f(\bar{a}) = \overline{f(a)}$ and $f(\tau) = \tau$. This operator (applied on a process p) is written $p[f]$ and called *relabelling*. Also there is an encapsulation operator ∂_H only when $H = A \cup \bar{A}$ with $A \subseteq \mathcal{A}$. This operator is called *restriction* and is written $p\backslash A$. Parallel composition is written $|$ instead of $\|$, but there is no further difference between CCS and $\mathrm{aprACP}_F(Act, |)$.

3.2 ACP

There are many methodological differences between the ACP approach to *process algebra* and the CCS approach. I will not address these here. As a language, ACP can be regarded as a modification of CCS in four directions.

- First of all, ACP makes a distinction between deadlock and successful termination. As a consequence, action prefixing can be replaced by action constants and a general sequential composition.
- ACP adds two auxiliary operators, the *left merge* and the *communication merge*, denoted $\lfloor\!\lfloor$ and $|$, to enable a finite equational axiomatization of the parallel composition.
- Whereas CCS combines communication and abstraction from internal actions in one operator, in ACP these activities are separated. In CCS the result of any communication is the unobservable action τ. In ACP it is an observable action, from which (in the extended language ACP_τ) one can abstract by applying an *abstraction* operator, renaming designated actions into τ.

- CCS adheres to a specific communication format, admitting only handshaking communication, whereas ACP allows a variety of communication paradigmas, including ternary communication, through the choice of the communication function $|$.

In this paper only the last feature of ACP is of importance. I don't distinguish observable and unobservable actions and therefore work with ACP rather than ACP_τ. As I also don't deal with the distinction between deadlock and successful termination, I restrict attention to the sublanguage aprACP of ACP that doesn't make this distinction and consequently supports prefixing only. Whereas in the original papers on ACP the set A of action constants was required to be finite, I allow it to be infinite. I add the subscript R (or F) to indicate the addition of (functional) renaming operators, which were not included in the syntax of ACP. Subsequently I drop the auxiliary operators $\lfloor\!\lfloor$ and $|$ from the language, since these can be expressed in the other operators of aprACP_F, as I will show in Section 3.5. The resulting language extends CCS in only one essential way, namely through the general communication format. This generality greatly enhances the expressiveness of the language.

3.3 Meije

Like CCS, also BOUDOL's language MEIJE [1, 4, 13] can be regarded as (a sublanguage of) an instantiation of aprACP with functional renaming. MEIJE is parametrised with a set $\mathcal{A}$ of *atomic actions* and a set $\mathcal{S}$ of *signals*. The set *Act* of *actions* is a commutative monoid, namely the free commutative product of the free commutative moniod generated by $\mathcal{A}$ and the free commutative group generated by $\mathcal{S}$. This means that the elements of *Act* are a kind of multisets over $\mathcal{A}$ and $\mathcal{S}$ with the stipulation that elements of $\mathcal{S}$ may also have negative multiplicities. These can also be seen as ordinary multisets over $\mathcal{A}$, $\mathcal{S}$ and $\mathcal{S}^{-1} = \{s^{-1} \mid s \in \mathcal{S}\}$ in which s and s^{-1} cancel. A typical element of *Act* is denoted as $a^5b^2s^3t^{-1}$. The product operation . on *Act* is such that $a^5b^2s^3t^{-1}.a^2c^4s^{-3}t^3u^{-3} = a^7b^2c^4t^2u^{-3}$. MEIJE can now be presented as a sublanguage of $\mathrm{aprACP}_F(Act, .)$.

In MEIJE there are two kind of renaming operators. There is a renaming operator ρ_Φ, written $\langle\Phi\rangle(p)$, for any *morphism* $\Phi : Act \rightarrow Act$. Here a morphism is a function satisfying $\Phi(a.b) = \Phi(a).\Phi(b)$ for $a, b \in Act$. In addition there is a renaming operator $s * p$, called *ticking*, for any signal $s \in S$. This operator renames any action a into $s.a$. The only type of encapsulation operators permitted in MEIJE are the restriction operators $p\backslash s$ for any signal $s \in S$. $p\backslash s$ is $\partial_H(p)$ for H the set of all actions containing s, i.e. all "multisets" in which s has a positive or negative multiplicity. MEIJE also has an operator *triggering* which is expressible in the others, and it lacks the operator $+$, because, as explained in Section 3.6, that operator is expressible in the others as well, but there is no further difference between MEIJE and $\mathrm{aprACP}_F(Act, .)$.

3.4 A Decidable Signature for aprACP

The language $\mathrm{aprACP}_R(A, |)$ defined so far has an uncountable signature if A is infinite. There are namely uncountably many encapsulation and renaming operators. Computationally it makes sense to restrict attention to a fragment of aprACP$_R$ with a decidable signature. This can be achieved by requiring the set of actions A to be decidable, and by restricting the permitted encapsulation and renaming operators ∂_H and ρ_R to the ones where $A - H$ and R are recursively enumerable sets. Such sets can be represented by the code of a turing machine that enumerates them, and it is decidable whether an arbitrary piece of text is the code of a turing machine enumerating a recursive enumarable set. This makes the signature decidable. As a consequence the set of recursion-free terms will be decidable too.

I could have chosen H to be enumerable instead of its complement $A - H$. However, the choice above has the advantage that the encapsulation operators can (in an obvious way) be regarded as special relational renamings, so that one has one kind of operator less to be concerned about. A more compelling argument will be presented in Section 4.4.

In order to ensure that the set of recursive terms is decidable as well, I have to require recursive specifications, seen as sets of equations, to be recursively enumerable at least. This makes the set of open terms decidable. However, it remains undecidable whether a term is closed. The set of closed terms is not even enumerable.

Therefore one may wish to insist that in terms of the form $\langle X|S\rangle$ the set of recursion variables V_S is decidable as well. This makes S computable. However, it is undecidable whether a piece of text is the code of a turing machine deciding membership of a set. Thus with computable recursion even the set of open terms becomes undecidable again.

Hence an even more restrictive requirement on the desired kind of recursion is in order. Here I require S to be primitive decidable. This means that there is a primitive recursive function deciding membership of V_S, and in case of a variable $X \in V_S$ returning the term S_X. It is decidable whether a piece of text is the source of a primitive recursive function, thus with this restriction the signature as well as the sets of open and closed terms are decidable.

If moreover the communication function is required to be partial recursive, the resulting variant of aprACP$_R$ will be denoted aprACP$_R^{\text{r.e.}}$. The other languages I mentioned can be adapted in the same way. In aprACP$_F^{\text{r.e.}}$ I have to allow partial recursive renaming functions. In the original version of aprACP$_F$, these where expressible in terms of total renamings and encapsulation.

3.5 Expressing the Left- and Communication Merge

The language ACP has two operators, $\mathbin{\|\!\_}$ and $|$, that I didn't include in the syntax of aprACP$_F$. The reason is that these operators can be expressed in the other operators of the language, and thus need not be introduced as primitives. Here I show how. Table 2 shows the action rules for the two operators. The

$\dfrac{x \xrightarrow{a} x'}{x \mathbin{\|\!\_} y \xrightarrow{a} x' \| y}$	$\dfrac{x \xrightarrow{a} x',\ y \xrightarrow{b} y',\ a\|b = c}{x \mid y \xrightarrow{c} x' \| y'}$

Table 2: The left- and communication merge of ACP

left merge, $\mathbin{\|\!\_}$, behaves exactly like the *merge* or parallel composition, $\|$, except that the first action is required to come from its leftmost argument. The communication merge, $|$, behaves exactly like $\|$, except that its first action is required to be a communication between its two arguments. The operators' most crucial use is in the axiom CM1

$$x\|y = x \mathbin{\|\!\_}\, y + y \mathbin{\|\!\_}\, x + x \mid y$$

that plays an essential rôle in axiomatising ACP with bisimulation semantics.

Note that the symbol $|$ is used for the communication function as well as the communication merge. This overloading is intentional, as the communication merge can be thought of as an extension of the communication function, which is defined on actions only. The vertical bar in the middle of an expression $\langle X|S\rangle$ is pronounced *where*—and sometimes even written that way—and has nothing to do with the communication function and merge. The vertical bar in a set expression like $\{n \in \mathbb{N} \mid n > 5\}$ is also pronounced *where* and constitutes a fourth use of this symbol. Finally $|$ is used to denote parallel composition in CCS. It is generally easy to determine from the context which $|$ is meant.

In order to express $\mathbin{\|\!\_}$ and $|$ in aprACP$_F(A, |)$ I assume that the set of actions A is divided into a set A_0 of actions that may be encountered in applications, and the remainder $H_0 = A - A_0$, which is used as a working space for implementing useful operators, such as $\mathbin{\|\!\_}$ and $|$. On A_0 the communication function is dictated by the applications, but on H_0 I can choose it in any way that suits me. The cardinality of A_0 should be infinite, and equal to the cardinality of A.

For today's implementation I assume that H_0 contains actions **skip**, **first**, **next**, $a_{\texttt{first}}$ and $a_{\texttt{next}}$ (for $a \in A_0$). The communication function given on A_0 is extended to these actions as indicated below (applying the convention that if $a|b$ is not defined it is undefined).

$a \mid \texttt{first} = a_{\texttt{first}}$	$(a \in A_0)$
$a \mid \texttt{next} = a_{\texttt{next}}$	$(a \in A_0)$
$a_{\texttt{first}} \mid b_{\texttt{first}} = a \mid b$	$(a, b \in A_0)$
$a \mid \texttt{skip} = a$	$(a \in A_0)$

Let $H_1 = A_0 \cup \{\texttt{first}, \texttt{next}\}$. I will use a renaming operator f_1 that satisfies $f_1(a_{\texttt{next}}) = a$, $f_1(a_{\texttt{first}}) = a_{\texttt{first}}$ and $f_1(a) = a$ for $a \in A_0$.

Let me first introduce the notation a^∞ to denote a process that perpetually performs the action a. This process is obtained as $a^\infty = \langle X \mid X = aX\rangle$. Now suppose p is a process that can do actions from A_0 only. Then

$$\partial_{H_1}(p\|\texttt{first}(\texttt{next}^\infty))$$

is a process that behaves exactly like p, except that every initial action has a tag (subscript) **first**, and every non-initial action has a tag **next**. Thus $\rho_{f_1}(\partial_{H_1}[p\|\texttt{first}(\texttt{next}^\infty)])$ is a process that behaves exactly like p, except that the initial actions are tagged **first**. It follows that for any two processes p and q with actions from A_0 only

$$p \mid q \leftrightarrows \partial_{H_0}\left(\rho_{f_1}(\partial_{H_1}[p\|\texttt{first}(\texttt{next}^\infty)])\middle\|\rho_{f_1}(\partial_{H_1}[q\|\texttt{first}(\texttt{next}^\infty)])\right).$$

In order to extend this result to processes with actions outside A_0 I use a bijective renaming $f_0 : A \to A_0$ and its inverse f_0^{-1}. The communication merge is expressed in $\mathrm{aprACP}_F(A, |)$, up to bisimulation equivalence, by

$$x \mid y \leftrightarrows \rho_{f_0^{-1}}\left(\partial_{H_0}\left(\rho_{f_1}(\partial_{H_1}[\rho_{f_0}(x)\|\mathtt{first}(\mathtt{next}^\infty)])\middle\|\rho_{f_1}(\partial_{H_1}[\rho_{f_0}(y)\|\cdots])\right)\right).$$

Finally the left merge is expressed in terms of the communication merge through

$$x \,\lfloor\!\lfloor\, y \leftrightarrows \rho_{f_0^{-1}}(\mathtt{skip}(\rho_{f_0}(y)) \mid \rho_{f_0}(x)).$$

3.6 Expressing Choice

As remarked in Section 3.3, the language MEIJE lacks the choice operator + of CCS and ACP. The reason this operator was omitted from the syntax of MEIJE was that it can be expressed in terms of the other operators. This is true in the setting of aprACP_F as well, so if one likes, this operator can be skipped from the signature.

For the implementation of choice, A is again divided in A_0 and H_0 and in H_0 we put the same actions as in the previous section, together with the action choose. The communication function also works as before, except that there is no communication possible between actions of the form $a_{\mathtt{first}}$ and $b_{\mathtt{first}}$. (So maybe one wants to use a different action first.) Instead one has the communication choose $\mid a_{\mathtt{first}} = a$ for $a \in A_0$. Recall that for p a process that can do actions from A_0 only, $\rho_{f_1}(\partial_{H_1}[p\|\mathtt{first}(\mathtt{next}^\infty)])$ is the same process in which the initial actions are tagged with a subscript first. This, by the way, is an implementation of the operator *triggering* of MEIJE. It follows that

$$p+q \leftrightarrows \partial_{H_0}\left(\rho_{f_1}(\partial_{H_1}[p\|\mathtt{first}(\mathtt{next}^\infty)])\middle\|\mathtt{choose}\middle\|\rho_{f_1}(\partial_{H_1}[q\|\mathtt{first}(\mathtt{next}^\infty)])\right)$$

since in the expression on the right choose can communicate with an initial action from only one of p or q, so that the other one is blocked forever. Using the renamings ρ_{f_0} and its inverse, just as in the previous section, + is expressed in the rest of aprACP_F.

3.7 Expressing the Communication Function

It may be felt as a drawback that the language (apr)ACP, unlike CCS, is parametrised by the choice of a communication function (besides the choice of a set of actions). This, one could argue, makes it into a collection of languages rather then a single one. Personally I do not share this concern. If in different applications different communication functions are used, they can, when desired, all be regarded as different fragments of the same communication function, each considered on only a small subset of the set of actions A. At any given time there is no need to know all actions and the entire communication function to be used in all further applications.

Alternatively one may argue that there are many parallel compositions possible in ACP, namely one for every choice of a communication function. Here I will present one instantiation of $\mathrm{aprACP}_F(A, |)$, such that for every other choice of a communication function the resulting parallel composition is expressible in this language.

Let, as in the previous section, $A_0 \subseteq A$ be the actions that are used in applications and $H_0 = A - A_0$ the working space. I fix a bijection $f_0 : A \rightarrow A_0$ and abbreviate ρ_{f_0} by ρ_0. For this implementation, H_0 should contain the actions (a, b) for every $a, b \in A_0$, as well as an action δ denoting deadlock. The communication function $|$ is defined by $a|b = (a, b)$ for $a, b \in A_0$ (thus undefined outside A_0). Now for any other communication function $\gamma : A^2 \rightarrow A$ let $\bar{\gamma} : A \rightarrow A$ be a renaming satisfying $\bar{\gamma}((\rho_0(a), \rho_0(b))) = c$ for those $a, b, c \in A$ with $\gamma(a, b) = c$, and $\bar{\gamma}(a) = f_0^{-1}(a)$ for $a \in A_0$. Furthermore, let H be $\{(a, b) \in A_0 \times A_0 \mid \gamma(a, b) \text{ undefined}\}$. Then the associated parallel composition $\|_\gamma$ is expressible in $\text{aprACP}_F(A, |)$ by

$$x\|_\gamma y \leftrightarrows \rho_{\bar{\gamma}}(\partial_H(\rho_0(x)\|\rho_0(y))).$$

3.8 Expressing Renaming and Encapsulation

The syntax of $\text{aprACP}_F^{\text{r.e.}}$ allows for a multitude of renaming operators. Here I show that one needs only two, namely the operator ρ_0, introduced earlier, which bijectively maps every action to one in the subset A_0 of A, and the *universal renaming operator* ρ_F. Every other functional renaming is then expressible.

To this end I introduce an action $\bar{f} \in H_0 = A - A_0$ for every partial recursive renaming function $f : A \rightarrow A$. I also introduce an action $(\bar{f}, a) \in H_0$ for every such f and every $a \in A_0$. The communication function is enriched by $\bar{f} \mid a = (\bar{f}, a)$ for $a \in A_0$ and the universal renaming F should satisfy $F((\bar{f}, \rho_0(a))) = f(a)$. Let $H_1 = A_0 \cup \{\bar{f} \mid f : A \rightarrow A\}$. Then $\partial_{H_1}(\bar{f}^\infty\|\rho_0(x))$ is a process that behaves exactly like x, except that every action a is renamed in $(\bar{f}, \rho_0(a))$. Hence ρ_f is expressible through $\rho_f(x) \leftrightarrows \rho_F(\partial_{H_1}(\bar{f}^\infty\|\rho_0(x)))$.

Note that every co-enumerable encapsulation operator can be regarded as a partial recursive renaming, so also all encapsulation operators can be expressed in aprACP with ρ_0 and ρ_F. The encapsulation ∂_{H_1} used in the construction can be incorporated in ρ_F as well.

In exactly the same way every enumerable relational renaming operator is expressible in aprACP with only ρ_0 and a *universal relational renaming.*

In the preceding section I showed how all operators $\|_\gamma$ with γ a communication function could be expressed in a particular instantiation of aprACP_F, using a multitude of renamings. Here I showed how all renaming operators of $\text{aprACP}_F^{\text{r.e.}}$ can be expressed in only two of them, using a particular communication function, and similarly for the encapsulations. It is an easy exercise to combine these results, and express all partial recursive renaming operators, all co-enumerable encapsulations, and all parallel compositions $\|_\gamma$ with γ a partial recursive communication function in a particular instantiation of $\text{aprACP}_F^{\text{r.e.}}$, with only one communication function and two renamings.

One may wonder whether *all* renaming operators can be expressed in aprACP, i.e. if it is possible to get rid of the last two. In general this is not possible. However, if one only cares about the behaviour of all the derived operators on the relevant subset A_0 of A, it is possible to omit the use of ρ_0 from all the constructions, encode the universal renaming in the communication function, and find for any derived operator (such as $\|\!\_$ or a renaming) an aprACP-expression with the same behaviour on A_0.

3.9 A Finite Signature for aprACP

Here I show how the syntax of aprACP$_R^{\text{r.e.}}$ can be reduced from a decidable one to a finite one. In the previous section the set of renaming and encapsulation operators was cut down to two elements, so we are left with an infinity of actions to get rid of. As in aprACP$_R^{\text{r.e.}}$ the set of actions is decidable, they can be numbered $a_0, a_1, a_2, \ldots$, such that the function $\texttt{succ} : A \to A$ given by $\texttt{succ}(a_i) = a_{i+1}$ is partial recursive (even computable). Hence every action is expressible in terms of a_0 and the renaming operator $\rho_{\texttt{succ}}$.

3.10 Expressing Relational Renaming

In order to express relational renamings in aprACP$_F$ one needs to add just one constant (or a third renaming) to the signature. One has to assume the existence of actions $[a,b]$ in H_0 for $a,b \in A_0$. The desired constant is $\texttt{all} = \Sigma_{a,b\in A_0}[a,b]0$. Here $\Sigma_{i\in I}$ is an infinite version of choice, to be formally introduced in Section 4.1. $\Sigma_{i\in I}$ is not a standard ingredient in the syntax of aprACP—if it were there would be no reason to add all as a constant. all can be expressed as $\rho_R(c0)$ where c is an action chosen from A and R is the relation $\{(c,[a,b]) \mid a,b \in A_0\}$.

Now $\texttt{allever} = \langle X \mid X = \texttt{all} \,\|\!\_\, X\rangle$ is a process that perpetually performs an action of the form $[a,b]$, and at each step has the choice between all such actions. For any relation $R \subseteq A \times A$ the process $\partial_{A\times A-R}(\texttt{allever})$ has at each step the choice between executing one the actions $[a,b]$ with $R(a,b)$. Let $\texttt{copy} : A_0 \to A$ be a renaming that sends each action $a \in A_0$ to the (new) action $a_{\texttt{copy}}$. Define the communication function on the new actions by $a_{\texttt{copy}} \mid [a,b] = b$. Then for p a process that does actions from A_0 only

$$\rho_R(p) \leftrightarrows \partial_{H_0}(\rho_{\texttt{copy}}(p) \| \partial_{A\times A-R}(\texttt{allever}))$$

Thus, by means of ρ_0 and its inverse to deal with action from outside A_0, all relational renaming operators are expressible in aprACP$_F$ with all.

all can also be expressed in aprACP$_F$ using so-called *unguarded recursion* (Section 4.3) as $\texttt{all} = \langle X \mid X = [a_0,a_0]0 + \texttt{succ2}(X)\rangle$ where succ2 is a renaming function enumerating the elements of $A \times A$. Thus, as long as unguarded recursion is permitted, aprACP$_F$ is equally expressive as aprACP$_R$. When unguarded recursion is banned, however, aprACP$_R$ turns out to be more expressive.

4 Specifying Process Graphs

In this section I will isolate, for each of the classes of process graphs mentioned in the introduction, a corresponding class of process expressions that denote only graphs from that class. When possible, I make these classes of expressions so large that every graph of the appropriate kind can be denoted by an expression in the corresponding class.

Definition 11 (*Kinds of graphs*). A process graph $G = (S,T,I) \in \mathbb{G}(A)$ is

- *κ-bounded* (for an uncountable cardinal κ) if for every state $s \in S$ there are less than κ outgoing transitions $s \xrightarrow{a} s'$,

- *countable* if for every state $s \in S$ there are at most countably many outgoing transitions $s \xrightarrow{a} s'$,
- *finitely branching* if for every state $s \in S$ there are only finitely many outgoing transitions $s \xrightarrow{a} s'$,
- *recursively enumerable* if there exists an algorithm enumerating all transitions $s \xrightarrow{a} s'$,
- *decidable* if there exists an algorithm that, when given a triple $(s, a, s') \in S \times A \times S$, determines whether this is a transition from T,
- *computable* if there exists an algorithm that, when given a state $s \in S$, returns the complete finite list of outgoing transitions $s \xrightarrow{a} s'$ and indicates when the list is complete,
- *primitive recursive* if there is such an algorithm that is primitive recursive,
- and *regular* if it has only finitely many states and transitions.

The class of all expressions defined so far will be the class of countable process expressions. The κ-bounded process expressions are obtained by enlarging the signature, whereas the other classes of Figure 1 are obtained by means of restriction.

4.1 The κ-bounded Process Expressions

In order to define the κ-bounded process expressions I have to generalise the syntax of aprACP$_R$, even beyond the boundaries imposed by Definition 1. Whenever I is an index set and p_i are process expressions for $i \in I$, $\Sigma_{i \in I} p_i$ is now a process expression too. It represents a choice between the processes p_i ($i \in I$). Since choice is associative and commutative for virtually every semantics proposed in the literature, I may be chosen to range over sets rather than sequences. The corresponding action rules (one abstract rule for every index set I and index $j \in I$) are

$$\frac{x_j \xrightarrow{a} y}{\Sigma_{i \in I} x_i \xrightarrow{a} y}.$$

For this purpose the set of variables should at least have cardinality κ. The expression $p_1 + p_2$ can now be regarded as an abbreviation for $\Sigma_{i=1,2} p_i$ and 0 as the summation over an empty index set. In the same fashion it is possible to introduce infinitary parallel compositions.

Now the *κ-bounded process expressions* can be defined as the ones in which all index sets, as well as the sets V_S in recursive specifications S, have cardinality less than κ. Also, for every relational renaming ρ_R and $a \in A$, the set $\{b \mid R(a, b)\}$ should have less than κ elements. It is straightforward to prove that process graphs associated to κ-bounded process expressions are κ-bounded. The converse is true as well:

Proposition 1 Every κ-bounded process graph can up to isomorphism be denoted by a κ-bounded process expression.

Proof: Let $G = (S, T, I)$ be a κ-bounded process graph. Take a variable X_s for every state $s \in S$ and let $\widetilde{G}$ be the recursive specification $\{X_s = \Sigma_{(s \xrightarrow{a} t) \in T} aX_t \mid s \in S\}$. Now $\langle X_I | \widetilde{G} \rangle$ is a closed process expression with $[\![\langle X_I | \widetilde{G} \rangle]\!] \cong G$. □

4.2 The Countable Process Expressions

In the special case of the countable process expressions ($\kappa = \aleph_1$) one allows countable alternative and parallel compositions and countable recursion. However, countable compositions can be expressed in terms of binary compositions and countable unguarded recursion. Namely

$$\Sigma_{i \in \mathbb{N}} p_i = \langle X_0 \mid \{X_i = p_i + X_{i+1} \mid i \in \mathbb{N}\}\rangle.$$

Thus the countable process expressions can be redefined to be the ones with countable recursion but only binary alternative and parallel composition.

Continuing from that perspective it can be observed that the addition of arbitrary infinite recursion does not add to the expressive power of the language, since in any expression $\langle X|S\rangle$ only countably many variables are reachable from X. Thus the countable process expressions can again be redefined to be exactly the ones introduced in Section 3. It follows that

Proposition 2 Countable process expressions yield countable process graphs and every countable process graph is denoted by a countable process expression.

4.3 The Bounded Process Expressions

The notion of guardedness was proposed in MILNER [8] to syntactically isolate a class of recursive specifications that have unique solutions. The definition below stems from BAETEN, BERGSTRA & KLOP [2].

Definition 12 (*Guardedness*). A free occurrence of a variable in a process expression t is *unguarded* if it does not occur in a subterm of the form at'. Let S be a recursive specification. The relation $\stackrel{u}{\longrightarrow} \subseteq V_S \times V_S$ is given by $X \stackrel{u}{\longrightarrow} Y$ iff Y occurs unguarded in S_X. S is *guarded* if the relation $\stackrel{u}{\longrightarrow}$ is well-founded.

Besides ensuring unique solutions, the same requirement also helps to keep the denoted process graphs finitely branching. Let a process expression be *guarded* if in all its subexpressions $\langle X|S\rangle$ the recursive specification S is guarded.

Proposition 3 Guarded expressions in the languages (apr)ACP$_F$, CCS, SCCS and MEIJE denote finitely branching process graphs. Moreover, every finitely branching process graph is denoted by a guarded process expression.

Proof: The first statement follows with a straightforward induction on the structure of terms, with in the case of recursion a nested induction on the length of chains $X_1 \stackrel{u}{\longrightarrow} X_2 \stackrel{u}{\longrightarrow} \cdots$.

The second statement follows immediately from the proof of Proposition 1, considering that unguarded recursion wasn't used there. In Proposition 2 unguarded recursion has to be used to replace infinite alternative compositions, but in order to denote finitely branching graphs this is unnecessary. □

Due to the relational renaming (or inverse image) operator this proposition does not hold for aprACP$_R$ and CSP. In case the set $\{b \mid R(a,b)\}$ is infinite, $\rho_R(a)$ denotes an infinitely branching process graph.

Let a relation $R \subseteq A \times A$ be *image-finite* if $\forall a \in A : \{b \mid R(a,b)\}$ is finite. Then the *bounded* process expressions can be defined as the ones with only guarded recursion and image-finite renaming operators. It follows that the bounded process expressions denote only finitely branching graphs.

4.4 The Recursive Enumerable Process Expressions

The *recursively enumerable* process expressions are the ones that appear in the language $\mathrm{aprACP}_R^{\mathrm{r.e.}}(A,|)$. This is the variant of aprACP_R with a decidable signature, introduced in Section 3.4. A has to be decidable and $|$ a partial recursive function. Moreover, only encapsulation operators ∂_H with H co-r.e. are allowed (i.e. the complement of H should be enumerable), and only renaming operators ρ_R with R enumerable. Finally recursive specifications S are required to be primitive recursive.

Proposition 4 Any process $\Sigma_{i\in I}p_i$ with $\{p_i \mid i \in I\}$ a recursive enumerable set of recursive enumerable process expressions is expressible in $\mathrm{aprACP}_R^{\mathrm{r.e.}}$.

Proof: A classic recursion theoretic theorem states that any non-empty r.e. set can be obtained as the image of a primitive recursive function. See e.g. Corollary 4.18 in MANIN [7]. It follows that $[\![\Sigma_{i\in I}p_i]\!] \leftrightarrow [\![\Sigma_{i\in\mathbb{N}}p(n)]\!]$ for certain primitive recursive function $p : \mathbb{N} \rightarrow$ (the closed $\mathrm{aprACP}_R^{\mathrm{r.e.}}$-expressions). Now use the construction from Section 4.2. □

Proposition 5 The r.e. process expressions denote exactly the r.e. graphs.

Proof: It is straightforward to enumerate (recursively) the valid proofs of transitions between $\mathrm{aprACP}_R^{\mathrm{r.e.}}$ expressions, and hence the transitions themselves. Note that for this to be true one needs the complement of H in ∂_H to be enumerable rather than H itself. This is the argument promised in Section 3.4.

The other direction follows immediately from the proof of Proposition 1, using Proposition 4. □

DE SIMONE [13] proved that in order to denote every r.e. process graph it is sufficient to use finite recursion only. However, whereas the Propositions 1–5 are rather trivial and only use inaction, action prefix, choice and recursion, De Simone's construction is more intricate and uses the entire syntax of MEIJE.

Theorem 1 Let A be an decidable set of actions. There exists an decidable set of signals S, such that for every r.e. process graph $G \in \mathbb{G}(A)$ there exists a closed r.e. expression t with finite recursion in MEIJE(A,S) for which $[\![t]\!] \leftrightarrow G$.

As MEIJE can be implemented in aprACP_F, this theorem implies that for every decidable set A there is a decidable set $A' \supseteq A$ and a r.e. communication function $|$ on A', such that every r.e. process graph can, up to bisimulation, be denoted by an expression in $\mathrm{aprACP}_F^{\mathrm{r.e.}}(A',|)$ with only finite recursion.

It follows immediately from Proposition 3 that the use of unguarded recursion is unavoidable in De Simone's result. Still, it is possible to limit such use to a minimum. Suppose A contains actions a_i and b_i for $i \in \mathbb{N}$. The process $\Sigma_{i\in\mathbb{N}}a_ib_i0$ can be obtained with unguarded recursion as $\langle X | X = a_0b_00 + \rho_f(X)\rangle$ in which f is the renaming with $f(a_i) = a_{i+1}$ and $f(b_i) = b_{i+1}$ for $i \in \mathbb{N}$. Adding this process as a constant U to the language aprACP_F—thereby obtaining aprACP_U—makes it possible to recreate De Simone's result without using unguarded recursion.

Theorem 2 Let A be a decidable set of actions. There exists an decidable set of actions A' and a partial recursive communication function $|$ on A', such that for every r.e. process graph $G \in \mathbb{G}(A)$ there exists a closed expression t with finite guarded recursion in the language $\mathrm{aprACP}_U^{\mathrm{r.e.}}(A',|)$ for which $[\![t]\!] \leftrightarrow G$.

Proof: The proof is a variation on the one of De Simone. Consider the r.e. process graphs over A with as nodes the natural numbers. Since I consider graphs up to bisimulation equivalence, I may assume that there is at most one edge between every two nodes. Such a graph G can be represented by an algorithm g that, when given a pair of nodes (i, j), runs for some time and returns a in case there is a transition (i, a, j). In case there is no transition from i to j it returns the value δ or runs forever.

Now let A' be the set of all such algorithms g, together with A, a special symbol δ denoting (dead)lock, and the actions $\mathtt{to}_i$ and $\mathtt{from}_i$ for $i \in \mathbb{N}$. Note that the set of algorithms (in a particular form, such as Turing machine code) of partial recursive functions is decidable, i.e. it is decidable whether an arbitrary piece of text constitutes such an algorithm, and hence an element of A'. Let the communication function $|$ be given by $g \mid \mathtt{from}_i \mid \mathtt{to}_j = g(i, j)$. [To be precise, in order to implement this in the ACP communication format I also need actions of the form $\mathtt{fromto}_{ij}$, $g(i, \cdot)$ and $g(\cdot, j)$].

The next thing I need is the *left merge* operator $⌊\!⌊$ from ACP, which, as we saw in Section 3.5, can be expressed in aprACP$_F$. By means of a renaming, the new constant U can be turned into $\mathtt{flow} = \Sigma_{i\in\mathbb{N}}\mathtt{to}_i\mathtt{from}_i 0$. This process describes the flow of control through an arbitrary state. It says that when one enters state i, the next thing to do is leaving the same state. Using only guarded recursion I subsequently define the process $\mathtt{control} = \mathtt{flow} ⌊\!⌊ \mathtt{control}$. This process will be put in a context where in each step a $\mathtt{from}$ action and a $\mathtt{to}$ action synchronise. As a result, when the nth synchronisation involves a $\mathtt{to}_i$ action, denoting the arrival in state i, the next synchronisation involves a $\mathtt{from}_i$, denoting departure from the same state. The choice of the $\mathtt{to}$ action is not restricted (by the $\mathtt{control}$ process). In order to initialise the process properly, and to allow a first synchronisation, I use the initialised control $C = \mathtt{from}_0 \| \mathtt{control}$, in which 0 is supposed to be the initial state.

Now a graph G is represented by the expression $\partial_H(g^\infty \| C)$. Here g^∞ is a shorthand for $\langle X \mid X = gX\rangle$, i.e. the process that repeatedly performs the action g, and $H = A' - A$. It is easy to see that $[\![\partial_H(g^\infty \| C)]\!] \cong G$. □

Corollary 1 Let A be a countably infinite decidable set of actions. There exists a partial recursive communication function $|$ on A, such that for every r.e. process graph $G \in \mathbb{G}(A)$ there exists a closed expression t with finite guarded recursion in the language aprACP$^{\text{r.e.}}_U(A, |)$ for which $[\![t]\!] \leftrightarrows G$.

Proof: Partition A into two infinite decidable subsets A_0 and H_0. It suffices to prove the statement for $G \in \mathbb{G}(A_0)$, since an arbitrary r.e. process graph can be obtained as $\rho_f(G)$ for such a G with $f: A_0 \rightarrow A$ a bijective renaming.

By Theorem 2 there is an extension A' of A_0 and a partial recursive communication function $|'$ on A', such that for every r.e. process graph $G \in \mathbb{G}(A_0)$ there exists a closed expression t with finite guarded recursion in the language aprACP$^{\text{r.e.}}_U(A', |')$ for which $[\![t]\!] \leftrightarrows G$. Let $h: A' \rightarrow A$ be a recursive bijection with $h(a) = a$ for $a \in A_0$. Define the partial recursive communication function $| : A^2 \rightarrow A$ by $h(a) \mid h(b) = h(c)$. Let $\tilde{t}$ be the closed aprACP$^{\text{r.e.}}_U(A, |)$-expression obtained from t by replacing all action names a by $h(a)$, including the action names in the subscripts of encapsulation and renaming operators. Then $[\![t]\!] \leftrightarrows G$ immediately implies $[\![\tilde{t}]\!] \leftrightarrows G$. □

4.5 The Decidable and the Effective Expressions

De Simone [12] shows that with unguarded recursion it is easy to specify undecidable processes. Therefore the first requirement of a *decidable* process expression is that only guarded recursion is permitted. In this setting there are already three different variants of aprACP to consider: the language aprACP_F with functional renaming, the language aprACP_R with relational renaming, and the language aprACP_U with the constant U, introduced in the previous section. The subscript U reminds of *universal* and is inspired by Theorem 2. Note that the guarded version of $\text{aprACP}_U^{\text{r.e.}}$ is at least as expressive as the guarded version of $\text{aprACP}_R^{\text{r.e.}}$. Namely, as A is decidable, the constant `all` of Section 3.10 can be obtained from U by means of encapsulation and renaming, and $\text{aprACP}_R^{\text{r.e.}}$ turned out to be guardedly expressible in terms of $\text{aprACP}_F^{\text{r.e.}}$ with `all`. As long as unguarded recursion was allowed, the three languages were equally expressive, but this is here no longer the case.

A second requirement for decidable process expressions has to do with the computable nature of the operators of the language. For the relational renaming operators ρ_R it is not sufficient to require the relations R to be decidable, as any recursively enumerable relation can be obtained as the composition of two decidable ones. In particular, the process $\Sigma_{a\in A}a0$ is surely decidable, and can be obtained as the image of a single action under a suitable decidable relational renaming. However, for any nonempty recursive enumerable set of actions $B \subseteq A$, the (generally undecidable) process $\Sigma_{b\in B}b0$ can be obtained as $\rho_f(\Sigma_{a\in A}a0)$ with f a primitive recursive function (recalling that a primitive recursive function is a special total recursive function, and any total recursive function, seen as a relation, is decidable). This is Corollary 4.18 in Manin [7].

A similar problem arises for the communication function. There are two ways in which to strengthen the decidability requirement so as to avoid these problems. The renaming operators as well as the communication function should be either *effective* or *coeffective*. The requirement of effectivity comes, in the more general setting of De Simone languages, from Vaandrager [14].

Definition 13 (*Decidable terms*) An $\text{aprACP}_R(A, |)$-expression is *effective* if
- A is a decidable set,
- $|$ is given as a total recursive function $| : A^2 \rightarrow A \dot{\cup} \{\delta\}$ —$a|b = \delta$ means that a and b do not communicate,
- it contains only computable guarded recursion,
- it contains only encapsulation operators ∂_H for which H is decidable
- and only renamings ρ_R for R such that $\forall a \in A : (\{b \mid R(a,b)\}$ is finite), and the total function which yields for any $a \in A$ this finite set is recursive.

A $\text{aprACP}_U(A, |)$-expression is *coeffective* if
- A is a decidable set,
- $|$ satisfies $\forall c \in A : (\{(a,b) \mid a|b = c\}$ is finite), and the total function which yields for any $c \in A$ this finite set is recursive,
- it contains only computable guarded recursion,
- it contains only encapsulation operators ∂_H for which H is decidable
- and only renamings ρ_R for R such that $\forall b \in A : (\{a \mid R(a,b)\}$ is finite), and the total function which yields for any $b \in A$ this finite set is recursive.

A $\text{aprACP}_U(A, |)$ expression is *decidable* if it is either effective and without the constant U or coeffective.

Note that, due to the use of countable recursion, the (co)effective expressions are not a subclass of the enumerable ones. Using only primitive recursive recursion would be a more serious restriction than in the the previous section, as Proposition 4 crucially depends on the use of unguarded recursion.

Mixing effective and coeffective ingredients in one process expression leads in general to undecidable processes. As indicated above, adding U to the effective processes is already catastrophic.

Proposition 6 Effective process expressions denote only computable graphs. Decidable expressions denote only decidable graphs.

Proof: Two straightforward structural inductions. □

I have no idea how to represent *all* decidable graphs by decidable process expressions. The proof strategy adopted for Theorem 2 makes use of a highly non-coeffective communication function as well as the non-effective constant U. Similarly I don't know if it is possible to represent all computable graphs by effective process expressions with finite recursion. However, the same recipe as used in the previous sections yields

Proposition 7 Every computable process graph is denoted by an effective process expression, using only choice, (in)action and recursion.

Proof: If $G \in \mathbb{G}(A)$ is computable it follows immediately that the recursive specification $\widetilde{G}$ constructed in the proof of Proposition 1 is computable. □

In PONSE [11], every computable graph is denoted by an effective expression in the language μCRL. The trivial proof above could be seen as a considerable simplification of his construction. However, Ponse uses finite recursion schemata, parametrised with recursive data parameters, whereas I use plain infinite recursion.

It is interesting to compare this positive result with the following negative one.

Definition 14 (*Effectivity*) An interpretation of a decidable language in a graph domain is *effective* if it induces a total recursive function from the closed terms to (descriptions of) computable process graphs.

The concept of an effective interpretation is due to VAANDRAGER [14]. Let a language be *decidable* if its set of closed terms is decidable (as in $\text{aprACP}_R^{\text{r.e.}}$). The following theorem stems from BAETEN, BERGSTRA & KLOP [2]. Is has been sharpened in VAANDRAGER [14], who established it for the even coarser notion of trace equivalence.

Theorem 3 Let A be a set of at least two actions. No decidable language with an effective interpretation in $\mathbb{G}(A)$ is able to denote all computable process graphs, up to bisimulation equivalence.

If follows that the set of effective process expressions is not decidable. This is due to the presence of computable recursion, which was observed to make the language undecidable already in Section 3.4. Similarly, the set of expressions in Ponse's language is undecidable. It also follows that the decidability requirement in the above theorem is essential.

4.6 The Primitive Effective Process Expressions

In Section 3.4 I proposed a variant $\text{aprACP}_R^{\text{r.e.}}$ of aprACP_R with a decidable signature. In Section 4.4 this language turned out to denote only enumerable process graphs. Furthermore in Section 3.7 there turned out to be a *canonical* instantiation of this language, such that any other instantiation (with a different communication function) is expressible in it. Here I search for a similar variant of aprACP in which only decidable graphs can be denoted.

The first idea could be to take the language of all effective process expressions in $\text{aprACP}_U^{\text{r.e.}}$ (or the coeffective ones or both). However, such a language has an undecidable signature. To be precise, as it is undecidable whether a piece of text is the code of a Turing machine representing a total recursive function, the set of renaming (and encapsulation) operators is undecidable.

Thus the collection of permitted encapsulation and renaming operators has to be cut down until their codes form a decidable set. It is tempting to think that Section 3.8 offers a solution, as it allows all these operators to be expressed in only two of them. However, the construction enabling this requires an action to be introduced in A for any renaming operator. This only shifts the undecidable signature problem from the renaming operators to the set of actions.

Another idea is allow any decidable selection of (co)effective renaming operators to constitute a valid instantiation of the desired language. This is basically what is done in VAANDRAGER [14] for the recursive specifications. There only one computable recursive specification is allowed. However, this violates the desired property of canonicity, as one cannot express all (co)effective renamings (or computable specifications) in one decidable selection.

Hence a (complexity) class of such operators has to be found for which it is decidable whether (a description of) an operator is in this class. As in Section 3.4 I take the class of *primitive recursive* operators. It should be admitted that many other (complexity) classes would serve the purpose equally well.

In order to maintain the canonicity result of Section 3.7, I also have to require primitive recursion for the communication function, which in turn requires A to be primitive decidable.

Definition 15 (*Primitive*) An $\text{aprACP}_R(A, |)$-term is *primitive effective* if
- A is a primitive decidable set,
- $|$ is given as a primitive recursive function $| : A^2 \to A \dot{\cup} \{\delta\}$,
- it contains only primitive recursive guarded recursion,
- only encapsulation operators ∂_H for which H is primitive decidable
- and only renamings ρ_R for R such that $\forall a \in A : (\{b \mid R(a,b)\}$ is finite), and the function which yields for any $a \in A$ this finite set is primitive recursive.

A $\text{aprACP}_U(A, |)$-expression is *primitive coeffective* if
- A is a primitive decidable set,
- $|$ satisfies $\forall c \in A : (\{(a,b) \mid a|b = c\}$ is finite), and the total function which yields for any $c \in A$ this finite set is primitive recursive,
- it contains only primitive recursive guarded recursion,
- only encapsulation operators ∂_H for which H is primitive decidable
- and only renamings ρ_R for R such that $\forall b \in A : (\{a \mid R(a,b)\}$ is finite), and the function which yields for any $b \in A$ this finite set is primitive recursive.

A $\text{aprACP}_U(A, |)$ expression is *primitive decidable* if it is either primitive effective and without the constant U or primitive coeffective.

The languages of primitive (co)effective aprACP expressions will be denoted $\text{aprACP}_R^{\text{p.e.}}(A, |)$ and $\text{aprACP}_U^{\text{p.c.}}(A, |)$. It is easy to see that the signatures and sets of open and closed terms of these languages are decidable. As it is very easy to recognise the sources of primitive recursive functions, they are even primitive decidable. Also the canonicity results of Section 3.7, as well as the expressibility results of Sections 3.5 and 3.6 apply to these variants of aprACP.

The *primitive decidable* process graphs are defined just like the decidable ones (Definition 11), but with the additional requirement that the involved algorithm uses only primitive recursion. Exactly as before it follows that

Proposition 8 Primitive effective process expressions denote only primitive recursive process graphs. Primitive decidable expressions denote only primitive decidable graphs.

Every primitive recursive process graph is denoted by a primitive effective process expression. □

4.7 The Regular and the Recursion-free Expressions

Finally the *regular* process expressions are the ones with finite guarded recursion, and no other operators than inaction, action prefix and choice occurring in recursive specifications, whereas the *recursion-free* ones obviously have no recursion at all. It is easy to show that regular process expressions denote only regular process graphs, and recursion-free expressions only finite and acyclic process graphs. Conversely, every regular process graph is denotable by a regular process expression, and every every finite and acyclic process graph is, up to bisimulation equivalence, denotable by a recursion-free expression.

5 De Simone Languages

A *De Simone language* is a language of which the syntax is given as an annotated signature, and the semantics as a TSS over that signature of a particular form, known as the *De Simone* format.

Definition 16 (*The De Simone format*). A TSS is in the *De Simone* format if for every recursive specification S and $X \in V_S$ it has a rule

$$\frac{S_x[\langle Y|S\rangle/Y]_{Y\in V_S} \xrightarrow{a} z}{\langle X|S\rangle \xrightarrow{a} z}$$

and each of its other rules (the *De Simone rules*) has the form

$$\frac{\{x_i \xrightarrow{a_i} y_i \mid i \in I\}}{f(x_1,\ldots,x_n) \xrightarrow{a} t}$$

where $(f,n) \in \Sigma$, $I \subseteq \{1,\ldots,n\}$ and $t \in \mathbb{T}(\Sigma)$ is univariate recursion-free term containing no other variables than x_i $(1 \leq x \leq n$ and $i \notin I)$ and y_i $(i \in I)$. Here *univariate* means that each variables occurs at most once.

In a rule of the above form, (f,n) is the *type*, a the *action*, t the *target*, and the tuple $(l_1,\ldots,l_n)$ with $l_i = a_i$ if $i \in I$ and $l_i = *$ otherwise, the *trigger* [14].

Most process description languages encountered in the literature, including CCS, SCCS, CSP, ACP and MEIJE, are De Simone languages. De Simone languages are known to satisfy all the sanity requirements of Section 2.2 up to bisimulation equivalence. Below I will generalise the classification of process expressions in aprACP$_R$ to a classification of De Simone languages. I will not consider the classes of finite, regular and κ-bounded expressions. The De Simone languages from Definition 16 are the countable ones.

Definition 17 (*Guarded*). Let $P = (\Sigma, A, R)$ be a TSS in De Simone format. For $(f,n) \in \Sigma$ and $1 \leq i \leq n$, the ith argument of (f,n) is *awake* if there is a rule in R of type (f,n) with i in its index set. For $t \in \mathbb{T}^r(\Sigma)$ a free occurrence of a variable in t is *awake* or *unguarded* if for every subterm $f(t_1, ..., t_n)$ of t such that the occurrence is in t_i, the ith argument of f is awake.

Let S be a recursive specification. The relation $\stackrel{u}{\longrightarrow} \subseteq V_S \times V_S$ is given by $X \stackrel{u}{\longrightarrow} Y$ iff Y is awake in S_X. S is *guarded* if the relation $\stackrel{u}{\longrightarrow}$ is well-founded.

This notion of guardedness is due to VAANDRAGER [14]. Guarded recursive specifications in any De Simone language have unique solutions.

Definition 18 A TSS P in De Simone format is said to be

- *(recursively) enumerable* if Σ is decidable, only primitive recursive recursion is allowed, and the set of De Simone rules is r.e.
- *bounded* [14] if only guarded recursion is allowed and for each type and trigger the set of rules involving that type and trigger is finite.
- *effective* [14] if Σ is decidable, only computable guarded recursion is allowed, and there exists a total recursive function associating with each type and trigger the finite set of rules with that type and trigger.
- *coeffective* if Σ is decidable, only computable guarded recursion is allowed, and there exists a total recursive function associating with each type, action and target the finite set of corresponding rules.
- *primitive effective* if Σ is primitive decidable, only primitive recursive guarded recursion is allowed, and there exists a primitive recursive function associating with each type and trigger the finite set of corresponding rules.
- *primitive coeffective* if Σ is primitive decidable, only primitive recursive guarded recursion is used, and there is a primitive recursive function giving for each type, action and target the finite set of corresponding rules.

It is not difficult to apply these requirements to the De Simone language aprACP$_U$ and verify that they coincide exactly with the ones from Section 4.

Proposition 9 Terms in a De Simone language with a property on the left

countable	countable
bounded	finitely branching
recursively enumerable	recursively enumerable
effective	computable
coeffective	decidable
primitive effective	primitive recursive
primitive coeffective	primitive decidable

denote only process graphs satisfying the corresponding property on the right.
Proof: Straightforward. □

The main results announced in this extended abstract concern the expressibility of arbitrary De Simone languages in aprACP$_R$. In order to state which expressiveness results have been obtained, more properties of De Simone languages need to be defined.

Definition 19 (*Dependence of operators*). Let $P = (\Sigma, A, R)$ be a TSS in De Simone format, then *dependence* is the smallest transitive binary relation on Σ such that (f, n) *is dependent on* (g, m) if there is a rule with type (f, n) and with (g, m) occurring in its target.

Definition 20 A TSS P in De Simone format is said to be

- *width-finitary* if for each type there are only finitely many targets (such that there is a rule with that type and target).
- *(primitive) width-effective* if there exists a (primitive) recursive function giving for each type the finite set of corresponding targets.
- *finitary* if
  - (*depth*:) each type is dependent on only finitely many other types,
  - (*width*:) and for each type there are only finitely many targets.
- *image-finite* if for each type and trigger the matching set of rules is finite.
- *functional* if there exists a finite upperbound for the number of rules with any given type and trigger.

The first two properties can best be understood when viewing a De Simone language as an abstract TSS. Width-finitariness is then the property that for every type there are only finitely many abstract rules. (Primitive) width-effectiveness moreover requires that there is a (primitive) recursive function associating with each type the finite set of abstract rules of that type. A language is finitary if the behaviour of a finite term can be deduced by considering only finitely many abstract rules. Thus a finitary De Simone language can be obtained as the combination of a number of De Simone languages with finitely many abstract rules, each of which is trivially primitive width-effective. Boundedness is the combination of guarded- and image-finiteness. As seen in Section 3, aprACP$_R$ is width-finitary, and even primitive width-effective. As in aprACP$_R$ every type is dependent only on itself (at most), the language is finitary as well. Image-finiteness for aprACP$_R$ reduces to image-finiteness of the relational renamings, and functionality corresponds with the restriction to aprACP$_F$.

The following table lists a number of *translatable properties* of De Simone languages. Every translatable property consist of a full row in the table, thus being the conjunction of a left and a right side. For aprACP$_R$, in each property the left side is either always true or implied by the right side.

Theorem 4 Any De Simone language satisfying certain translatable properties of Table 3 is expressible in the version of aprACP$_R$ with the same properties.

Proof: To be supplied in the full version of this paper. My proof is an adaptation of De Simone's construction, but avoids, when possible, the use of unguarded recursion, by resorting to a richer synchronisation algebra $(A, |)$. There is essentially only one construction, translating any open term in any (countable) De Simone language into aprACP$_R$. For any of the properties on the

width-finitary	with guarded recursion
width-effective	with computable guarded recursion
primitive width-effective	with prim. rec. guarded recursion
finitary	with finite guarded recursion
image-finite	with image-finite renaming
functional	with functional renaming
–	recursively enumerable
primitive width-effective	primitive effective

Table 3: Properties of De Simone languages preserved under translation to aprACP_R

right in Table 3, I followed the construction backwards to see which additional requirements on De Simone languages are needed to ensure that the version of aprACP they are translated into meets that property. This yielded the properties on the left. □

By taking the dependencies between the properties in Table 3 into account—finite recursion is surely primitive recursive, primitive effectivity entails both image-finiteness and primitive recursive guarded recursion, and with unguarded recursion aprACP_F is as good as aprACP_R—Theorem 4 establishes 30 expressibility results. Since virtually all De Simone languages encountered in practice are finitary, the most significant results are

1. Any finitary De Simone language is expressible in aprACP_R with finite guarded recursion.
2. Any finitary image-finite De Simone language is expressible in aprACP_R with finite guarded recursion and image-finite renamings.
3. Any finitary functional De Simone language is expressible in aprACP_F with finite guarded recursion.
4. Any finitary enumerable De Simone language is expressible in $\text{aprACP}_R^{\text{r.e.}}$ with finite guarded recursion.
5. Any finitary enumerable image-finite De Simone language is expressible in $\text{aprACP}_R^{\text{r.e.}}$ with finite guarded recursion and image-finite renamings.
6. Any finitary enumerable functional De Simone language is expressible in $\text{aprACP}_F^{\text{r.e.}}$ with finite guarded recursion.
7. Any finitary primitive effective De Simone language is expressible in $\text{aprACP}_R^{\text{p.e.}}$ with finite guarded recursion.
8. Any finitary primitive effective functional De Simone language is expressible in $\text{aprACP}_F^{\text{p.e.}}$ with finite guarded recursion.

In each of these results the De Simone languages are also assumed to have finite guarded recursion only, but, by the compositionality of recursion, the same results hold without requiring or getting guardedness, finiteness or both.

Result 4 generalises the original theorem by De Simone, saying that any finitary recursively enumerable De Simone language with finite recursion is expressible in the recursively enumerable version of MEIJE with finite recursion. The generalisation is that, under the assumption that the source languages have only guarded recursion, the target language (now aprACP_R) can be required to use only guarded recursion as well.

Using the constant U yields an even stronger result for recursive enumerable De Simone languages, namely by dispensing with finitariness. This result has no effective counterpart.

Theorem 5 Any recursively enumerable De Simone language is expressible in aprACP_U with finite guarded recursion.

Acknowledgments Many thanks to Hanna Walińka and Anna Patterson for proofreading, and to the editors and publisher of this proceeding for delaying publication until my contribution was ready.

References

[1] D. AUSTRY & G. BOUDOL (1984): *Algèbre de processus et synchronisations.* *Theoretical Computer Science* 30(1), pp. 91–131. See also [4].

[2] J.C.M. BAETEN, J.A. BERGSTRA & J.W. KLOP (1987): *On the consistency of Koomen's fair abstraction rule.* Theoretical Computer Science 51(1/2), pp. 129–176.

[3] J.A. BERGSTRA & J.W. KLOP (1984): *The algebra of recursively defined processes and the algebra of regular processes.* This volume.

[4] G. BOUDOL (1985): *Notes on algebraic calculi of processes.* In K. Apt, editor: *Logics and Models of Concurrent Systems*, Springer-Verlag, pp. 261–303. NATO ASI Series F13.

[5] S.D. BROOKES, C.A.R. HOARE & A.W. ROSCOE (1984): *A theory of communicating sequential processes.* *JACM* 31(3), pp. 560–599.

[6] J.F. GROOTE & F.W. VAANDRAGER (1992): *Structured operational semantics and bisimulation as a congruence.* Information and Computation 100(2), pp. 202–260.

[7] YU.I. MANIN (1977): *A Course in Mathematical Logic*, Graduate Texts in *Mathematics* 53. Springer-Verlag.

[8] R. MILNER (1980): *A Calculus of Communicating Systems*, LNCS 92. Springer-Verlag.

[9] R. MILNER (1983): *Calculi for synchrony and asynchrony.* Theoretical *Computer Science* 25, pp. 267–310.

[10] G.D. PLOTKIN (1981): *A structural approach to operational semantics.* Report DAIMI FN-19, Computer Science Department, Aarhus University.

[11] A. PONSE (1992): *Computable processes and bisimulation equivalence.* Report CS-R9207, CWI, Amsterdam.

[12] R. DE SIMONE (1984): *On* MEIJE *and SCCS: infinite sum operators vs. non-guarded definitions.* Theoretical Computer Science 30, pp. 133–138.

[13] R. DE SIMONE (1985): *Higher-level synchronising devices in* MEIJE–*SCCS.* Theoretical Computer Science 37, pp. 245–267. For more details see [12] and: *Calculabilité et Expressivité dans l'Algebra de Processus Parallèles* MEIJE, Thèse de 3ᵉ cycle, Univ. Paris 7, 1984.

[14] F.W. VAANDRAGER (1993): *Expressiveness results for process algebras.* In J.W. de Bakker, W.P. de Roever & G. Rozenberg, editors: *Proceedings REX Workshop on Semantics: Foundations and Applications*, Beekbergen, The Netherlands, June 1992, LNCS 666, Springer-Verlag, pp. 609–638.

Definability with the State Operator in Process Algebra

Javier Blanco
Department of Computing Science, Eindhoven University of Technology
Eindhoven, The Netherlands. E-mail: javier@win.tue.nl

Abstract

The defining power of some process algebras can be enlarged if a state operator is added to their signature. Such an operator models a (local) state in which a process can act. The state space is always assumed to be finite. In this work we obtain some results about extensions of process algebras with the state operator. Furthermore, we show that if only linear equations are considered then every process definable using this operator can be defined in ACP with renamings.

1 Introduction

The state operator is used to model the idea of a process working in a (local) environment. It has been used (e.g. [Vaa90]) for the translation of a programming language into process algebra. In this work this operator is studied from the point of view of its defining power. Given the definition of the operator based on two functions over a set of states and a set of atomic actions, some restriction should be imposed in order to avoid a trivial answer for the question of whether a process is definable or not. For example, if every computable function is allowed then all processes with uniformly bounded non-determinism, i.e. with a constant bound for the outdegree of each node in a representing graph, are definable. The restriction we make is a strong one, we take only finite sets of atomic actions and states.

We study the two systems that already appear in [BB91a]. The first consists of processes defined only by sequential and alternative composition that run in a (global) context. This class of processes is closely related to the graphs of pushdown automata [Cau90]. The second class allows the occurrence of the state operator inside the recursive specification (in a restricted way) and constitutes an interesting subclass (as we will show) of the class of processes definable in ACP with renamings. In [BB91a] some examples are presented, for instance a bag and a queue.

In [Pon91] an implicit form of the state operator appears, in order to use Hoare logic in the context of process algebra. In this work the calculus is based on operational semantics (there are no equations) and the version of bisimulation equivalence is stronger, in the sense that only terms with the same state can be related. The operator definition principle of [Ver92] allows the same set of processes as those obtained by using the state operators to be defined (besides giving the possibility of proving equality of operators).

Acknowledgements. To Jos Baeten, Jan Bergstra and Alban Ponse for their helpful comments about this work.

2 Preliminaries

In this section we present a quick description of process algebra. We refer the reader to [BW90] for further information.

Several systems will be used. The largest signature considered is the one of ACP (Algebra of Communicating Processes), which still can be enlarged with projections, renamings and the state operator presented in the next section. The signature of ACP has

constants a finite set **A** of atomic actions and a special constant $\delta \in \mathbf{A}$ indicating inaction.

unary operators given $H \subset \mathbf{A} \setminus \{\delta\}$ the encapsulation operator ∂_H.

binary operators $+, \cdot, \|, ⌊\!⌊, |$. $+$ represents alternative composition, $\cdot$ sequential composition, and $\|$ parallel composition (merge). The auxiliary operators ⌊⌊ (left merge) and $|$ (communication merge) are used to define the merge.

2.1 Basic Process Algebra

The theory BPA_δ has a restricted signature with only **A**, + and $\cdot$. The axioms are given in table 1.

A1	$x + y = y + x$
A2	$x + (y + z) = (x + y) + z$
A3	$x + x = x$
A4	$(x + y) \cdot z = x \cdot z + y \cdot z$
A5	$(x \cdot y) \cdot z = x \cdot (y \cdot z)$
A6	$x + \delta = x$
A7	$\delta \cdot x = \delta$

Table 1: Axioms of BPA_δ

2.2 Process Algebra

The signature of the theory PA_δ contains $\|$ and ⌊⌊ besides the elements of the signature of BPA_δ. The $\|$ represents the free merge. The additional axioms are presented in table 2 (a ranges over **A**).

CM1	$x \parallel y = x \lfloor\!\lfloor y + y \lfloor\!\lfloor x$
CM2	$a \lfloor\!\lfloor x = a \cdot x$
CM3	$a \cdot x \lfloor\!\lfloor y = a \cdot (x \parallel y)$
CM4	$(x + y) \lfloor\!\lfloor z = x \lfloor\!\lfloor z + y \lfloor\!\lfloor z$

Table 2: Additional axioms of PA_δ

2.3 Algebra of Communicating Processes

The theory called ACP is presented. The theory is parameterized by a communication function γ that indicates which atomic actions communicate. This function is assumed to be commutative and associative, and it satisfies the equation $\gamma(\delta, a) = \delta$ for all $a \in \mathbf{A}$. The axioms of table 1 should be extended with the axioms of table 3 ($a, b \in \mathbf{A}$).

CM1	$x \parallel y = x \lfloor\!\lfloor y + y \lfloor\!\lfloor x + x \mid y$	
CM2	$a \lfloor\!\lfloor x = a \cdot x$	
CM3	$a \cdot x \lfloor\!\lfloor y = a \cdot (x \parallel y)$	
CM4	$(x + y) \lfloor\!\lfloor z = x \lfloor\!\lfloor z + y \lfloor\!\lfloor z$	
CF1	$a \mid b = \gamma(a, b)$	
CM5	$a \cdot x \mid b = (a \mid b) \cdot x$	
CM6	$a \mid b \cdot x = (a \mid b) \cdot x$	
CM7	$a \cdot x \mid b \cdot y = (a \mid b) \cdot (x \parallel y)$	
CM8	$(x + y) \mid z = x \mid z + y \mid z$	
CM9	$x \mid (y + z) = x \mid y + x \mid z$	
D1	$\partial_H(a) = a$	if $a \notin H$
D2	$\partial_H(a) = \delta$	if $a \in H$
D3	$\partial_H(x + y) = \partial_H(x) + \partial_H(y)$	
D4	$\partial_H(x \cdot y) = \partial_H(x) \cdot \partial_H(y)$	

Table 3: Additional axioms of ACP

2.4 Renamings

A feature that can be added to the previous algebras is the possibility of renaming atomic actions, given a function $f : \mathbf{A} \setminus \{\delta\} \to \mathbf{A}$. The operator ρ_f is defined in table 4 ($a \in \mathbf{A}$).

2.5 Projections

Any of the signatures defined above can be extended by an infinite set of unary operators π_n with n a natural number greater than 0. The intended meaning of $\pi_n(P)$ (in some appropriate model) is the process that behaves as P but stops

RN0	$\rho_f(\delta) = \delta$	
RN1	$\rho_f(a) = f(a)$	if $a \neq \delta$
RN2	$\rho_f(x \cdot y) = \rho_f(x) \cdot \rho_f(y)$	
RN3	$\rho_f(x + y) = \rho_f(x) + \rho_f(y)$	

Table 4: Renamings

after executing n steps. The axioms for the projection operators are given below in table 5 ($a \in \mathbf{A}$).

PR1	$\pi_n(a) = a$
PR2	$\pi_1(a \cdot x) = a$
PR3	$\pi_{n+1}(a \cdot x) = a \cdot \pi_n(x)$
PR4	$\pi_n(x + y) = \pi_n(x) + \pi_n(y)$

Table 5: Projections

2.6 Recursive definitions

The processes we wish to consider are defined by using a set of recursive equations. Properly speaking, we want processes that are solutions of such a set in a suitable model. However, all the work can be done without explicit reference to the model. This implies one can use any model which satisfies a few principles stated below. The definability results given below are interesting only if one only considers to finite sets of equations. However, in order to give some counterexamples we sometimes use infinite sets of equations.

Definition 2.6.1.

1. A system of *recursion equations* (over $\mathrm{BPA}_\delta, \mathrm{PA}_\delta$, ACP, etc) is a finite set of the form

$$E = \{X_i = s_i(X_0, \ldots, X_n); i = 0, \ldots, n\}$$

where the $s_i(X_0, \ldots, X_n)$ are expressions in the required signature, and the variables of s_i are among $X_0, \ldots, X_n$.

2. An infinite system is defined similarly

$$E = \{X_i = s_i(\vec{X}); i \in \mathbf{N}\}$$

3. A *solution* (in a certain model) of a recursive specification is a set of processes, one for each equation, such that the equations become true statements when the variables are interpreted as the corresponding process. We use sometimes the word solution for the process in that set that corresponds to the main variable of the specification

4. A specification is *guarded* if every occurrence of a variable in the right hand side of an equation is, modulo A4 A5, in a term of the form $a \cdot s$, for some atom a. Guardedness is a sufficient condition to guarantee uniqueness of solutions in many models.

5. A recursive specification is linear if each equation is either of the form

$$X = \sum_{i<n} a_i \cdot X_i + \sum_{j<m} b_j$$

where $m + n \geq 1$, or

$$X = \delta$$

6. A process is *definable* if it is a solution of a guarded recursive specification.

7. A recursive specification is in *restricted Greibach Normal Form (GNF)* if every equation is of the form

$$X_i = \delta$$

or

$$X_i = \sum_j a_{ij} \cdot X_{p(i,j)} \cdot X_{q(i,j)} + \sum_k b_{ik} \cdot X_{r(i,k)} + \sum_l c_{il}$$

It is a well known fact (see [BBK87b]) that any definable process over BPA admits a specification in restricted GNF.

□

Lemma 2.6.2. *Let P be a definable process. Then*

a *P has a head normal form, i.e. we can write P as*

$$\sum_{i<n} a_i \cdot Q_i + \sum_{j<m} b_j$$

where all Q_i are definable.

b *For every n, $\pi_n(P)$ equals a closed term.*

Proof. see [BW90] □

We work independently of a particular model, all the results apply for any model that satisfies the following principles.

1. The *Recursive Definition Principle (RDP^-)* says that every guarded specification has a solution.

2. The *Approximation Induction principle (AIP)* is the following

$$\frac{\forall n \geq 1. \pi_n(x) = \pi_n(y)}{x = y}$$

3. The *Recursive Specification Principle (RSP)* states that every guarded recursive specification has at most one solution. This principle is a consequence of AIP.

2.7 Operational semantics

In this section the operational semantics (sometimes referred to as action rules) is defined. The last two rules of table 6 are parameterized by a recursive specification E, given in one of the signatures. In the following we assume that s_i is the right hand side of the variable X_i in E. Let T be the set of terms in the corresponding signature generated by the set of variables of the recursive specification E. Two relations are defined,

- $\longrightarrow \subseteq T \times A \times T$, where $x \stackrel{a}{\longrightarrow} y$ means that x can do an a-step and transform into y.
- $\longrightarrow \surd \subseteq T \times A$, where $x \stackrel{a}{\longrightarrow} \surd$ means that x can do an a-step and terminate

Both relations are defined inductively, i.e. they are the least relations satisfying the rules in tables 6, 7 and 8.

$\overline{a \stackrel{a}{\longrightarrow} \surd}$	
$\dfrac{x \stackrel{a}{\longrightarrow} \surd}{x + y \stackrel{a}{\longrightarrow} \surd}$	$\dfrac{x \stackrel{a}{\longrightarrow} x'}{x + y \stackrel{a}{\longrightarrow} x'}$
$\dfrac{y \stackrel{a}{\longrightarrow} \surd}{x + y \stackrel{a}{\longrightarrow} \surd}$	$\dfrac{y \stackrel{a}{\longrightarrow} y'}{x + y \stackrel{a}{\longrightarrow} y'}$
$\dfrac{x \stackrel{a}{\longrightarrow} \surd}{x \cdot y \stackrel{a}{\longrightarrow} y}$	$\dfrac{x \stackrel{a}{\longrightarrow} x'}{x \cdot y \stackrel{a}{\longrightarrow} x' \cdot y}$
$\dfrac{s_i \stackrel{a}{\longrightarrow} \surd}{X_i \stackrel{a}{\longrightarrow} \surd}$	$\dfrac{s_i \stackrel{a}{\longrightarrow} y}{X_i \stackrel{a}{\longrightarrow} y}$

Table 6: Operational Semantics for BPA_δ

In order to determine a reasonable semantic for processes they must be divided out by some congruence. The action rules are used for that purpose.

$$\frac{x \xrightarrow{a} \surd}{x \parallel y \xrightarrow{a} y} \qquad \frac{x \xrightarrow{a} x'}{x \parallel y \xrightarrow{a} x' \parallel y}$$

$$\frac{y \xrightarrow{a} \surd}{x \parallel y \xrightarrow{a} x} \qquad \frac{y \xrightarrow{a} y'}{x \parallel y \xrightarrow{a} x \parallel y'}$$

$$\frac{x \xrightarrow{a} \surd}{x \mathbin{⌊\!⌊} y \xrightarrow{a} y} \qquad \frac{x \xrightarrow{a} x'}{x \mathbin{⌊\!⌊} y \xrightarrow{a} x' \parallel y}$$

Table 7: Additional rules for PA_δ

$$\frac{x \xrightarrow{a} \surd}{x \parallel y \xrightarrow{a} y} \qquad \frac{x \xrightarrow{a} x'}{x \parallel y \xrightarrow{a} x' \parallel y}$$

$$\frac{y \xrightarrow{a} \surd}{x \parallel y \xrightarrow{a} x} \qquad \frac{y \xrightarrow{a} y'}{x \parallel y \xrightarrow{a} x \parallel y'}$$

$$\frac{x \xrightarrow{a} \surd}{x \mathbin{⌊\!⌊} y \xrightarrow{a} y} \qquad \frac{x \xrightarrow{a} x'}{x \mathbin{⌊\!⌊} y \xrightarrow{a} x' \parallel y}$$

assume now $\gamma(a, b) = c$

$$\frac{x \xrightarrow{a} \surd, y \xrightarrow{b} \surd}{x \parallel y \xrightarrow{c} \surd} \qquad \frac{x \xrightarrow{a} \surd, y \xrightarrow{b} \surd}{x \mid y \xrightarrow{c} \surd}$$

$$\frac{x \xrightarrow{a} x', y \xrightarrow{b} \surd}{x \parallel y \xrightarrow{c} x'} \qquad \frac{x \xrightarrow{a} x', y \xrightarrow{b} \surd}{x \mid y \xrightarrow{c} x'}$$

$$\frac{x \xrightarrow{a} \surd, y \xrightarrow{b} y'}{x \parallel y \xrightarrow{c} y'} \qquad \frac{x \xrightarrow{a} \surd, y \xrightarrow{b} y'}{x \mid y \xrightarrow{c} y'}$$

$$\frac{x \xrightarrow{a} x', y \xrightarrow{b} y'}{x \parallel y \xrightarrow{c} x' \parallel y'} \qquad \frac{x \xrightarrow{a} x', y \xrightarrow{b} y'}{x \mid y \xrightarrow{c} x' \parallel y'}$$

$$\frac{x \xrightarrow{a} \surd, a \notin H}{\partial_H(x) \xrightarrow{a} \surd} \qquad \frac{x \xrightarrow{a} x', a \notin H}{\partial_H(x) \xrightarrow{a} x'}$$

Table 8: Additional rules for ACP

There exist many different notions of equivalence, one of the most used is bisimulation.

Definition 2.7.3. A relation $R \subseteq T \times T$ is called a *bisimulation* if whenever $s \xrightarrow{a} s'$ and sRt then there exist t' such that $t \xrightarrow{a} t'$ and $s'Rt'$ and whenever $s \xrightarrow{a} \surd$ and sRt then $t \xrightarrow{a} \surd$ and vice versa in both cases.

Two processes are *bisimilar* if there exists a bisimulation between them. □

Bisimulation appears originally in [Par81]. In [BW90] there is a extensive treatment of equivalences based on bisimulation.

3 The state operator

3.1 Introduction

The state operator λ is a generalization of the renaming operators, and is used to describe processes whose actions may have a side effect on a state space. Depending on the state, actions can be renamed or blocked, and in addition transform the state. This operator is useful in process verification and in the semantics of programming languages.(See [BB88, Vaa90]).

The state operator is defined by a finite set of states $\mathbf{S}$ and two functions

$$act : \mathbf{A} \times \mathbf{S} \longrightarrow \mathbf{A}$$

$$\mathit{eff} : \mathbf{A} \times \mathbf{S} \longrightarrow \mathbf{S}$$

The function *act* acts as the renaming part of the state operator while *eff* gives the new state after the execution of an action.

It is also required that δ is inert, i.e. $act(\delta, s) = \delta$ and $\mathit{eff}(\delta, s) = s$ for all states s. We say that *act* and *eff* are trivial when $act(a, s) = a$ and $\mathit{eff}(a, s) = s$ respectively. The axioms and operational semantics are given below.

SO1	$\lambda_s(a) = act(a, s)$
SO2	$\lambda_s(a \cdot x) = act(a, s) \cdot \lambda_{\mathsf{eff}(a,s)}(x)$
SO3	$\lambda_s(x + y) = \lambda_s(x) + \lambda_s(y)$

Table 9: Axioms for the state operator

A state I is called *inert* if for all actions a, $act(a, I) = a$ and $\mathit{eff}(a, I) = I$ Obviously, for all definable processes P, $\lambda_I(P) = P$. In the following we assume the existence of an inert state called I in every state space.

Example 3.1.1. A small example of how this operator works is the following clock that gives a signal each hour telling it.

$$X = tick \cdot X$$

$$\frac{x \xrightarrow{a} x', \mathrm{act}(a,s) \neq \delta}{\lambda_s(x) \xrightarrow{\mathrm{act}(a,s)} \lambda_{\mathrm{eff}(a,s)}(x')}$$

$$\frac{x \xrightarrow{a} \surd, \mathrm{act}(a,s) \neq \delta}{\lambda_s(x) \xrightarrow{\mathrm{act}(a,s)} \surd}$$

Table 10: Operational Semantics for the state operator

$$Clock = \lambda_{12}(X)$$

where $\mathbf{S} = \{1, \ldots, 12\}$ and

$$act(tick, n) = n$$

$$eff(tick, n) = (n \bmod 12) + 1$$

□

Lemma 3.1.2. *For every definable process P, state s and* $n > 0$

$$\pi_n(\lambda_s(P)) = \lambda_s(\pi_n(P))$$

Proof. Straightforward, by induction on n. □

Definition 3.1.3. Let λ be defined over $\mathbf{S}$ by *act* and *eff*. A new state operator λ^2 over $\mathbf{S} \times \mathbf{S}$ is defined by

$$act^2(a, < s, t >) = act(act(a,t), s)$$

$$eff^2(a, < s, t >) =< eff(act(a,t), s), eff(a,t) >$$

□

In spite of the different name, λ^2 is a state operator. The state operator is parameterized by $\mathbf{S}$, *act* and *eff*, so one has a different operator for each value of that set and functions. The change in the name was done in order to avoid confusion. The justification for the square in its name is given in the following lemma. In [BW90] a set of objects is used as an extra parameter of the state operator. In that framework the two different state operators can be described using different object names.

Lemma 3.1.4. *For all definable processes P, states s,t*

$$\lambda^2_{<s,t>}(P) = \lambda_s(\lambda_t(P))$$

Proof.

(i) For closed terms. By straightforward induction. For example, if $P = a \cdot Q$ and we have proven the lemma for Q, then

$$\begin{aligned}
\lambda^2_{<s,t>}(P) &= \lambda^2_{<s,t>}(a \cdot Q) \\
&= act^2(a, <s,t>) \cdot \lambda^2_{eff^2(a,<s,t>)}(Q) \\
&= act(act(a,t),s) \cdot \lambda^2_{<eff(act(a,t),s),eff(a,t)>}(Q) \\
&= act(act(a,t),s) \cdot \lambda_{eff(act(a,t),s)}(\lambda_{eff(a,t)}(Q)) \\
&= \lambda_s(act(a,t) \cdot \lambda_{eff(a,t)}(Q)) \\
&= \lambda_s(\lambda_t(P))
\end{aligned}$$

(ii) For definable processes. Using AIP, lemma 2.6.2 and lemma 3.1.2.

$$\begin{aligned}
\pi_n(\lambda^2_{<s,t>}(P)) &= \lambda^2_{<s,t>}(\pi_n(P)) \\
&= \lambda_s(\lambda_t(\pi_n(P))) \\
&= \pi_n(\lambda_s(\lambda_t(P)))
\end{aligned}$$

□

3.2 Recursive definitions with the state operator

In this section some results of [BB91a] are revised and new results are obtained.

The concepts defined in definition 2.6.1 can be used also in a context with the state operator. It may not be immediate how to extend the property of linearity. If the definition above is used without modification then the state operator would not appear at all in a linear specification. The most obvious extension, the one we will adopt is to allow equations of the form

$$X = \sum a_i \cdot \lambda_{\sigma_i}(X_i) + \sum b_j$$

or

$$X = \delta$$

This includes the form given in definition 2.6.1 due to the possibility of having an inert state.

The classes of processes considered here are:

- BPA_δrec (PA_δrec, ACPrec): processes defined by an guarded recursive specification over BPA_δ (PA_δ, ACP).
- λ(BPA_δrec): processes obtained by an application of λ to a process in BPA_δrec.
- (BPA_δ + λ)lin: processes defined by a linear recursive specification over BPA_δ + λ.

The main result of this section is that $\lambda(\text{BPA}_\delta\text{rec}) \subset (\text{BPA}_\delta + \lambda)\text{lin}$. As a corollary $\text{BPA}_\delta\text{rec} \subset (\text{BPA}_\delta + \lambda)\text{lin}$. This last inclusion was stated in [BB91a] but the proof was not correct.

Lemma 3.2.5. *Let X be a process in* $(\text{BPA}_\delta + \lambda)\text{lin}$, *then for every state s* $\lambda_s(X) \in (\text{BPA}_\delta + \lambda)\text{lin}$.

Proof. If X satisfies the premises of the lemma, then we can write it in the form

$$X = \sum_{i<n} a_i \cdot \lambda_{\sigma_i}(X_i) + \sum_{j<m} b_j$$

where X_i are defined also by a linear specification over $\text{BPA}_\delta + \lambda$ (the case $X = \delta$ is trivial).

Let Y be defined by

$$Y = \sum_{i<n} act(a_i, s) \cdot \lambda^2_{<s,\sigma_i>}(X_i') + \sum_{j<m} act(b_j, s)$$

where X_i' is defined as X_i where λ_s is replaced by $\lambda^2_{<I,s>}$. It is immediate to see that $\lambda_s(X)$ is a (unique) solution of Y. □

Using the notation of the beginning of this section we can paraphrase this lemma as $(\text{BPA}_\delta + \lambda)\text{lin} = \lambda((\text{BPA}_\delta + \lambda)\text{lin})$.

Lemma 3.2.6. *Let X be a process in* BPA_δ*rec. Then X is in* $(\text{BPA}_\delta + \lambda)\text{lin}$ *(with an enriched alphabet).*

Proof. Let a recursive specification E over BPA_δ be given that uses variables from $X_0, \ldots, X_n$ and has X_0 as solution. Assume the specification of each X_i is in restricted GNF, i.e.

$$X_i = \sum_j a_{ij} \cdot X_{p(i,j)} \cdot X_{q(i,j)} + \sum_k b_{ik} \cdot X_{r(i,k)} + \sum_l c_{il}$$

Define (using another set of atomic actions)

$$\begin{aligned} X \;=\; & \sum_{i,j} < a_{ij}, i, q(i,j) > \cdot \lambda_{p(i,j)}(X) + \sum_{i,k} < b_{ik}, i, r(i,k) > \cdot X + \\ + \; & \sum_{i,l} < c_{il}, i > \cdot X + \sum_{i,l} < c_{il}, i >_t \end{aligned}$$

The state space $\mathbf{S}$ is $\{0, \ldots, n, \text{init}_0, \ldots, \text{init}_n, I\}$. The functions *act* and *eff* are trivial except in the following cases:

$$\begin{aligned}
& act(< a, i >, m) = act(< a, i >, init_m) = \delta \text{ if } i \neq m \\
& act(< a, i >, init_i) = act(< a, i >_t, i) = \delta \\
& act(< a, i, k >, i) = act(< a, i, k >, init_i) = a \\
& act(< a, i >, i) = act(< a, i >_t, init_i) = a \\
& eff(< a, i, k >, i) = k \\
& eff(< a, i, k >, init_i) = init_k \\
& eff(< a, i >, i) = I
\end{aligned}$$

Now we claim that for each sequence $n_1, \ldots, n_m, s$ the following equality holds

$$X_{n_1} \cdots X_{n_m} \cdot X_s = \lambda_{init_s}(\lambda_{n_m}(\cdots(\lambda_{n_1}(X)\cdots)$$

The intuition underlying this construction is that the state operators indicate what the process still has to do. There are two kinds of terminating actions: If the process is in the initial state, i.e. there is nothing remaining, then the $<>_t$ actions are allowed, that actually terminate, otherwise the termination is indicated by the disappearance of this state (evolving into the inert state).

We prove the claim by showing that both sides of the equality satisfy the same (infinite) recursive specification.
We abbreviate $\lambda_{s_1}(\cdots(\lambda_{s_n}(X)\cdots)by\lambda_{s_1\ldots s_n}(X)$

$$\lambda_{init_i}(X) = \sum_j a_{ij} \cdot \lambda_{init_{q(i,j)}}(\lambda_{p(i,j)}(X)) + \sum_k a_{ik} \cdot \lambda_{init_{r(i,k)}}(X) + \sum_l c_{il}$$

Let σ be a sequence of states.

$$\begin{aligned}
\lambda_\sigma(\lambda_i(X)) &= \lambda_\sigma(\sum_j a_{ij} \cdot \lambda_{q(i,j)}(\lambda_{p(i,j)}(X)) + \\
&\quad + \sum_k b_{ik} \cdot \lambda_{r(i,k)}(X) + \sum_l c_{il} \cdot \lambda_I(X)) \\
&= \sum_j a_{ij} \cdot \lambda_\sigma(\lambda_{q(i,j)}(\lambda_{p(i,j)}(X))) + \\
&\quad + \sum_k b_{ik} \cdot \lambda_\sigma(\lambda_{r(i,k)}(X)) + \sum_l c_{il}.\lambda_\sigma(X)
\end{aligned}$$

It is immediate that if we replace $\lambda_{init_s}(\lambda_{n_m}(\cdots(\lambda_{n_1}(X)\cdots)$ by $X_{n_1} \cdots X_{n_m} \cdot X_s$ the equations above are valid.

□

We are now able to prove the main result of this section

Theorem 3.2.7. *Let X be a process in* $\lambda(\text{BPA}_\delta\text{rec})$. *Then X is in* $(\text{BPA}_\delta + \lambda)\text{lin}$.

Proof. Immediate from lemmas 3.2.5 and 3.2.6. □

The next and last result of this section shows that the application of a state operator outside of a BPA_δ specification adds to the expressive power of PA_δ (and to that of ACP if we require that the alphabet of the process is disjoint with the image of the communication function, see [BB88]). The converse of this, i.e. that there exist processes definable in PA_δ and not in $\lambda(\text{BPA}_\delta\text{rec})$ (even not in $(\text{BPA}_\delta + \lambda)\text{lin}$), was already answered affirmatively in [BB91a].

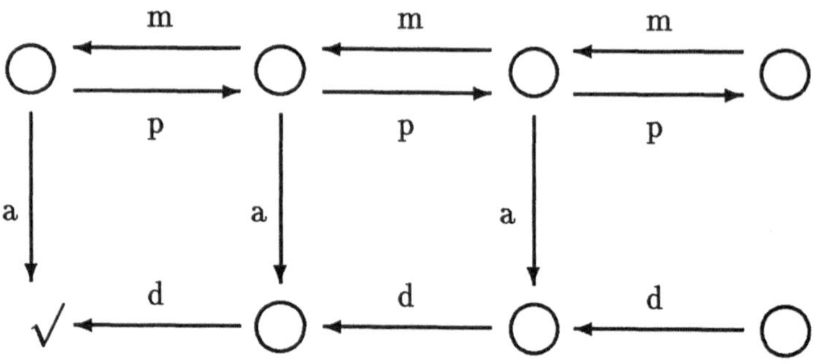

Figure 1: counter with an extra action

Example 3.2.8. Consider the following (infinite) recursive specification

$$\begin{aligned} X_0 &= p \cdot X_1 + a \\ X_{i+1} &= p \cdot X_{i+2} + m \cdot X_i + a \cdot D_i \\ D_0 &= d \\ D_{i+1} &= d \cdot D_i \end{aligned}$$

The intuitive meaning of this process is a counter that once in its existence can do an a-action and then do as many d-actions as the current value of the counter. The graph of the process is indicated in figure 3.2.8. A similar example appears in [Cau90]. Note that if the a-actions and the D_i are removed then the usual counter is obtained.

We see that this process is definable in $\lambda(\mathrm{BPA}_\delta\mathrm{rec})$ as follows. Let $\mathbf{S} = \{0, 1\}$ and

$$\begin{aligned} act(a, 0) &= d \\ act(p, 0) &= \delta \\ act(m, 0) &= \delta \\ eff(a, 1) &= 0 \end{aligned}$$

and trivial otherwise. Now

$$X_0 = \lambda_1(C)$$

where

$$\begin{aligned} C &= p \cdot T \cdot C + a \\ T &= p \cdot T \cdot T + m + a \end{aligned}$$

If the a-action is removed in both equations, then C defines a counter. The proof that $X_0 = \lambda_1(C)$ is in the appendix. □

Lemma 3.2.9. *The counter of 3.2.8 cannot be defined in* PA_δ.

Proof. See Appendix. □

As an immediate consequence

Theorem 3.2.10. *There is a process in* $\lambda(\mathrm{BPA}_\delta\mathrm{rec})$ *that is not definable in* PA_δ.

Example 3.2.11. Another interesting example of a process that is not definable in PA_δ but it is in $\lambda(\mathrm{BPA}_\delta\mathrm{rec})$ is the following.

Let S be defined by the following (infinite) specification, where ϵ is the empty string and $*$ means concatenation.

Let $d \in \{a, b\}$

$$\begin{aligned}
S_\epsilon &= push(a) \cdot S_a + push(b) \cdot S_b \\
S_{d*\sigma} &= push(a) \cdot S_{a*d*\sigma} + push(b) \cdot S_{b*d*\sigma} + pop(d) \cdot S_\sigma + top(d) \cdot S'_{d*\sigma} \\
S'_\epsilon &= push(a) \cdot S'_a + push(b) \cdot S'_b \\
S'_{d*\sigma} &= push(a) \cdot S'_{a*d*\sigma} + push(b) \cdot S'_{b*d*\sigma} + pop(d) \cdot S'_\sigma
\end{aligned}$$

This "stack" can once in its existence show its top. Indeed it is definable in $\lambda(\mathrm{BPA}_\delta\mathrm{rec})$ as follows.

Let $\mathbf{S} = \{0, 1\}$. Let $d \in \{a, b\}$, define *eff*(top(d),0) = 1 and *act*(top(d),1) = δ and act and eff be trivial otherwise.

Furthermore, let

$$\begin{aligned}
S &= (push(a) \cdot T_a + push(b) \cdot T_b).S \\
T_d &= push(a) \cdot T_a \cdot T_d + push(b) \cdot T_b \cdot T_d + pop(d) + top(d) \cdot T_d
\end{aligned}$$

(note that by removing the *top* action one obtains a stack).

Then

$$S_\lambda = \lambda_1(S)$$

The proof of this last fact is analogous to the one in [BB91a], claim 3.5. The fact that it is not definable in PA_δ is not easy to prove, but the ideas of [BT87] are used in a straightforward way. □

3.3 Limits to the expressive power of the state operator

The main result of this section is the fact that all processes of $(\mathrm{BPA}_\delta + \lambda)$lin are definable in ACP with renamings. Some other interesting results about the defining power of the state operator are stated.

Definition 3.3.12. Let **A** be the set of atomic actions parametrizing the signature of ACP. Let P be a process definable in such a theory. The *alphabet* of P is the set of atomic actions that P can perform, so it is a subset of **A**. For a formal definition see [BBK87a]. □

Definition 3.3.13. Let **S**, *act*, *eff* be the parameters for a state operator and let $s \in \mathbf{S}$. Let $\mathbf{A}_1$ be the set of atomics actions. and γ_1 a communication function over $\mathbf{A}_1$. Then we define an operator κ_s over ACP + RN with a larger set of atomic actions $\mathbf{A}$ (i.e. $\mathbf{A}_1 \subset \mathbf{A}$) and using a communication function γ that extends γ_1 by

$$\kappa_s(X) = \rho_f \circ \partial_H(X \parallel P_s)$$

Here

- $\mathbf{A} = \{a_s | a \in \mathbf{A}_1, s \in \mathbf{S}\} \cup \{a' | a \in \mathbf{A}_1\} \cup \mathbf{A}_1$
- $\gamma(a, a_s) = act(a, s)'$
- $f(a') = a$
- $H = \{a_s | a \in \mathbf{A}_1\} \cup \mathbf{A}_1$
- $P_s = \sum_{a \in \mathbf{A}} a_s \cdot P_{\mathit{eff}(a,s)}$

□

Note that given a process P definable in ACP + RN with a set $\mathbf{A}$ of atomic actions and a state operator, then the alphabet of $\kappa_s(P)$ will still be a subset of $\mathbf{A}$.

Lemma 3.3.14. *Let P be a definable process over* ACP + RN, *then*

$$\kappa_s(P) = \lambda_s(P) \cdot \delta$$

Proof.

(i) For closed terms.

$P = a$

$$\begin{aligned} \kappa_s(a) &= \rho_f \circ \partial_H(a \parallel P_s) \\ &= \rho_f \circ \partial_H(a \mid P_s) \\ &= act(a,s) \cdot \rho_f \circ \partial_H(P_{\mathit{eff}(a,s)}) \\ &= act(a,s) \cdot \delta \\ &= \lambda_s(a) \cdot \delta \end{aligned}$$

$P = a \cdot Q$

$$\begin{aligned} \kappa_s(a.Q) &= \rho_f \circ \partial_H(a.Q \parallel P_s) \\ &= \rho_f \circ \partial_H(a.Q \mid P_s) \\ &= act(a,s) \cdot \rho_f \circ \partial_H(Q \parallel P_{\mathit{eff}(a,s)}) \\ &= act(a,s) \cdot \lambda_{\mathit{eff}(a,s)}(Q) \cdot \delta \\ &= \lambda_s(a.Q) \cdot \delta \end{aligned}$$

$P = Q + R$ straightforward.

(ii) For definable processes. Let Q be a definable process. Q has a normal form $\sum a_i \cdot Q_i + \sum b_j$. We prove by induction over n that

$$\pi_n(\kappa_s(Q)) = \pi_n(\lambda_s(Q) \cdot \delta)$$

Now we use lemma 2.6.2

$$\begin{aligned}\pi_1(\kappa_s(Q)) &= \sum act(a_i, s) + \sum act(b_j, s)\\ &= \pi_1(\lambda_s(Q) \cdot \delta)\end{aligned}$$

$$\begin{aligned}\pi_{n+1}(\kappa_s(Q)) &= \sum act(a_i, s) \cdot \pi_n(\kappa_s(Q_i)) + \sum act(b_j, s) \cdot \delta\\ &= \sum act(a_i, s) \cdot \pi_n(\lambda_s(Q_i) \cdot \delta) + \sum act(b_j, s).\pi_n(\delta)\\ &= \pi_{n+1}(\sum act(a_i, s) \cdot \lambda_s(Q_i) \cdot \delta + \sum act(b_j, s) \cdot \delta)\\ &= \pi_{n+1}(\lambda_s(Q) \cdot \delta)\end{aligned}$$

□

This lemma implies that in contexts where successful and unsuccessful termination cannot be distinguished (e.g. aprACP in [BB91b]) the state operator is definable in terms of the operators of ACP + RN. This also implies that it can be eliminated when dealing with perpetual processes as in the next example. Note that this is stronger than saying that every process definable with the state operator is also definable without it.

Example 3.3.15. In [BB91a] the following specification due to Vaandrager of a FIFO queue over $\{a, b\}$ in $(\text{BPA}_\delta + \lambda)$lin is given.

- S $= \{0, A, B, I\}$ is the state space.
- $act(\text{out}, A) = s(a)$, $act(\text{out}, B) = s(b)$, $e\!f\!f(\text{out}, A) = I$, $e\!f\!f(\text{out}, B) = I$;
- $act(s(a), B) = s(b)$, $act(s(b), A) = s(a)$, $e\!f\!f(s(a), B) = A$, $e\!f\!f(s(b), A) = B$;
- $act(\text{out}, 0) = \delta$

In the cases not mentioned above *act* and *eff* are trivial.

Now, a queue is given by the following equations

$$Q = \lambda_0(X)$$

$$X = r(a) \cdot \lambda_A(X) + r(b) \cdot \lambda_B(X) + out \cdot X$$

□

As an application of lemma 3.3.14 a recursive specification of a queue in ACP + RN can be obtained since Q is perpetual, i.e. a termination state (successful or not) is never reached. In [BW90] there is already a specification of a queue in ACP + RN, but the one obtained by the application of lemma 3.3.14 is different and interesting in its own right. For example, this specification can be extended in a straightforward way to one of a queue that additionally can do an action that deletes all a-elements of a queue. It seems difficult to do the same with the specification of [BW90]. The extension is done by adding an action $d(a)$ such that $\mathit{eff}(d(a), A) = I$ and is trivial for *act*. In the specification for X above a summand $d(a) \cdot X$ is also added. When it is translated to ACP + RN this delete action synchronizes with all the "A-states" and makes them inert.

Theorem 3.3.16. *Let X be a process in* $(\mathrm{BPA}_\delta + \lambda)$lin, *then X is definable in* ACP + RN.

Proof. Let X be a solution of a linear recursive specification over variables $X_0, \ldots, X_n$ where for every $i \in \{0, \ldots, n\}$

$$X_i = \sum_{c_{ij} \in C_i} c_{ij} \cdot \lambda_{\sigma_{ij}}(X_{p(i,j)}) + \sum_{b_{ik} \in B_i} b_{ik}$$

Define

$$P_{s,i} = \sum_{c_{ij} \in C_i} (c_{ij})_s \cdot P_{\mathit{eff}(c_{ij},s),p(i,j)} + \sum_{b_{ik} \in B_i} (b_{ik})_s$$

Let f, H, γ be as in definition 3.3.13, then

$$\lambda_s(X_i) = \rho_f \circ \partial_H(X_i \parallel P_{s,i})$$

This last equality follows by RSP since the left and right hand side satisfy the same recursive specification. □

3.4 Some remarks about decidability of bisimulation

The problem of whether bisimulation equivalence is decidable for certain classes of processes has received many partial answers. It is well known that bisimulation is decidable for BPA (see [BBK87b, Cau90, SCS92]). Bisimulation is undecidable for ACP (see [BK84]) and therefore for all systems containing it. For PA it is still an open problem (some work in that direction is [Chr92]). One of the most interesting open problems, that is also related with the decidability of deterministic languages, is whether bisimulation is decidable for processes in $\lambda(\mathrm{BPA}_\delta\mathrm{rec})$. Theorem 3.4.18 shows that if we move to $(\mathrm{BPA}_\delta + \lambda)$lin then the problem of decidability of bisimulation has a negative answer. The construction is based on one in [BK84].

We call pushdown automata graphs (following [Cau90]) the graphs of left derivation (prefix rewriting) where the transitions have the form (in GNF) $sX \xrightarrow{a} t\alpha$, s, t are states, α is a sequence of variables. It is immediate to see that any graph given by the operational semantics of $\lambda(\text{BPA}_\delta\text{rec})$ is a pushdown automata graph. On the other hand given a pushdown automata (a set of rules as above), one can obtain a specification in $\lambda(\text{BPA}_\delta\text{rec})$ with the same graph. This can be accomplished as in the following example taken from [Cau90].

Example 3.4.17. Take the following pushdown automata.

$$\begin{array}{lll} sX \xrightarrow{a} sXX & sX \xrightarrow{c} t & sX \xrightarrow{d} u \\ tX \xrightarrow{b} t & uX \xrightarrow{b} vX & vX \xrightarrow{b} u \end{array}$$

The corresponding $\lambda(\text{BPA}_\delta\text{rec})$ specification is the following.

$$X = a_1 \cdot X \cdot X + c_1 + d_1 + b_1 + b_2 \cdot X + b_3$$

then the pushdown automata is the graph of the process $\lambda_s(X)$, where

- $act(s, b_i) = \delta$
- $act(t, b_2) = act(t, b_3) = act(t, a_1) = act(t, c_1) = act(t, d_1) = \delta$
- $act(u, b_1) = act(u, b_3) = act(u, a_1) = act(u, c_1) = act(u, d_1) = \delta$
- $act(v, b_1) = act(v, b_2) = act(v, a_1) = act(v, c_1) = act(v, d_1) = \delta$
- $act(r, e_n) = e$ otherwise
- $\mathit{eff}(s, c_1) = t$
- $\mathit{eff}(s, d_1) = u$
- $\mathit{eff}(u, b_2) = v$
- $\mathit{eff}(v, b_3) = u$

□

Theorem 3.4.18. *Given two processes in* $(\text{BPA}_\delta + \lambda)$lin *there is no algorithm which decides if they are bisimilar.*

Proof. Let K be a r.e. set that is not recursive. In [HU79] it is proven that K can be recognized by a three counter machine. Such a machine has three counters a, b, c (ranged over by a metavariable α) and a sequence of instructions. The instructions have one of the following forms,

i $\alpha := \alpha + 1;$ goto j

ii $\alpha := \alpha \dot{-} 1$; goto j

iii if $\alpha = 0$ then goto j else goto j'

iv stop

The meaning of each instruction is obvious ($\dot{-}$ means subtract except if it is zero) The sequence of instructions is numbered and it is assumed that the labels of the goto instructions are among these numbers.

Now, let P be a program that recognizes K, i.e. $P(n,0,0) \longrightarrow$stop, where $P(x,y,z)$ means that x, y, and z are the initial values of the counters. We will give a recursive specification in (BPA_δ + λ)lin such that it can do an infinite number of actions if and only if the machine does not finish.

First we translate each instruction I_i into a linear equation in the following way (following the pattern above)

i $X_i = b \cdot \lambda_\alpha(X_j)$

ii $X_i = (O_\alpha + p_\alpha) \cdot X_j$

iii $X_i = O_\alpha \cdot X_j + p_\alpha \cdot \lambda_\alpha(X_{j'})$

iv $X_i = \delta$

The state operator will model the state of the counters of the machine. An expression of the form $\lambda_s(\lambda_\sigma(X_i))$ where $\sigma \in \{a,b,c,I\}^*$ means that the machine is executing instruction i and that counter α has value $\#_\alpha(\sigma)$, that is the number of occurrences of α in σ.

The *act* and *eff* functions are trivial except in the following cases

- $act(\alpha, 0_\alpha) = \delta$ (the counter has a positive value)
- $act(s, 0_\alpha) = b$ (the counter is empty)
- $act(\alpha, p_\alpha) = b$ (the counter has a positive value)
- $eff(\alpha, p_\alpha) = I$ (decrease the counter)
- $act(s, p_\alpha) = \delta$ (don't decrease an empty counter)

Given B defined by $B = b \cdot B$, then B is bisimilar to $\lambda_s(\lambda_\sigma(X_1))$ if and only if $P(n,0,0)$ diverges. □

4 Summary

The next picture illustrates all the (proper) inclusions among the different systems.

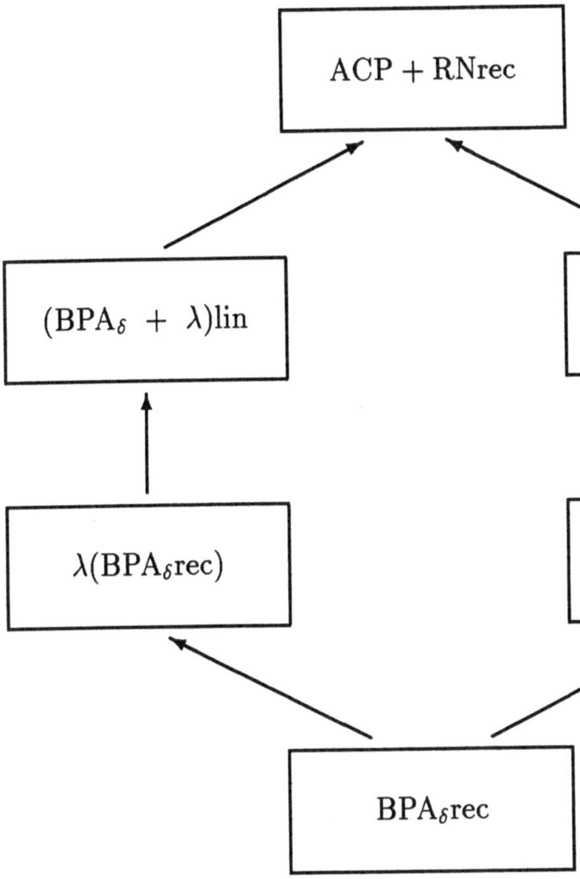
ACP + RNrec
(BPA$_\delta$ + λ)lin
ACPrec
λ(BPA$_\delta$rec)
PA$_\delta$rec
BPA$_\delta$rec

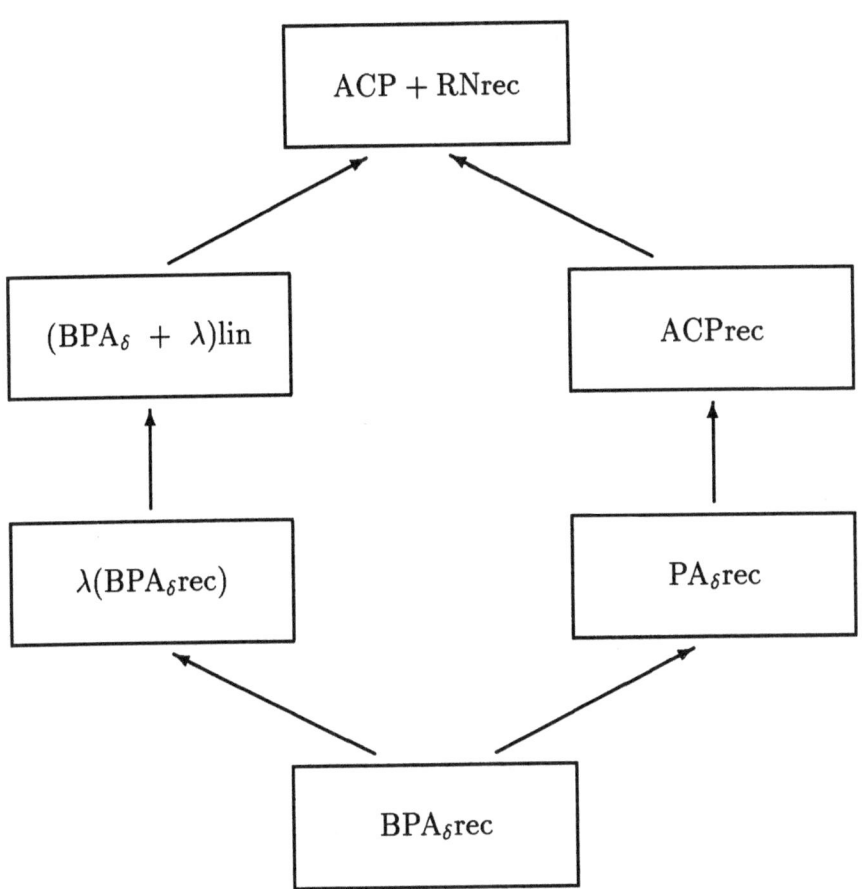

A Proofs

First we will show that the following specification defines the counter of 3.2.8

$$\begin{aligned} C &= p\cdot T\cdot C + a \\ T &= p\cdot T\cdot T + m + a \end{aligned}$$

We show that $\lambda_1(C) = X_0$ where

$$\begin{aligned} X_0 &= p\cdot X_1 + a \\ X_{i+1} &= p\cdot X_{i+2} + m\cdot X_i + a\cdot D_i \\ D_0 &= d \\ D_{i+1} &= d\cdot D_i \end{aligned}$$

Proof. They satisfy the same set of defining equations

$D_n = \lambda_0(T^n\cdot C)$

$$\begin{aligned} \lambda_0(C) &= \lambda_0(p\cdot T\cdot C) + \lambda_0(a) \\ &= \delta + d \\ &= d \end{aligned}$$

$$\begin{aligned} \lambda_0(T^{n+1}\cdot C) &= \lambda_0(p\cdot T\cdot T^{n+1}\cdot C) + \\ &\quad \lambda_0(m\cdot T^n\cdot C) + \lambda_0(a\cdot T^n\cdot C) \\ &= \delta + \delta + d\cdot\lambda_0(T^n\cdot C) \\ &= d\cdot\lambda_0(T^n\cdot C) \end{aligned}$$

$X_n = \lambda_1(T^n\cdot C)$

$$\begin{aligned} \lambda_1(C) &= \lambda_1(p\cdot T\cdot C) + \lambda_1(a) \\ &= p\cdot\lambda_1(T\cdot C) + a \end{aligned}$$

$$\begin{aligned} \lambda_1(T^{n+1}\cdot C) &= \lambda_1(p\cdot T\cdot T^{n+1}\cdot C) + \\ &\quad \lambda_1(m\cdot T^n\cdot C) + \lambda_1(a\cdot T^n\cdot C) \\ &= p\cdot\lambda_1(T^{n+2}\cdot C) + m\cdot\lambda_1(T^n\cdot C) + a\cdot\lambda_0(T^n\cdot C) \\ &= p\cdot\lambda_1(T^{n+2}\cdot C) + m\cdot\lambda_1(T^n\cdot C) + a\cdot D_n \end{aligned}$$

□

Now we will prove that this process is not definable in $\mathbf{PA}_\delta$ looking at the term model (or isomorphically at the graph model, see [BW90]).

The following definitions are adaptations of some appearing in [Vaa91].

Definition A.0.1.

1. Let s be a term in PA_δ. An occurrence of a subterm t of s is *sleeping* in s if it occurs in r in a subterm of s of the form $r' \lfloor\!\lfloor r$ or $r' \cdot r$.
2. A subterm t of s is sleeping in s if every occurrence of t in s is sleeping.
3. A subterm t of s is *dead* if for every s' such that $s \xrightarrow{\sigma} s'$ t is sleeping in s'
4. Let E be a recursive specification. A term s is dead in E if it is dead in every right hand side of E.

□

Fact A.0.2. If a term t is dead in a recursive specification E with root X, then X is bisimilar to X', where X' is the root of the specification E' which equals E except that every occurrence of t is replaced for t' (t' any term). □

Fact A.0.3. If $p \| q = d^n, d \in \mathbf{A}$ then $\exists m, p$ such that $p = d^m, q = d^p$ and $n = p + m$ □

As a consequence of fact A.0.3, if $p \| q = d^n$ then $p \| q = \bar{p} \cdot \bar{q}$ where $\bar{x}$ is x where all $\|$ and $\lfloor\!\lfloor$ are replaced by $\cdot$.

Fact A.0.4. Let E be a specification over PA_δ with root X. Then there exists a specification E' with root X' such that $X = X'$ and all occurrences of $\lfloor\!\lfloor$ appear inside the scope of a $\|$. □

Proof. Rewrite every $p \lfloor\!\lfloor q$ that is outside the scope of a $\|$ into

$$\sum a_i \cdot (t_i \| q) + \sum b_j \cdot q$$

where $\sum a_i \cdot t_i + \sum b_j$ is the head normal form of t. □

Theorem A.0.5. *Let E be a specification with root X of the process X_0 of example 3.2.8 over* PA_δ, *then there exist a specification E' of the same process over* BPA_δ.

Proof. Let E' be E where all $\|$ and $\lfloor\!\lfloor$ are replaced by $\cdot$.

Let $p \| q$ be a subterm of (a right hand side of) E not in the scope of a $\|$ or a $\lfloor\!\lfloor$. We will show that $p \| q$ is dead or equal to d^n.

Suppose $X \xrightarrow{\sigma} t$ and $p \| q$ is a subterm of t that is awake and not equal to d^n for any n. Since t cannot be d^n for some n, then t should be able to perform an a-action.

As $p \| q$ is awake then

$$t = p \| q \cdot r + s$$

where eventually r missing or $s = \delta$. Suppose that $p \xrightarrow{a} p'$ or $p \xrightarrow{a} \surd$. Then $t \xrightarrow{a} p' \| q$ or $t \xrightarrow{a} q$, in any case q should be able to do a d-action.

But then t can do a d-action and then an a-action which is impossible, so we conclude that the a-action is done by s. But then p is able to do a p-action or a m-action. In both cases $t \xrightarrow{p} p' \parallel q \cdot r$ or $t \xrightarrow{p} q \cdot r$. But now $p' \parallel q$ or q should be able to do an a-action and the same argument as before applies. □

Now, Lemma 3.2.9 follows since the process of 3.2.8 is not definable in BPA_δ (see [Cau90]).

References

[BB88] J.C.M. Baeten and J.A. Bergstra. Global renaming operators in concrete process algebra. *Information and Computation*, 78(3):205–245, 1988.

[BB91a] J.C.M. Baeten and J.A. Bergstra. Recursive process definitions with the state operator. *Theoretical Computer Science*, 82:285–302, 1991.

[BB91b] J.C.M. Baeten and J.A. Bergstra. A survey of axiom systems for process algebras. Report P9111, Programming Research Group, University of Amsterdam, 1991.

[BBK87a] J.C.M. Baeten, J.A. Bergstra, and J.W. Klop. Conditional axioms and α/β calculus in process algebra. In M. Wirsing, editor, *Formal Description of Programming Concepts – III, Proceedings of the 3th IFIP WG 2.2 working conference,* Ebberup 1986, pages 53–75, Amsterdam, 1987. North-Holland.

[BBK87b] J.C.M. Baeten, J.A. Bergstra, and J.W. Klop. Decidability of bisimulation equivalence for processes generating context-free languages. In J.W. de Bakker, A.J. Nijman, and P.C. Treleaven, editors, *Proceedings PARLE conference,* Eindhoven, *Vol. II (Parallel Languages),* volume 259 of *Lecture Notes in Computer Science*, pages 94–113. Springer-Verlag, 1987.

[BK84] J.A. Bergstra and J.W. Klop. The algebra of recursively defined processes and the algebra of regular processes. In J. Paredaens, editor, *Proceedings 11th ICALP,* Antwerp, volume 172 of *Lecture Notes in Computer Science*, pages 82–95. Springer-Verlag, 1984.

[BT87] J.A. Bergstra and J. Tiuryn. Process algebra semantics for queues. *Fundamenta Informaticae*, X:213–224, 1987. Also appeared as: Report IW 241, Mathematisch Centrum, Amsterdam, 1983.

[BW90] J.C.M. Baeten and W.P. Weijland. *Process Algebra.* Cambridge Tracts in Theoretical Computer Science 18. Cambridge University Press, 1990.

[Cau90] D. Caucal. On the transition graphs of automata and grammars. Report 1318, INRIA, 1990.

[Chr92] S. Christensen. Distributed bisimilarity is decidable for a class of infinite state-space processes. In W.R. Cleaveland, editor, *CONCUR '92*, volume 630 of *Lecture Notes in Computer Science*, pages 148–161, 1992.

[HU79] J.E. Hopcroft and J.D. Ullman. *Introduction to Automata Theory, Languages and Computation.* Addison-Wesley, 1979.

[Par81] D.M.R. Park. Concurrency and automata on infinite sequences. In P. Deussen, editor, 5th *GI Conference*, volume 104 of *Lecture Notes in Computer Science*, pages 167–183. Springer-Verlag, 1981.

[Pon91] A. Ponse. Process expressions and Hoare's logic. *Information and Computation*, 95(2):192–217, 1991.

[SCS92] H. Hüttel S. Christensen and C. Stirling. Bisimulation is decidable for all context-free processes. In W.R. Cleaveland, editor, *CONCUR '92*, volume 630 of *Lecture Notes in Computer Science*, pages 138–147, 1992.

[Vaa90] F.W. Vaandrager. Process algebra semantics of POOL. In J.C.M. Baeten, editor, *Applications of process algebra*, Cambridge Tracts in Theoretical Computer Science 17, pages 173–236. Cambridge University Press, 1990.

[Vaa91] F.W. Vaandrager. On the relationship between process algebra and I/O automata. In *Proceedings LICS 91*, pages 387–398. Proceedings of the IEEE, 1991.

[Ver92] C. Verhoef. *Linear unary operators in proecess algebra.* PhD thesis, University of Amsterdam, 1992.

Normed BPP and BPA

Javier Blanco
Department of Computing Science, Eindhoven University of Technology
Eindhoven, The Netherlands. E-mail: javier@win.tue.nl

Abstract

Some properties of normed processes that belong both to BPP (Basic Parallel Processes) and to BPA (Basic Process Algebra) are established. Also, the intersection of normed BPP and λ(BPA) is examined. They are used to solve certain problems concerning decidability of bisimulation equivalence.

1 Introduction

There exist algorithms for deciding bisimulation equivalence in BPP (Basic Parallel Processes) and in BPA (Basic Process Algebra) (see [BBK93, Cau90a, Chr93, HS91]). An interesting question is whether one can decide if a BPP process is bisimilar with a BPA process.

In both cases an important subset for which the problem of deciding bisimulation is easier, is the set of normed processes. A process graph is normed if from any node, there exists a finite sequence of steps to the termination state. In this article the previously mentioned problem for normed processes is solved. The strategy to solve the problem is the following: given a normed BPP process we can decide if it is also a BPA process, in which case we can construct a BPA specification for it and therefore use the algorithm to decide bisimulation in BPA.

Also, given a normed BPP process, we can decide whether it is a λ(BPA) process. The decidability problem of bisimulation for λ(BPA), even for the normed case, is an open problem.

2 λ(BPA)

There are plenty of references about the state operator, hence we are not going to repeat them all here ([BB88, BB91, Bla92]).

The state operator is defined by a finite set $\mathbf{S}$ and two functions

$$act : \mathbf{A} \times \mathbf{S} \longrightarrow \mathbf{A}$$

$$\mathit{eff} : \mathbf{A} \times \mathbf{S} \longrightarrow \mathbf{S}$$

The function *act* acts as the renaming part of the state operator, while *eff* gives the new state after the execution of an action.

It is also required that δ is inert, i.e. *act*(δ,s) = δ and *eff*(δ,s) = s for all states s. We say that *act* and *eff* are trivial when *act*(a,s) = a and *eff*(a,s) = s, respectively. The axioms and operational semantics are given below.

SO1	$\lambda_s(a) = act(a,s)$
SO2	$\lambda_s(a \cdot x) = act(a,s) \cdot \lambda_{\mathrm{eff}(a,s)}(x)$
SO3	$\lambda_s(x+y) = \lambda_s(x) + \lambda_s(y)$

Table 1: Axioms for the state operator

$$\frac{x \xrightarrow{a} x', \mathrm{act}(a,s) \neq \delta}{\lambda_s(x) \xrightarrow{\mathrm{act}(a,s)} \lambda_{\mathrm{eff}(a,s)}(x')}$$

$$\frac{x \xrightarrow{a} \surd, \mathrm{act}(a,s) \neq \delta}{\lambda_s(x) \xrightarrow{\mathrm{act}(a,s)} \surd}$$

Table 2: Operational Semantics for the state operator

The class of processes λ(BPA) is defined as the processes obtained by an application of the state operator to a recursive definition in BPA. The graphs of processes in λ(BPA) coincide with the push-down automaton graphs ([Bla92, Cau90b]).

We list now some results about λ(BPA) graphs form [Cau90b, MS85].

Definition 2.1. Let P be a normed λ(BPA) graph. For any vertex v with norm $|v|$ we define $P(v)$ as the connected component containing v in the subgraph of P generated by the vertices with norm greater or equal than $|v|$

The vertices of $P(v)$ with norm equal to $|v|$ are called *frontier points* □

Definition 2.2. An *end-isomorphism* is a label preserving graph isomorphism that maps frontier points onto frontier points □

Theorem 2.3. *Let P be a normed $\lambda(BPA)$ process. The set*

$$\{P(v) | v \textit{ is a vertex of } P\}$$

has only finitely many isomorphism classes under end-isomorphism.

Corollary 2.4. *For any n, the number of frontier points of norm n is bounded.*

3 Basic Parallel Processes

The BPP specifications use alternative composition, parallel composition and action prefix. The general sequential composition is not allowed. In [Chr93] a normal form was found for specifications in BPP. Throughout this section we will assume BPP specification E with variables $\mathrm{Var}(E) = \{X_1 \dots X_n\}$, such that each variable is defined as

$$X_i = \sum_{j=1}^{m_i} a_{ij} \cdot \alpha_{ij}$$

where α_{ij} is a parallel composition of variables which will be identified with a multiset (eventually empty), when it is convenient. We also assume that every variable appears in at least one multiset reachable from the root (otherwise it can be removed from the specification).

Definition 3.1. The following relation on $\mathrm{Var}(E)$ is defined:

$X_i \to X_k$ iff there exists an α_{ij} with $X_k \in \alpha_{ij}$

Furthermore, we define the set

$R(X) = \{Y \in \mathrm{Var}(E) : X \to^+ Y\}$

□

Definition 3.2.

- A variable X is *bounded* if there exists a number n such that in any state of the graph of E the multiplicity of X is at most n.
- a variable is *unbounded* if it is not bounded.

□

Fact 3.3. An infinite state BPP process has at least one unbounded variable.

□

The following theorem will be useful in finding a necessary condition for a BPP process to belong to $\lambda(\mathrm{BPA})$.

Proposition 3.4. *Let Y be an unbounded variable. If $X \in R(Y)$, then X is also unbounded*

Lemma 3.5. *Given a specification E, a variable Y is unbounded if and only if at least one of the following properties holds:*

1. *there is another unbounded variable X such that $Y \in R(X)$*
2. $Y \xrightarrow{\sigma}{}^{*} Y^n, n \geq 2$
3. *there exist a variable X and a path $X \xrightarrow{\sigma}{}^{*} X \parallel Y$.*

Proof. It is clear that if one of the conditions holds then Y is unbounded. Now, suppose that Y is unbounded and that the first two conditions are not true. Since Y is unbounded, there is an infinite path

$$\alpha_1 \xrightarrow{\sigma_1}{}^* \alpha_2 \parallel Y^{\parallel k_1} \xrightarrow{\sigma_2}{}^* \dots \xrightarrow{\sigma_n}{}^* \alpha_n \parallel Y^{\parallel k_{n-1}} \xrightarrow{\sigma_{n+1}}{}^* \dots$$

where for all $i, k_i < k_{i+1}$ and $Y \notin \alpha_i$.

Moreover, since condition 1 is not true we can assume that for all i, all variables in α_i are bounded. Otherwise we can take another infinite path where we take out all the unbounded variables from the α_i. Note that we can do this because every variable is normed.

For cardinality reasons we know then that there exist α, σ such that $\alpha \parallel Y^{\parallel k} \xrightarrow{\sigma}{}^* \alpha \parallel Y^{\parallel k+r}, r > 0$. We can assume that $r = 1$ since Y is normed. From the fact that condition 2 is not true it follows that there exists σ such that $\alpha \xrightarrow{\sigma}{}^* \alpha \parallel Y$, because Y cannot reduce to a variable in α since they are all normed. Again we can assume that all variables of α are used.

Assume that for any $X \in \alpha, X \xrightarrow{\sigma_i}{}^* \beta_i$, such that $\bigcup \beta_i = \alpha \cup \{Y\}$. This can be done because one can split the computation σ according to the variable in α that originates each step. If some of the β_i is empty, then we can remove the variable from α (again because it is normed). If one of the $\beta_i = \{Y\}$, say $X \xrightarrow{\sigma_i}{}^* Y$ then there is another variable X' such that $X' \xrightarrow{\sigma_{i'}}{}^* X$. Then we can get rid of the X. In this way we can reduce the α until we have a variable X such that $X \xrightarrow{\sigma_i}{}^* X' \parallel Y$. Since all other variables reduce through their correspondent σ_j to exactly one variable in α, then it follows that $X \in R(X')$, since X must be also in α. □

Remark 3.6. We use $Y^{\parallel k}$ to indicate $Y \parallel Y \parallel \dots \parallel Y$ (k- times) and Y^k for $Y \cdot Y \cdots Y$. □

4 BPP ∩ λ(BPA)

Theorem 4.1. *Let E be a canonical BPP specification and X, Y two variables from E such that in the graph defined by E there is an infinite number of p's and of q's such that $X^{\parallel p} \parallel Y^{\parallel q}$ is the label of a node, then there is no specification in λ(BPA) of the same process*

Proof. Assume that the norms of X and Y are given by

$$|X| = m$$

$$|Y| = n$$

then, it is immediate that for any p, q

$$|X^{\parallel p} \parallel Y^{\parallel q}| = pm + qn$$

Since all variables are normed then $X^{\|p} \parallel Y^{\|q}$ is the label of a node for *any* p, q. Moreover there is always a path from $X^{\|(p+p1)} \parallel Y^{\|(q+q1)}$ to $X^{\|p} \parallel Y^{\|q}$ for any $p1, q1$. This implies that if we take only the nodes which norm is greater than a constant, then two nodes of this form will belong to the same connected component. Now we will show that the number of frontier points is unbounded.

For any number k the number of nodes with norm equal to kmn is at least $k+1$.

$$|X^{\|kn}| =$$

$$= |X^{\|(k-1)n} \parallel Y^{\|m}| =$$

$$\ldots$$

$$= |X^{\|(k-i)n} \parallel Y^{\|im}| =$$

$$\ldots$$

$$= |Y^{\|km}| =$$

$$= kmn$$

□

Finally, the following theorem illustrates that this condition is not only necessary, but sufficient as well.

Theorem 4.2. *A BPP process that does not have a pair of unboundded variables as in theorem 4.1 is also a $\lambda(BPA)$ process.*

Proof. We prove it for the case when there is only one unbound variable. The general case is similar. Let

$$E = \{X_i = \sum a_{ij} \cdot \alpha_{ij} : 1 \leq i \leq n\} \cup \{Y = \sum b_j \cdot \beta_j\}$$

Define $\#_X(\alpha)$ as the multiplicity of X in the multiset α. Assume Y is the unbound variable and for all i let m_i be the bound corresponding to X_i. Now define

$$\begin{aligned} B \;=\; & \sum_i \sum_j (i, a_{ij}) \cdot B'^{\#_Y(\alpha_{ij})} \cdot B + \sum_{\#_Y(\beta_j) \geq 1} b_j \cdot B'^{\#_Y(\beta_j)-1} \cdot B + \\ & \sum_{\#_Y(\beta_j)=0} (b'_j \cdot C + b''_j) \end{aligned}$$

$$\begin{aligned} B' \;=\; & \sum_i \sum_j (i, a_{ij}) \cdot B'^{\#_Y(\alpha_{ij})+1} + \sum_{\#_Y(\beta_j) \geq 1} b_j \cdot B'^{\#_Y(\beta_j)} + \\ & \sum_{\#_Y(\beta_j)=0} + b_j \end{aligned}$$

$$C = \sum \sum_{\#_Y(\alpha_{ij}) \geq 1} (i, a_{ij}) \cdot B'^{\#_Y(\alpha_{ij})-1} \cdot B +$$
$$\sum \sum_{\#_Y(\alpha_{ij})=0} (i, a_{ij})' \cdot C + \sum \sum_{\#_Y(\alpha_{ij})=0} (i, a_{ij})''$$

The state space is defined as

$$S = \{X_1^{k_1} \parallel \cdots \parallel X_n^{k_n} : k_i \leq m_i\}$$

The functions *act* and *eff* are defined as follows:

- $act((i, a_{ij}), X_1^{k_1} \parallel \cdots \parallel X_n^{k_n}) = \delta$, if $k_i = 0$
- $act((i, a_{ij}), X_1^{k_1} \parallel \cdots \parallel X_n^{k_n}) = a_{ij}$, if $k_i \geq 0$
- $act((i, a_{ij})', X_1^{k_1} \parallel \cdots \parallel X_n^{k_n}) = \delta$, if $k_i = 0$ or for all $j \neq i, k_j = 0$
- $act((i, a_{ij})'', X_1^{k_1} \parallel \cdots \parallel X_n^{k_n}) = \delta$, if $k_i = 0$ or there exist a $j \neq i$ such that $k_j \geq 0$
- $act((i, a_{ij})', X_1^{k_1} \parallel \cdots \parallel X_n^{k_n}) = a_{ij}$, if $k_i \geq 0$ and there exist a $j \neq i$ such that $k_j \geq 0$
- $act((i, a_{ij})'', X_1^{k_1} \parallel \cdots \parallel X_n^{k_n}) = a_{ij}$, if $k_i \geq 0$ and for all $j \neq i, k_j = 0$
- $act(b'_j, X_1^{k_1} \parallel \cdots \parallel X_n^{k_n}) = \delta$, if for all $j \neq i, k_j = 0$
- $act(b''_j, X_1^{k_1} \parallel \cdots \parallel X_n^{k_n}) = \delta$, if there exist a $j \neq i$ such that $k_j \geq 0$
- $act(b'_j, X_1^{k_1} \parallel \cdots \parallel X_n^{k_n}) = b_j$, if there exist a i such that $k_i \geq 0$
- $act(b''_j, X_1^{k_1} \parallel \cdots \parallel X_n^{k_n}) = b_j$, if for all $i, k_i = 0$
- $eff((i, a_{ij}, X_1^{k_1} \parallel \cdots \parallel X_n^{k_n}) = X_1^{k_1+(\#_{X_1}(\alpha_{ij}))} \parallel \cdots \parallel X_i^{k_i+(\#_{X_i}(\alpha_{ij})-1)} \parallel \cdots \parallel X_n^{k_n+(\#_{X_n}(\alpha_{ij}))}$

Accordingly, it is easy to prove that

$$\lambda_{X_1^{k_1} \parallel \cdots \parallel X_n^{k_n}}(C) = X_1^{\parallel k_1} \parallel \cdots \parallel X_n^{\parallel k_n}$$

and

$$\lambda_{X_1^{k_1}\|\cdots\|X_n^{k_n}}(B'^r \cdot B) = X_1^{\|k_1} \| \cdots \| X_n^{k_n} \| B^{\|r+1}$$

□

Example 4.3. We provide an example in which there is more than one unbounded variable

$$Z = a \cdot X \| M_2 + b \cdot Y \| M_1$$

$$X = p_1 \cdot X \| M_1 + t_1$$

$$Y = p_2 \cdot Y \| M_2 + t_2$$

$$M_1 = m_1$$

$$M_2 = m_2$$

The corresponding specification in λ(BPA) is given by:

$$B_1 = (X, p_1) \cdot B_1' \cdot B_1 + t_1 \cdot B_1 + m_2 \cdot B_1 + m_1' \cdot C_1 + m_1''$$

$$B_1' = (X, p_1) \cdot B_1' \cdot B_1' + t_1 \cdot B_1' + m_2 \cdot B_1' + m_1$$

$$C_1 = (X, p_1) \cdot B_1 + (X, t_1)' \cdot C_1 + (X, t_1)'' + (X, m_2)' \cdot C_2 + (X, m_2)''$$

$$B_2 = (X, p_2) \cdot B_2' \cdot B_2 + t_2 \cdot B_2 + m_1 \cdot B_2 + m_2' \cdot C_2 + m_2''$$

$$B_1' = (X, p_2) \cdot B_2' \cdot B_2' + t_2 \cdot B_2' + m_1 \cdot B_2' + m_2'$$

$$C_1 = (X, p_2) \cdot B_2 + (X, t_2)' \cdot C_2 + (X, t_2)'' + (X, m_1)' \cdot C_1 + (X, m_1)''$$

$$A = a \cdot B_1 + b \cdot B_2$$

Where the *act* and *eff* functions are defined in the obvious way, for example:

$$\lambda_{\emptyset}(A) = a \cdot \lambda_{M_2}(C_1) + b \cdot \lambda_{M_1}(C_2)$$

$$\lambda_{Y\|M_1}(B_2') = p_2 \cdot \lambda_{Y\|M_1}(B_2' \cdot B_2') + m_2 + t_2 \cdot \lambda_{M_1}(B_2') + m_1 \cdot \lambda_Y(B_2')$$

□

5 BPA

Some results concerning BPA processes are recalled in this section.

Definition 5.1. Given a graph of a process, a vertex s is a multiple start for a vertex t, if there exist two paths from s to t such that their only common vertices are s and t. □

The following proposition is taken from [Cau90b].

Proposition 5.2. *The set of multiple starts for any state of a BPA graph is finite.*

Theorem 5.3. *Let E specify a process in BPP $\cap$ BPA with an unbound variable Y. Then Y cannot appear an infinite number of times in a state with another variable.*

Proof. Suppose that Y appears an infinite number of times beside X. Given that the process is normed we know then that $X \parallel Y^{\parallel p}$ is a state for any natural number n. There exists a path

$$X \xrightarrow{a_1} \alpha_1 \xrightarrow{a_2} \cdots \xrightarrow{a_n} Y^{\parallel r}$$

where $r \geq 0$ and all α_i are different than $Y^{\parallel s}$ for any s. (This is easy, take the first state of the form $Y^{\parallel r}$ in a path from X to the termination state or take $r = 0$ if no Y is encountered) Now the set

$$\{X \parallel Y^{\parallel j} : j > r\}$$

is a set of multiple starts for Y^r. We can take the path from $X \parallel Y^{\parallel j}$ to X and then from there to $Y^{\parallel r}$ or take the path from $X \parallel Y^{\parallel j}$ to $Y^{\parallel j+r}$ and then reduce it in j steps to $Y^{\parallel r}$. It is clear that the paths satisfy the properties of a multiple start. (Note that because Y cannot do a step to another variable and it is normed, then it must be able to terminate in just one step.) □

Corollary 5.4. *Given an infinite process in BPP $\cap$ BPA with an unbound variable Y, there is no variable X (different than Y) such that $X \xrightarrow{\sigma}{}^{*} X \parallel Y$.*

Corollary 5.5. *An unbounded variable cannot appear with a different variable in a state at all.*

Proof. From corollary 5.4, the only possibility for Y to be unbound is that it can do a step to $Y^{\parallel k}$ with $k \geq 2$. Yet, if there is a state of the form $X \parallel Y$ then there is also a $X \parallel Y^{\parallel j}$ for any j. □

Finally, all this gives us an algorithm to decide whether a BPP process is bisimilar to a BPA process.

Theorem 5.6. *Given a BPP specification we can decide if the process defined belongs to BPA, in which case we can construct a BPA specification for it.*

Proof. Using the algorithm in [Chr93] for normed processes based on unique decomposition, we can find a specification of the canonical graph of that process. One can easily calculate whether a variable is unbounded using lemma 3.5. If more than one unbounded variable appears together, then the process does not belong to BPA. Otherwise we know that it already belongs to λ(BPA). If it is the case that for every unbounded variable Y no other variable X can perform a sequence of actions to $X \parallel Y$, then the process is in BPA. To obtain a specification in BPA for this process, first find a linear specification (as in [BW90]) for the regular process obtained by removing all occurrences of the unbounded variables. For every unbounded variable Y, if a variable in the original BPP specification has a summand of the form $a \cdot Y^{\parallel r}$, then add a sumand $a \cdot Y^{\parallel r}$ to the corresponding variable in the new specification. Finally, add an equation for every unbouded variable where all $\parallel$ are replaced by $\cdot$. □

Example 5.7. The BPP specification

$$X = a \cdot (Z \parallel W) + b \cdot (Y \parallel Y \parallel Y) + b \cdot Y'$$

$$Y = c \cdot (Y \parallel Y) + d$$

$$Y' = c' \cdot (Y' \parallel Y' \parallel Y') + d'$$

$$Z = e \cdot Z + f$$

$$W = g$$

is transformed into

$$A = a \cdot B + b \cdot Y \cdot Y \cdot Y + b \cdot Y'$$

$$B = e \cdot B + f \cdot D + g \cdot C$$

$$C = e \cdot C + f$$

$$D = g$$

$$Y = c \cdot Y \cdot Y + d$$

$$Y' = c' \cdot (Y' \cdot Y' \cdot Y') + d'$$

□

6 Conclusion

In this article we have shown another use for the known algorithms deciding bisimulation equivalence. The techniques introduced may be helpful in extending these results to certain sets of processes for which the decidability of bisimulation is still unknown.

References

[BB88] J.C.M. Baeten and J.A. Bergstra. Global renaming operators in concrete process algebra. *Information and Computation*, 78(3):205–245, 1988.

[BB91] J.C.M. Baeten and J.A. Bergstra. Recursive process definitions with the state operator. *Theoretical Computer Science*, 82:285–302, 1991.

[BBK93] J.C.M. Baeten, J.A. Bergstra, and J.W. Klop. Decidability of bisimulation equivalence for processes generating context-free languages. *Journal of the ACM*, 40(3):653–682, 1993.

[Bla92] J. Blanco. Definability with the state operator in process algebra. Report P9221, University of Amsterdam, Amsterdam, 1992.

[BW90] J.C.M. Baeten and W.P. Weijland. *Process Algebra*. Cambridge Tracts in Theoretical Computer Science 18. Cambridge University Press, 1990.

[Cau90a] D. Caucal. Graphes canoniques de graphes algébriques. *Theoretical Informatics and Applications*, 24(4):339–352, 1990.

[Cau90b] D. Caucal. On the transition graphs of automata and grammars. Report 1318, INRIA, 1990.

[Chr93] S. Christensen. *Decidability and Decomposition in Process Algebra*. PhD thesis, Department of Computer Science, University of Edinburgh, Edinburgh, 1993.

[HS91] H. Hüttel and C. Stirling. Actions speak louder than words: Proving bisimilarity for context-free processes. In *Proceedings* 6th *Annual Symposium on Logic in Computer Science*, Amsterdam, The Netherlands, pages 376–386. IEEE Computer Society Press, 1991.

[MS85] D.E. Muller and P.E. Schupp. The theory of ends, pushdown automata, and second order logic. *Theoretical Computer Science*, 37:51–75, 1985.

A Real Time μCRL Specification of a System for Traffic Regulation at Signalized Intersections

M.J. Koens
University of Amsterdam
The Netherlands

L.H. Oei
University of Amsterdam
The Netherlands

Abstract

This paper takes up the challenge advanced in Report P9313 of the Programming Research Group of the University of Amsterdam [VW93]. It provides an alternative way of specifying an existing traffic regulation system at signalized intersections of the firm **Nederland Haarlem**[1]. The specification proposed in this paper features the major differences with the previous attempt, which are threefold:

1. it is process oriented rather than data oriented, i.e. each phase is modeled as a process that runs in parallel with the others. Furthermore there is one central control process that decides which phases are realized (turn green).
2. it uses the $\rho\mu$CRL (Real Time Micro Common Represention Language), a Real Time extension of μCRL, instead of PSF (Process Specification Formalism);
3. it features Prefix Integrated Real Time Algebra of Communicating Processes (ACP).

Key words & phrases: algebraic specification, real time process algebra, $\rho\mu$CRL, traffic regulation.

1 Introduction

This paper has been written in the context of the course Real Time Process Algebra at the University of Amsterdam and brings a real time specification of a Traffic Regulation System. As specification language we chose μCRL as described in [GP90]. We did not use the time extended version $r\mu$CRL as found in [Fok92], because it uses absolute time instead of relative time, and it has a notion of integration different from the kind we were familiar with. μCRL features data, but is untimed, so we added prefixed integration and relative time and named this extension $\rho\mu$CRL. The theory on which these additions are based can be found in [Klu93], which is based on [BB91].

We would like to stress here that as far as we know our attempt to specify such a large domain in $\rho\mu$CRL has no precedence and brings about all the obvious experimental hazards one can think of. Our specification of a traffic regulation system is based upon the work of Bas van Vlijmen and Arjan van Waveren [VW93]. They in turn have based their work on Nederland Haarlem documents written by Jan Kroone [Kro90].

[1]Nederland Haarlem, P.O. Box 665, 2003 RR, Haarlem, the Netherlands

In this paper we first will give a short description of the problem to be specified. After a brief intermezzo of some theoretical backgrounds we then give a thorough explanation of the specification. The next chapter describes the major differences with the specification as found in [VW93]. Finally the specification itself is appended.

2 Acknowledgments

We would like to express our gratitude to the following persons : Jan Bergstra for his provocative manner of teaching and his interest for reading our draft, Steven Klusener for his dedication to his assistantship and his innumerable tips and hints, Bas van Vlijmen for being the in between with Nederland Haarlem and Willem Jan Fokkink for his draft of rμCRL syntax in ASF-SDF, which has motivated us in using Real Time μCRL. We would like to thank Nederland Haarlem especially for giving us the opportunity to present and communicate our ideas.

3 Dynamic Traffic Regulation System at Signalized Intersections

Within this paper we will maintain the jargon proposed by Nederland Haarlem as to maintain downward compatibility with the origins of this specification. In our strive for conciseness we have decided to describe the problem by a list of definitions, which is essentially a summary of the exposition given by [VW93] and indirectly also of [Kro90]:

3.1 Basic Traffic Jargon

3.1.1 Objects

Streams are subsections of the influx of traffic participants into the intersection scene from one particular road branch, e.g., one queuing section for cars wanting to turn left of a T-intersection would be one indivisible stream. To each stream a constant weight factor is attributed.

Signalized intersections are traffic intersections where every stream is regulated with a signal; typically a traffic light.

Phases are collections of streams synchronized by effectively the same signal. This signal could be split into a special car, bicycle or pedestrian signal, but as the behavior of these signals are completely synchronous, these various streams would make one phase. The meaning of *phase* is therefore idiomatic for Nederland Haarlem and should not be confused with the colloquial semantics of changing states.
The following constants are attributed to each phase:

- a minimal green time,
- a clearance duration,
- a yellow time,

- and an absolute priority, to be able to order phases when their weighed waiting times are the same.

In our specification each phase has been modeled as a process that accounts for its own weighed waiting time, which depends linearly on the number of traffic participants waiting and their weight factors. (see below: *Dynamic priorities accounting*).

Detection devices are associated with each stream. In our specification each stream has one detection device. Nederland Haarlem specifies detectors for pedestrians to be able to detect one pedestrian only. They also carry detection devices that are sensitive to special traffic like ambulances. Since the internal behavior of detection devices is not our goal of study, we have modeled detectors as atomic actions without communications. The detection of a traffic participant then becomes just an event.

3.1.2 Objectives

Realization of a phase means that its signal turned green. A phase that has been cleared (see definition below) and has traffic waiting is requesting for realization.

Clearance duration of a phase is the time constant associated with the phase in which it is expected to have 'dispatched' its realized traffic participants and has effectively forced all incoming traffic to a standstill. Popularly speaking: the phase has been red long enough after being green and yellow so as not to endanger further development on the critical area of the intersection.

3.2 Realization paraphernalia

3.2.1 The four order regulation system

The four order regulation system is a proprietary dynamic scheduling algorithm developed by Nederland Haarlem, a company that is leading the market of traffic regulation systems. The sophistication of this algorithm is expressed by not only achieving the primary goal of a traffic regulation system, namely: the absence of collisions, but also in the optimalization of throughput without frustrating the sense of fairness associated with dynamic configuration of waiting traffic participants with respect to time.

The algorithm of the four order regulation system resides in the control process. The control process loops over a time constant that in real life situations amounts to one second. The control process has been modeled by a single process within our specification.

3.2.2 The conflict matrix

A conflict matrix, the matrix of pairs of conflicting phases (phases that physically cannot be realized at the same time — since they would simply collide — form pairs), is associated with the control process as its main datastructure. [2]

[2] Note that the specified conflict matrix here is like in [VW93] a simplification of the real conflict matrix which contains a lot more data.

3.2.3 Dynamic priority accounting

Like in the previous study the dynamic priority of each phase is calculated each second. It is determined by the weighed waiting time of the phases and their absolute priorities in case some waiting times are equal.

The weighed waiting time is calculated each accounting cycle (in our specific case: one tenth of a second). In every cycle it is increased with the product of the number of traffic participants waiting and their associated weight factor.

3.2.4 First order realization

A phase will be realized first order if:

1. it is requesting realization,
2. it has higher priority than any other conflicting phase requesting realization,
3. and is not in conflict with non-cleared phases (read: moving traffic).

3.2.5 Second order realization

A phase will be realized second order if:

1. it is requesting realization, but cannot be realized first order,
2. and it does not frustrate any waiting phases of higher priority,
3. and is not in conflict with non-cleared phases.

Intuitively: a phase can be realized second order, because the phases with higher priority it could block, are already blocked by another phase. A phase realized in second order is just traffic that is 'covered' by a phase that will get a higher order realization or a phase that is non-cleared.

3.2.6 Third order realization

A phase will be realized third order if:

1. it is cleared and it has no traffic waiting,
2. and it can be realized without frustrating a cleared phase with a higher priority that has no traffic waiting,
3. and it can be realized without frustrating a phase with a request for realization pending,
4. and it is not in conflict with any of the non-cleared phases.

Within our specification the third order realization algorithm is immersed in the second order realization function: phases with no traffic waiting get the lowest priorities, so they are treated last automatically in our specification, without frustrating others.

3.2.7 Fourth order realization

Fourth order realizations are realizations granted to vehicles owning special privileges like: fire patrol, police, ambulance, public transportation and the like. This feature does not figurate in our specification, but we feel justified in mentioning it since the nomenclature of the algorithm would otherwise be very puzzling.

3.2.8 Green time accounting

The green time granted to a realized phase into first order is taken as the maximum of the minimal greentime and an estimate of the requested greentime. As a simplification the latter is equated to the number of traffic participants waiting.

Since a second order realization is not allowed to frustrate other waiting phases, it must be cleared again at the moment the first of the conflicting waiting phases with a higher priority is no longer frustrated by another realized phase.

4 Some Theoretical Backgrounds

4.1 Syntax: Actions, Relative Time, Prefixed Integration

Time stamped actions take no time and can be performed immediately after each other. $a[t]$ denotes an action a to be performed t time steps after the previous action. [3]

An integral preceding a time stamped action is used to give the possibility to perform an action at *some* time given by a condition, typically a time interval.

A silent step (or internal step) denoted by the greek letter τ, is the abstraction from internal behaviour. Communications are renamed to τ's.

4.2 Examples of Process Algebra Semantics

- $a[2] \cdot b[1]$ means: after waiting two time steps perform action a and after waiting another time step perform action b; the $\cdot$ is the sequential operator.
- $a[2] \cdot b[0]$ means: perform after two time steps a then immediately b.
- $a[2] + b[4]$ means: perform either a after two time steps or b after four time steps. The $+$ is the alternative operator; the choice is nondeterministic. After waiting two time steps this process has been evolved into $a[0] + b[2]$,

[3] In this document actions are implicitly *urgent* actions as described in [Klu93], e.g. $a[0]$ means that a must be executed *immediately* by which we allow consecutive actions at the same point in time.

The square brackets explicitly determines the relative time setting. An alternative would be absolute time denoted by $a(t)$.

idling forth determines a choice for the b; by idling another time step $a[0]+b[2]$ evolves to $b[1]$. Note that $a[0]+b[2]$ can also execute the action a immediately.

- $P\|Q$ means: perform P in parallel with Q.
- $\int_{v>2}(a[v]\cdot b[v+1])$ means: perform action a after waiting some period $v>2$ and then perform b after waiting another period $v+1$.
- $a(p:Pid)[t]$ means: action a parametrized with the data type Pid. $a(1)[t]$ would be in this context an action a instantiated by the process-id '1'.
- $\sum_{p\in Pid} Pcleared(p)$ means: the generalized summation over the alternative compositions over the set of process identifiers, it rewrites to: $Pcleared(1)+Pcleared(2)+...+Pcleared(12)$ in case $Pid=1,\ldots,12$.
- $P(w:Time)=\int_{v\in[0,w\rangle} detect[v]\cdot P(w-v)+\tau[w]\cdot P(cycletime)$ means: the process P can either detect traffic participants between now and time w and then call itself with the remaining part of w, or wait until w and perform the internal action τ after which it calls itself, where w is set to $cycletime$.

5 The Specification

5.1 The Process Equations

5.1.1 The Phase Processes

The P processes describe the behavior of a phase.

- $P_{red}^{cleared}$ describes the behavior of a phase when it is cleared and turned red.

$$P_{red}^{cleared}(p:Pid, wt:WeighedTimeTable, w:Time)=$$

1. Its first parameter gives its identification number p, its Pid.
2. The next parameter is a table wt, containing information about the number of traffic participants waiting at each of its stream, the total number of traffic participants waiting and a weighed waiting time.
3. The last parameter is w, denoting the time it has to wait to recalculate its weighed waiting time.

This process can perform four actions. The first three actions can be performed during an interval from immediately to the time given by w.

$$\int_{v\in[0,w\rangle}\sum_{s\in Sid} detect(s)[v]\cdot P_{red}^{cleared}(p, update(wt,s), w-v)$$

In this interval it can detect a traffic participant at any of its streams, after which it calls itself with an updated wt and the remaining part of w.

$$\int_{v\in[0,w\rangle} send_{wt}(p, weighedTime(wt), totalTp(wt))[v]$$

$$\cdot P^{cleared}_{red}(p, wt, w - v)$$

It can also send the control process a tuple, holding the number of waiting traffic participants and the weighed waiting time. After this action the process calls itself as well.

$$\int_{v\in[0,w\rangle} \sum_{t\in Time} rec_{realization}(p,t)[v] \cdot set_{green}(p)[0] \cdot P_{green}(p,t)$$

The third possibility is to receive a realization message from the control process, holding the length of the period it has to be green. After receiving such a message the process performs a set_{green}, turning the lights green, and changes its behavior by calling P_{green}. P_{green} needs its Pid and the greentime, so these are given as parameters.

$$\tau[w] \cdot P^{cleared}_{red}(p, calculate_{wt}(p, wt), cycletime)$$

The last alternative of $P^{cleared}_{red}$ is to do nothing until the end of the aforementioned period (by performing a $\tau[w]$) and call itself again with a table wt with updated weighed waiting time and a new time ($cycletime$) to denote when to update the weighed waiting time again.

$$\begin{aligned} & P^{\neg cleared}_{red}(p : Pid, wt : WeighedTimeTable, w : Time) \\ = \\ & \int_{v\in[0,w\rangle} \sum_{s\in Sid} detect(s)[v] \cdot P^{\neg cleared}_{red}(p, update(wt,s), w - v) \\ & + \tau[w] \cdot P^{cleared}_{red}(p, calculate_{wt}(wt), cycletime) \end{aligned}$$

- $P^{\neg cleared}_{red}$ has a similar behavior as $P^{cleared}_{red}$, but it will never receive a message to turn green from the control process, nor will it be asked to send its waiting information. The w denotes the clearance duration. After this period the phase is cleared and $P^{cleared}_{red}$ is called.

$$P_{green}(p : Pid, t : Time) = set_{yellow}(p)[t] \cdot P_{yellow}(p)$$

- P_{green} is called immediately after the lights at a phase are turned green. The second parameter gives the time after which it must turn the lights to yellow and call P_{yellow}.

$$\begin{aligned} P_{yellow}(p : Pid) &= set_{red}(p)[yellowtime(p)] \\ &\quad \cdot P_{red}^{\neg cleared}(p, initstreams(p), clearance(p)) \end{aligned}$$

- P_{yellow} turns the lights from yellow to red again. The period after which this must happen is a constant given by the function *yellowtime*.

5.1.2 The Control Process

The CP processes describe the behavior of the control process.

- CP_{rec} (which abbreviates $CP_{receive}$) has three parameters.

  1. A set of cleared phases $\langle p, cs \rangle$ (first element is p) [4],
  2. a set of non-cleared phases ncs and
  3. a set ws of cleared phases which have sent their waiting information.

$$\begin{aligned} & CP_{rec}(\langle p, cs \rangle : PidSet, ncs : PendingSet, ws : WaitingSet) \\ = & \\ & \sum_{t \times tp \in Time \times Int} rec_{wt}(p, t, tp)[0] \cdot CP_{rec}(cs, ncs, \langle p, t, tp, ws \rangle) \end{aligned}$$

  The first parameter gives the set of phases that may be realized. CP_{rec} asks with rec_{wt} a phase p in this set for its waiting information: its weighed waiting time (t) and the number of traffic participants waiting (tp).

  The received information is stored with the Pid of the phase in ws and CP_{rec} calls itself with the updated ws and a cs where the processed phase is left out.

$$\begin{aligned} & CP_{rec}(\emptyset : PidSet, ncs : PendingSet, ws : WaitingSet) \\ = & \\ & CP_{send}(peel(ws), ncs, realizations(ncs, ws)) \end{aligned}$$

  When the set of cleared phases is empty (all cleared phases have sent their waiting information), CP_{send} is called with the result of the function *realizations*. This function calculates the dynamic priority of the phases and then finds which phases can be realized for how long. The result of

[4] A set is implemented as a list, which has a first element, to be able to walk through the set.

realizations is a list of pairs of a phase and the time that phase must be turned green. To be able to prevent conflicts it needs the set of phases that are non-cleared as an extra parameter. This set can be derived by *peel*ing the time part and the traffic participants part from *ws*.

- CP_{send} is called with a set of cleared phases, a set of non-cleared phases and the set of phases that have to be realized together with their green-time.

$$CP_{send}(cs : PidSet, ncs : PendingSet, \langle p, t, ps\rangle : PendingSet)$$
$$=$$
$$send_{realization}(p,t)[0] \cdot CP_{send}(extract(p, cs),$$
$$\langle p, t + yellowtime(p) + clearance(p), ncs\rangle, ps)$$

CP_{send} takes a Pid p and a greentime from its third argument and sends the phase with Pid p the message to turn green for a period t.

It then calls itself with updated parameters: p is removed from the set of cleared phases cs and added to the set of non-cleared phases ncs and the processed phase is removed from the Pending-list denoting the set of phases that must set to green.

The time after which a phase is cleared is given by its greentime + its *yellowtime* + its *clearance* duration. This is the time that is stored with p in ncs.

$$CP_{send}(cs : PidSet, ncs : PendingSet, \emptyset : PendingSet)$$
$$=$$
$$\tau[controlcycletime] \cdot CP_{rec}(newcs(cs, ncs), newncs(ncs), \emptyset)$$

When all the phases that must turn green are informed (ps is empty) the control process waits a period *controlcycletime* and starts a new control cycle by calling CP_{rec} again. This is done with updated ncs and cs: the duration to clearance of every non-cleared phase as stored in ncs is decremented by *controlcycletime*.

When a duration to clearance is (less than or equal to) zero it means that this particular phase is cleared now and its Pid is moved from ncs to cs.

5.1.3 The Traffic Regulation System Process

Finally, process *RTRS* describes the overall process.

$$\begin{aligned} RTRS \;=\; & \tau_{\{comm_{realization}, comm_{wt}\}}(\\ & \partial_{\{send_{wt}, rec_{wt}, send_{realization}, rec_{realization}\}}(\\ & CP_{rec}(interPids, \emptyset, \emptyset) \| \\ & \|_{p \,\in\, interPids} \; (P^{cleared}_{red}(p, initstreams(p), cycletime)))) \end{aligned}$$

It starts a CP merged with one P for every phase in the particular intersection. All these processes are initialized with values specified in the actual intersection part of the specification. By renaming isolated send and receive actions to δ (*deadlock*) we force these actions to be performed in pairs, resulting in a communication.

These resulting communications are internal and are renamed to τ (*silent step*). We note that a hierarchy emerges because CP determines the time to communicate with P, while P accepts communication using an interval.

5.1.4 The Process Equations put together

$$
\begin{aligned}
& P_{red}^{cleared}(p : Pid, wt : WeighedTimeTable, w : Time) \\
= \\
& \int_{v \in [0,w\rangle} \sum_{s \in Sid} detect(s)[v] \cdot P_{red}^{cleared}(p, update(wt, s), w - v) \\
& + \int_{v \in [0,w\rangle} send_{wt}(p, weighedTime(wt), totalTp(wt))[v] \\
& \quad \cdot P_{red}^{cleared}(p, wt, w - v) \\
& + \int_{v \in [0,w\rangle} \sum_{t \in Time} rec_{realization}(p, t)[v] \cdot set_{green}(p) \cdot P_{green}(p, t) \\
& + \tau[w] \cdot P_{red}^{cleared}(p, calculate_{wt}(p, wt), cycletime)
\end{aligned}
$$

$$
\begin{aligned}
& P_{red}^{\neg cleared}(p : Pid, wt : WeighedTimeTable, w : Time) \\
= \\
& \int_{v \in [0,w\rangle} \sum_{s \in Sid} detect(s)[v] \\
& \cdot P_{red}^{\neg cleared}(p, update(wt, s), w - v) \\
& + \tau[w] \cdot P_{red}^{cleared}(p, calculate_{wt}(wt), cycletime)
\end{aligned}
$$

$$
P_{green}(p : Pid, t : Time) \quad = \quad set_{yellow}(p)[t] \cdot P_{yellow}(p)
$$

$$
\begin{aligned}
P_{yellow}(p : Pid) \quad = \quad & set_{red}(p)[yellowtime(p)] \\
& \cdot P_{red}^{\neg cleared}(p, initstreams(p), clearance(p))
\end{aligned}
$$

$$
\begin{aligned}
& CP_{rec}(\emptyset : PidSet, ncs : PendingSet, ws : WaitingSet) \\
= \\
& CP_{send}(peel(ws), ncs, realizations(ncs, ws))
\end{aligned}
$$

$$CP_{rec}(\langle p, cs\rangle : PidSet, ncs : PendingSet, ws : WaitingSet)$$
$$= \sum_{t \times tp \in Time \times Int} rec_{wt}(p,t,tp)[0] \cdot CP_{rec}(cs, ncs, \langle p,t,tp,ws\rangle)$$

$$CP_{send}(cs : PidSet, ncs : PendingSet, \emptyset : PendingSet)$$
$$=$$
$$\tau[controlcycletime] \cdot CP_{rec}(newcs(cs,ncs), newncs(ncs), \emptyset)$$

$$CP_{send}(cs : PidSet, ncs : PendingSet, \langle p,t,ps\rangle : PendingSet)$$
$$=$$
$$send_{realization}(p,t)[0] \cdot CP_{send}(extract(p,cs), \langle p, t + yellowtime(p) + clearance(p), ncs\rangle, ps)$$

$$\begin{aligned} RTRS \;=\; & \tau_{\{comm_{realization}, comm_{wt}\}}(\\ & \partial_{\{send_{wt}, rec_{wt}, send_{realization}, rec_{realization}\}}(\\ & CP_{rec}(interPids, \emptyset, \emptyset) \| \\ & \|_{p \in interPids} \; (P^{cleared}_{red}(p, initstreams(p), cycletime)))) \end{aligned}$$

5.2 Abstract Data Types and Operators

We will now explain the remaining part of the specification: the sorts and functions of the Abstract Datatypes

5.2.1 Generic Datatypes from Standard Libraries

As μCRL does not have "built-in" datatypes we have to provide specifications even for the simple datatypes as booleans and integers. The previous study of the Traffic Regulation System made use of a standard library of Abstract Data Type primitives: The PSF Library [Wam91]. Since μCRL does not allow us to import a library, we had to adapt these modules by hand.

- The sorts `Bool` and `Int` are self-explanatory. Integers, or numbers in general are especially handy for coding Object Id's. The sorts for instance `Pid` and `Sid` are used to identify phases and streams within phases. They are mapped to integers.

5.2.2 $\rho\mu$CRL Time Datatype

We specified sort `Time` as the Rational Numbers, because we were not able to specify Reals in a simple fashion. This means that our specification is not as 'Real' as the title might suggest. An action at time sqrt(2) cannot occur. On the other hand an action very near to sqrt(2) with any desired precision can be modeled.

The result of a calculation is not simplified, i.e. `plus(R(1,2),R(1,2))` $\rightarrow$ `R(2,2)` ($\frac{1}{2}+\frac{1}{2} \rightarrow \frac{2}{2}$), but this does not cause problems, as we have an equality function `eq` as well by which `eq(R(2,2), R(1,1))` reduces to true.

A sort `Bool` and a sort `Time` are essential in $\rho\mu$CRL and should be supplied with every specification [Fok92]. However, Jan Bergstra commented that this still leaves enough degrees of freedom.

5.2.3 Phase related datastructures and operators

- Sort `WeighedTimeTable` provides a table for all actual information a phase needs to keep. It holds the accounted weighed priority for the phase and its accumulated traffic participants for its associated set of streams. `calculateWeight` recalculates the weighed waiting time and `update` is used when a new traffic participant is detected and the list of streams must be updated.

5.2.4 Control Process related datastructures and operators

- Sort `ConflictMatrix` lets the user build a conflict matrix by adding conflicting-pairs to an `emptyConflictMatrix`. A `PidSet` is used to keep track of sets of `Pid`'s, for instance the set of cleared phases *cs* as used in the process equations.
- A `WaitingSet` is used to keep a set (*ws* in the equations) of phases together with their weighed waiting time and number of traffic participants.
- A 'sort'-function, using the insertion sort algorithm, is defined to tell `realizations` which phases in the list have the highest priority to be set to green first order.
- A `PendingSet` holds a set of (`Pid x Time`)-pairs. We have seen the `PendingSets` *ncs* and *ps* in the process equations.
- Functions `newCs` and `newNcs` are used after every cycle of the control process to determine which phases are cleared and moved from `ncs` to `cs`.

5.2.5 The Realization Algorithm

- The functions `realizations`, `rIterate`, `r1` and `r2` are used to find which phases for how long should be turned green. They are the core of this specification.
- `rIterate` finds phases that are not in conflict with non-cleared phases. If such a phase is not in conflict with a non-realizable phase with higher priority (a phase in `notUsed`) it is realized first order by `r1`, otherwise it is realized second order by `r2`, if the time given is long enough.
- The available time is calculated by `spareTime`. This function first finds the phases (say P) in `notUsed` it can frustrate. Then it finds the time the first of these phases can be realized; it takes the time the first phase in `ncs` that frustrate phases in P is cleared. This minimum can be found,

i.e. the set of phases is not empty, because the function `spareTime` is only called when a phase `p` conflicts with `notUsed` and every phase in `notUsed` has a conflict with one or more phases in `ncs`.

5.2.6 Actual Intersection Data

Finally, the data constants are given which determine the behavior of the example intersection to be specified, which in our case has been given the same values as in Appendix C of [VW93]. Unlike [VW93] we have not filled in an instance of a traffic situation complete with detected traffic participants for the simple reason that there are no tools yet available to simulate a $\rho\mu$CRL specification in Real Time.

6 A Comparison with the Previous Study

We started specifying using the prose of [VW93]. Their specification was only used for fine-tuning our specification.

The most important differences between the two specifications are described in this chapter.

6.1 Data versus Processes

A system with an inherent control process like the Traffic Regulation System can be naturally specified using the Algebra of Communicating Processes (see [BW90]). This writing considers phases as processes instead of records in a datastructure of a control process. This enhances the readability of the specification considerably.

The nestings in the original specification of the Control-Process disappeared because more functions are called in the loops of the process-definitions. To illustrate the difference we take the liberty to 'quote' a fragment of the previous study in which only one artificial Control-Process is defined which triggers the accounting function. Due the lack of other processes, no communication or parallelism is defined.

The greatest disadvantage of the previous study is that at the top level the *states* of the phases is hidden away into lower levels. We think phases of the Traffic Regulation System at the Intersection level are too important to be factored out at the top module by abstraction. The behavior of phases is the point of interest when we study the system on this intermediate level.

```
% fragment taken from Appendix B. from [VlijWa93] p. 63/64

process module Control-Process
begin
...
definitions
  CONTROL-PROCESS
  = Control-Process(init-status)
```

```
 Control-Process(status)
 = Poll-Detection-Devices(status)

 Poll-Detection-Devices(status)
 = Poll-Detection-Devices(poll-first,empty-detection-set,status)

 Poll-Detection-Devices(detector,ds,status)
 = [ eq(detector,poll-last) = false ]
   -> sum(b in BOOLEAN,
           detect(detector,b) .
           poll-Detection-Devices(next-to-poll(detector),
                                  add(detection(detector,b),ds),status)
   +
   [ eq(detector,poll-last) = true ]
   -> sum(b in BOOLEAN,
           detect(detector,b) .
           Normal-Procedure(
             realization-third-order(
               realization-second-order(
                 realization-first-order(
                   account-for-time(
                     account-weighed-priorities(
                       account-for-detection(
                         add(detection(detector,b),ds),status)))))))
         )

  Normal-Procedure(new-status)
  = Change-Signals(new-status) . Control-Process(new-status)

  Change-Signals(empty-phase-set)
  = skip

  Change-Signals(add(f, status))
  = change-signal(f) . ChangeSignals(status)
end Control-Process
```

The original specification carries phases with all information around in the determination of the phases to be realized. We have chosen to carry minimal information around, but this resulted in sorts `PidSet`, `WaitingSet` and `PendingSet` and the need for functions `weightFactor` and `cumulative` to instruct function `update`. Those different lists gave rise to different versions of much-used functions (`conflict`, `in`, `if`, etc.). An alternative would be to use one kind of list that can hold all information needed, and fill it with dummies when necessary.

6.2 PSF versus $\rho\mu$CRL

The advantages we tasted while using $\rho\mu$CRL are unsurprisingly the lack of overhead features that were eliminated from PSF at the birth of μCRL. Especially the absence of the import and export facility and a great deal of typical ASF-PSF syntactical sugaring contribute to the reduction of the number of lines in the specification. But the lack of these facilities makes it very questionable whether a *complete* specification, where all necessary details for the final implementation are included and the nestings of functions and sorts are deep, should be specified in $\rho\mu$CRL. [5]

The original specification is more explicit (meaning: things are more explicitly spelled out), which causes the specification to be larger and (maybe) less efficient. Our specification is more compact, despite the absence of pre-defined sorts and functions. Compare for instance the two methods to determine which phases to realize.

6.3 Timeless versus Real Time

The original specification of the Traffic Regulation System relies heavily on the notion of cycli. These are assumed to be discrete time steps that can be modeled in loop counters. That approach worked, but the use of time as in our specification has some advantages: the notion of time is explicit. This is especially important when a genuine time critical system is to be specified. Real Time μCRL is in one way more expressive than PSF. The use of integrals provides a continuous timing, which may be difficult to express by clock-simulating processes in PSF. Moreover, writing the processcycle as timestep implies that the timing of the system depends implicitly on the speed of the hardware.

The behavior of our specification is different, because the P-processes calculate their weighed waiting time ten times a second (the *cycletime* is $\frac{1}{10}$ time step) and are asked to send this information every second (the *controlcycletime* is 1 time step). We added this extra cycle to the specification, partly to show the expressiveness of $\rho\mu$CRL. It is mentioned on page 20 of [VW93] as a property of the real system by Nederland Haarlem.

Our proposed features for $\rho\mu$CRL, however, are used without taking too much into consideration what consequences they have on the operational semantics of μCRL. For instance, we have only looked at the operators we needed for this specific instance of a Traffic Regulation System. Even on this experimental level there are enough differences between our syntax definition and the one given by [Fok92], most notably our usage of prefixed integration and our decision to settle for relative time instead of absolute time. This should give rise for a comparison and is an area for further study.

7 Conclusion

Algebraic Specification forces one into an analytical thought process. A description of a solution of a particular problem in the form of an algebraic specification is therefore less error prone than one created with a not so rigorous

[5]The modular structure of a PSF specification enables the specification of a hierarchy of separate modules, each with their local definitions. Unfortunately *Completeness* of the associated term rewriting system is not a modular property. A counter example given by Barendregt and Klop (1986) is reproduced in [BKM89]. For proof theoretical reasons modularity is not supported by μCRL.

formal method. This sounds like an open door but only by 'doing it' we have learned to 'feel' that way.

We have learned to appreciate Algebraic Specification as a Formal Document describing the data definition and process behavior of a system whereupon equational reasoning can be applied. We have experienced that it is unwise to optimize during the specification activity since this seriously endangers the readability of the specification and very easily introduces ambiguities and ironically enough also redundancies.

We have shown that Real Time Process Algebra embedded in μCRL is a considerable enrichment of the expressivity of μCRL. It would be a very tedious exercise indeed to perform this reverse engineering project the other way around, namely by creating a separate process to synchronize with a Timer process for each time stamped process in our specification. Not to mention the communication overhead.

The introduction of Asynchronous Communicating Processes in our distributed Traffic Regulation System specification adds another level of functional decomposition to the original specification. This does not mean, however, that our specification dictates a possible implementation to be executed in parallel. The specification as given in the appendix could be implemented as one monolithic block as well. We would argue that this is the virtue of abstraction.

References

[BB91] J.C.M Baeten & J.A. Bergstra, Real time process algebra. *Journal of Formal Aspects of Computing Science*, 3(2) : 142-188, 1991.

[BW90] J.C.M. Baeten & W.P. Weijland, *Process algebra* Cambridge Tracts in Theoretical Computer Science 18. Cambridge University Press, 1990.

[BKM89] J.A. Bergstra, J.W. Klop & A. Middeldorp, *Termherschrijfsystemen*, PTT Research 1989, Kluwer Programmatuurkunde, in Dutch.

[Fok92] W.J. Fokkink, *A simple specification language combining processes, time and data.* Center for Mathematics and Computer Science, Technical Report CS-R9209, Department of Software Technology, Amsterdam 1992.

[GP90] J.F. Groote & A. Ponse, The Syntax and Semantics of μCRL. In this volume: A.Ponse, C.Verhoef and S.F.M. van Vlijmen, editors. *Proceedings of ACP94*. Workshops in Computing, Springer-Verlag, 1994.

[Kli93] P. Klint, A meta-environment for generating programming environments. *ACM Transactions on Software Engineering Methodology*, 2(2):176-201, 1993.

[Klu93] A.S. Klusener, *Models and axioms for a fragment of real time process algebra.* Ph.D thesis, Technical University of Eindhoven, December 1993.

[Kro90] J.J.M.T. Kroone, *UAP-26 Universele applicatieprogramma's voor de FR34 en FR80.* B.V. Nederland Haarlem Verkeerssystemen, 1990.

[MV91] S. Mauw & G.J. Veltink, *A Process Specification Formalism.* Fundamenta Informaticae XIII, pp. 85-139, 1990.

[Vel91] G.J. Veltink, *The PSF Toolkit.* Technical Report P9107, Programming Research Group, University of Amsterdam 1991.

[VW93] S.F.M. van Vlijmen & A. van Waveren, *Algebraic Specification of a System for Traffic Regulation at Signalized Intersections.* Technical Report P9313, Programming Research Group, University of Amsterdam June 1993.

[Wam91] J.J.van Wamel *A Library for PSF*, Technical Report P9301, Programming Research Group, University of Amsterdam, 1993.

A The Specification

```
%%%%%%%%%%%%%%%%%%%%%%%%%%%%%%%%%%%%%%%%%%%%%%%%%%%%%%%%%%%%%%%%%%%%%%%%
%                                                                      %
%  Real Time Micro Common Representation Language Specification  %
% of a System for Traffic Regulation at Signalized Intersections %
%                                                                      %
%%%%%%%%%%%%%%%%%%%%%%%%%%%%%%%%%%%%%%%%%%%%%%%%%%%%%%%%%%%%%%%%%%%%%%%%

% Begin Standard Sorts

sort Bool
func T, F         :                -> Bool
     or, and, xor: Bool # Bool -> Bool
     not          : Bool         -> Bool
     if : Bool # Bool # Bool    -> Bool
var  b, b0        : Bool
rew  or(T, b)   = T
     or(F, b)   = b
     and(T, b) = b
     and(F, b) = F
     xor(T, b) = not(b)
     xor(F, b) = b
     not(T)     = F
     not(F)     = T
     if(T, b, b0) = b
     if(F, b, b0) = b0

sort Nat % only used to specify Ints
func zero                :              -> Nat
     s                   : Nat          -> Nat
     plus, sbtr, mlty : Nat # Nat -> Nat
     leq, eq             : Nat # Nat -> Bool
var  nat, nat0           : Nat
rew  plus(nat, zero)          = nat
     plus(nat, s(nat0))       = s(plus(nat, nat0))
     sbtr(nat, zero)          = nat
     sbtr(zero, s(nat0))      = zero
     sbtr(s(nat),s(nat0))     = sbtr(nat, nat0)
     mlty(nat, zero)          = zero
     mlty(nat, s(nat0))       = plus(mlty(nat, nat0), nat)
     leq(zero, nat)           = T
     leq(s(nat), s(nat0))     = leq(nat, nat0)
     leq(s(nat), zero)        = F
     eq(zero, zero)           = T
     eq(s(nat), zero)         = F
     eq(zero, s(nat))         = F
     eq(s(nat), s(nat0))      = eq(nat, nat0)

sort Sign % only used to specify Ints
```

```
func Pos, Neg    :              -> Sign
     eq          : Sign # Sign -> Bool
var  sign, sign0 : Sign
rew  eq(Pos, Pos) = T
     eq(Neg, Neg) = T
     eq(Pos, Neg) = F
     eq(Neg, Pos) = F

sort Int
func 0, 1, 2, 3, 4, 5, 6, 7, 8, 9, 10, 11, 12: -> Int
     I                  : Sign # Nat           -> Int
     plus, sbtr, mlty : Int # Int              -> Int
     if                 : Bool # Int # Int     -> Int
     leq, eq            : Int # Int            -> Bool
     pos                : Int                  -> Bool
var  n, n0          : Int
     sign, sign0    : Sign
     nat, nat0      : Nat
rew  if(T, n, n0) = n
     if(F, n, n0) = n0
     plus(I(sign,nat),I(Neg,nat0)) = sbtr(I(sign,nat),I(Pos,nat0))
     plus(I(Neg, nat),I(Pos,nat0)) = sbtr(I(Pos,nat0),I(Pos,nat))
     plus(I(Pos, nat),I(Pos,nat0)) = I(Pos, plus(nat, nat0))
     sbtr(I(sign,nat),I(Neg,nat0)) = plus(I(sign,nat),I(Pos,nat0))
     sbtr(I(Neg, nat),I(Pos,nat0)) = I(Neg, plus(nat, nat0))
     sbtr(I(Pos, nat),I(Pos,nat0)) = if(leq(nat, nat0),
                                         I(Neg, sbtr(nat0, nat)),
                                         I(Pos, sbtr(nat, nat0)))
     mlty(I(sign,nat),I(sign0,nat0)) = if(eq(sign, sign0),
                                         I(Pos, mlty(nat, nat0)),
                                         I(Neg, mlty(nat, nat0)))

     eq(I(sign,nat), I(sign0,nat0))  =
       or(and(eq(sign,sign0), eq(nat,nat0)),
          and(eq(nat,zero), eq(nat0,zero)))

     leq(I(Pos,nat),I(Pos,nat0)) = leq(nat,nat0)
     leq(I(Neg,nat),I(Pos,nat0)) = T
     leq(I(Pos,nat),I(Neg,nat0)) = and(eq(nat,zero),eq(nat0,zero))
     leq(I(Neg,nat),I(Neg,nat0)) = leq(nat0,nat)

     pos(n)                      = leq(0, n)

      0 = I(Pos, zero)
      1 = I(Pos, s(zero))
      2 = I(Pos, s(s(zero)))
      3 = I(Pos, s(s(s(zero))))
      4 = I(Pos, s(s(s(s(zero)))))
      5 = I(Pos, s(s(s(s(s(zero))))))
      6 = I(Pos, s(s(s(s(s(s(zero)))))))
```

```
       7 = I(Pos, s(s(s(s(s(s(s(zero))))))))
       8 = I(Pos, s(s(s(s(s(s(s(s(zero)))))))))
       9 = I(Pos, s(s(s(s(s(s(s(s(s(zero))))))))))
      10 = I(Pos, s(s(s(s(s(s(s(s(s(s(zero)))))))))))
      11 = I(Pos, s(s(s(s(s(s(s(s(s(s(s(zero))))))))))))
      12 = I(Pos, s(s(s(s(s(s(s(s(s(s(s(s(zero)))))))))))))

sort Time % defined as the Rational numbers
func R                          : Int # Int           -> Time
     plus, sbtr, min, max       : Time # Time         -> Time
     leq, lt, eq                : Time # Time         -> Bool
     if                         : Bool # Time # Time  -> Time
     time                       : Int                 -> Time
var  t, t0                      : Time
     n, n0, n1, n2              : Int
     sign                       : Sign
     nat                        : Nat
rew  plus(R(n, n0), R(n1, n2)) =
       R(plus(mlty(n, n2), mlty(n1, n0)), mlty(n0, n2))
     sbtr(R(n, n0), R(n1, n2)) =
       R(sbtr(mlty(n, n2), mlty(n1, n0)), mlty(n0, n2))
     leq(R(n, n0), R(n1, n2))  = if(xor(pos(n0), pos(n2)),
                                   leq(mlty(n0, n1), mlty(n, n2)),
                                   leq(mlty(n, n2), mlty(n0, n1)))
     eq(t, t0)                 = and(leq(t, t0), leq(t0, t))
     lt(t, t0)                 = and(leq(t, t0), not(leq(t0, t)))
     min(t, t0)                = if(leq(t, t0), t, t0)
     max(t, t0)                = if(leq(t, t0), t0, t)
     if(T, t, t0)              = t
     if(F, t, t0)              = t0

% End Standard Sorts

% Begin Global Data Structures Definitions

sort Sid % Stream Id
func S               : Int        -> Sid % constructor
     int             : Sid        -> Int % typecast
     eq              : Sid # Sid -> Bool
var  s, s0           : Sid
     n               : Int
rew  int(S(n))       = n
     eq(s, s0)       = eq(int(s), int(s0))

sort Pid % Phase Id
func P                                           : Int        -> Pid
     int                                         : Pid        -> Int
     eq                                          : Pid # Pid -> Bool
var  p, p0                                       : Pid
     n                                           : Int
```

```
rew  int(P(n)) = n
     eq(p, p0) = eq(int(p), int(p0))

sort ConflictMatrix
func add         : Pid # Pid # ConflictMatrix     -> ConflictMatrix
     emptyConflictMatrix, conflictMatrix:         -> ConflictMatrix
     conflict    : Pid # Pid                              -> Bool
     in          : Pid # Pid # ConflictMatrix             -> Bool
     if : Bool # ConflictMatrix # ConflictMatrix -> ConflictMatrix
var  m, m0       : ConflictMatrix
     p, p0, p1, p2 : Pid
rew  if(T, m, m0) = m
     if(F, m, m0) = m0

     conflict(p, p0) = if(leq(int(p), int(p0)),
                         in(p, p0, conflictMatrix),
                         in(p0, p, conflictMatrix))

     in(p, p0, emptyConflictMatrix) = F
     in(p, p0, add(p1, p2, m))     =
       or(and(eq(p, p1), eq(p0, p2)), in(p, p0, m)),

% End Global Data Structures Definitions

% Begin Actual Intersection Data Set

func weightFactor                              : Pid # Sid -> Int
     priority                                  : Pid       -> Int
     cumulative                                : Pid # Sid -> Bool
     cycleTime, controlCycleTime               :           -> Time
     yellowTime, minGreenTime, clearanceDuration : Pid     -> Time
     actualIntersectionPidSet                  :           -> PidSet
     initStreams                           : Pid -> WeighedTimeTable
var  p : Pid
     s : Sid
rew              cycleTime  = R(1, 10)
          controlCycleTime  = R(1, 1)

               priority(p) = int(p)
             yellowTime(p) = R(1, 1)
           minGreenTime(p) = R(2, 1)
      clearanceDuration(p) = R(1, 1)

         weightFactor(p, s) = 1
           cumulative(p, s) = T    % COUNT number of detects

     conflictMatrix =
       add(P(1),  P(5), add(P(1), P(9),
       add(P(2),  P(5), add(P(2), P(6),
       add(P(2),  P(9), add(P(2), P(10),
```

```
      add(P(2), P(11), add(P(2), P(12),
      add(P(3),  P(5), add(P(3), P(6),
      add(P(3),  P(7), add(P(3), P(8),
      add(P(3), P(11), add(P(3), P(12),
      add(P(4),  P(8), add(P(4), P(12),
      add(P(5),  P(8), add(P(5), P(9),
      add(P(6),  P(8), add(P(6), P(9),
      add(P(6), P(10), add(P(6), P(11),
      add(P(7), P(11), add(P(5), P(12),
      add(P(8), P(11), add(P(8), P(12),
      add(P(9), P(11), add(P(9), P(12),
      emptyConflictMatrix))))))))))))))))))))))))))

    actualIntersectionPidSet =
      add(P(1), add(P(2), add(P(3), add(P(4), add(P(5), add(P(6),
      add(P(7), add(P(8), add(P(9), add(P(10), add(P(11),
      add(P(12), emptyPidSet))))))))))))

    initStreams(P(n)) =
          Wt(R(0,1), 0, add(S(1), 0, emptyStreamDataSet))
     % n not used since as a simplification we assume every phase
     % to have one stream as in [VlijWa93].

% End Actual Intersection Data Set

% Begin Dynamic Accounting Functions Definitions

sort PidSet
func add          : Pid # PidSet                       -> PidSet
     emptyPidSet:                                      -> PidSet
     extract, selectConflict : Pid # PidSet            -> PidSet
     conflict     : Pid # PidSet                       -> Bool
     in           : Pid # PidSet                       -> Bool
     if           : Bool # PidSet # PidSet             -> PidSet
var  pidset, pidset0    : PidSet
     p, p0                                  : Pid
rew  extract(p, emptyPidSet)      = emptyPidSet  % error
     extract(p, add(p0, pidset)) = if(eq(p, p0),
                                       extract(p, pidset),
                                       add(p0, extract(p, pidset)))

     conflict(p, emptyPidSet)      = F
     conflict(p, add(p0, pidset)) =
              or(conflict(p0, p), conflict(p, pidset))

     selectConflict(p, emptyPidSet)      = emptyPidSet
     selectConflict(p, add(p0, pidset)) =
       if(conflict(p, p0),
          add(p0, selectConflict(p, pidset)),
          selectConflict(p, pidset))
```

```
     in(p, emptyPidSet)      = F
     in(p, add(p0, pidset)) = or(eq(p, p0), in(p, pidset))

     if(T, pidset, pidset0) = pidset
     if(F, pidset, pidset0) = pidset0

sort StreamDataSet WeighedTimeTable
func add          : Sid # Int # StreamDataSet    -> StreamDataSet
     emptyStreamDataSet  :                       -> StreamDataSet
     Wt           : Time # Int # StreamDataSet   -> WeighedTimeTable
     weighedTime: WeighedTimeTable               -> Time
     totalTp      : WeighedTimeTable             -> Int
     calculateWeight : Pid # WeighedTimeTable    -> WeighedTimeTable
     update       : Pid # WeighedTimeTable # Sid -> WeighedTimeTable
     update       : Pid # StreamDataSet # Sid    -> StreamDataSet
     cwIterate    : Time # StreamDataSet # StreamDataSet
                                                 -> WeighedTimeTable
     cwIterate    : Pid # Time # Int # StreamDataSet # StreamDataSet
                                                 -> WeighedTimeTable
     if: Bool # WeighedTimeTable # WeighedTimeTable
                                                 -> WeighedTimeTable
     if: Bool # StreamDataSet # StreamDataSet    -> StreamDataSet
var  wt, wt0                   : WeighedTimeTable
     sset, sset0               : StreamDataSet
     n, n0                     : Int
     s, s0                     : Sid
     p, p0                     : Pid
     t, t0                     : Time
rew  weighedTime(Wt(t, n, sset)) = t
     totalTp(Wt(t, n, sset)) = n

     if(T, wt, wt0) = wt
     if(F, wt, wt0) = wt0
     if(T, sset, sset0) = sset
     if(F, sset, sset0) = sset0

     calculateWeight(p, Wt(t, n, sset)) =
       cwIterate(p, t, n, sset, emptyStreamDataSet)

     cwIterate(p, t, n, emptyStreamDataSet, sset) = Wt(t, n, sset)
     cwIterate(p, t, n, add(s, n0, sset0), sset)  =
       cwIterate(p, plus(t, R(mlty(weightFactor(p, s), n0) ,1))),
           plus(n, n0), sset0, add(s, n0, sset))

     update(p, Wt(t, n, sset), s) = Wt(t, n, update(p, sset, s))

     update(p, emptyStreamDataSet, s) = emptyStreamDataSet % error
     update(p, add(s, n, sset), s0)   =
       if(eq(s, s0),
```

```
   if(cumulative(p, s),
      add(s, plus(n, 1), sset), % car: every detect counts
      add(s, 1, sset)),         % pedestrian
   add(s, n, update(p, sset, s0)))

sort WaitingSet
func add             : Pid # Time # Int # WaitingSet -> WaitingSet
     emptyWaitingSet :                               -> WaitingSet
     insertionSort   : WaitingSet                    -> WaitingSet
     insert         : Pid #  Time # Int # WaitingSet -> WaitingSet
     in              : Pid # WaitingSet              -> Bool
     peel            : WaitingSet                    -> PidSet
     if             : Bool # WaitingSet # WaitingSet -> WaitingSet
var  wl, ws, wset, wset0: WaitingSet
     t, t0              : Time
     p, p0              : Pid
     n, n0              : Int
rew  if(T, wset, wset0) = wset
     if(F, wset, wset0) = wset0

     % insertion descending sort on dynamic time and priority
     insertionSort(emptyWaitingSet)    =
       emptyWaitingSet
     insertionSort(add(p, t, n, wset)) =
       insert(p, t, n, insertionSort(wset))

     insert(p, t, n, emptyWaitingSet)     =
       add(p, t, n, emptyWaitingSet)
     insert(p, t, n, add(p0, t0, n, wset))=
       if(or(leq(t, t0), and(eq(t, t0),
                          leq(priority(p), priority(p0)))),
         add(p0, t0, n, insert(p, t, n, wset)),
         add(p, t, n, add(p0, t0, n, wset)))

     peel(emptyWaitingSet) =  emptyPidSet
     peel(add(p, t, n, wset)) = add(p, peel(wset))

sort PendingSet
func add               : Pid # Time # PendingSet     -> PendingSet
     emptyPendingSet   :                             -> PendingSet
     newCs             : PidSet # PendingSet         -> PidSet
     newNcs            : PendingSet                  -> PendingSet
     realizations      : PendingSet # WaitingSet     -> PendingSet
     rIterate   : PendingSet # WaitingSet # PidSet   -> PendingSet
     r1 : PendingSet # Pid # WaitingSet # PidSet # Time
                                                     -> PendingSet
     r2 : PendingSet # Pid # WaitingSet # PidSet # Time
                    # Time # Time                    -> PendingSet
     conflict          : Pid # PendingSet            -> Bool
     spareTime         : Pid # PendingSet # PidSet   -> Time
```

```
    maxWaitingTime     : PendingSet # PidSet          -> Time
    maxWaitingTime     : PendingSet # PidSet # Time   -> Time
    maxTime            : Pid # PendingSet # Time      -> Time
    if               : Bool # PendingSet # PendingSet -> PendingSet
var ncs, pset, pset0   : PendingSet
    t, t0, spare, clear: Time
    p, p0              : Pid
    cs, notUsed, pidset: PidSet
    n, n0              : Int
    ws, wset           : WaitingSet
    s                  : Sid
rew if(T, pset, pset0) = pset
    if(F, pset, pset0) = pset0

    newCs(cs, emptyPendingSet) = cs
    newCs(cs, add(p, t, pset)) =
      if( leq(t, controlCycleTime),
        add(p, newCs(cs, pset)),
        newCs(cs, pset))

    newNcs(emptyPendingSet) = emptyPendingSet
    newNcs(add(p, t, pset))   =
      if( leq(t, controlCycleTime),
        newNcs(pset),
        add(p, sbtr(t, controlCycleTime), newNcs(pset)))

    realizations(ncs, ws) =
                rIterate(ncs, insertionSort(ws), emptyPidSet)

    rIterate(ncs, add(p, t, n, wset), notUsed) =
      if(conflict(p, ncs),
        rIterate(ncs, wset, add(p, notUsed)),
        if(conflict(p, notUsed),
          % second order realization
          r2(ncs, p, wset, notUsed, R(n,1),
              plus(plus(R(n,1), yellowTime(p)),
              clearanceDuration(p)), spareTime(p, ncs, notUsed)),
          % first order realization
          r1(ncs, p, wset, notUsed, max(R(n,1),
              minGreenTime(p)))))

    r1(ncs, p, wset, notUsed, t) =
      add(p, t, rIterate(
          add(p, plus(plus(t, yellowTime(p)),
              clearanceDuration(p)), ncs), wset, notUsed))

    r2(ncs, p, wset, notUsed, t, clear, spare) =
      if(lt(spare, minGreenTime(p)),
        % not enough time to be realized
        rIterate(ncs, wset, add(p, notUsed)),
```

```
    % time available
    if(lt(spare, clear),
      % realize as long as allowed
      add(p, sbtr(sbtr(spare, yellowTime(p)),
          clearanceDuration(p)),
        rIterate(add(p, spare, ncs), wset, notUsed)),
      % realize as long as wanted
      add(p, t,
        rIterate(add(p, clear, ncs), wset, notUsed))))

  conflict(p, emptyPendingSet)  = F
  conflict(p, add(p0, t, pset)) =
    or(conflict(p, p0), conflict(p, pset))

  % find time first phase of notUsed frustrated by p is no
  % longer frustrated by ncs
  spareTime(p, ncs, notUsed) =
    maxWaitingTime(ncs, selectConflict(p, notUsed))

  % find time first phase in pidset is no longer frustr. by ncs
  maxWaitingTime(ncs, emptyPidSet)  = R(0, 1)
    % error: regular use guarantees non-empty pidSet
  maxWaitingTime(ncs, add(p, pidset)) =
    maxWaitingTime(ncs, pidset, maxTime(p, ncs, R(0, 1)))

  maxWaitingTime(ncs, emptyPidSet, t)  = t
  maxWaitingTime(ncs, add(p, pidset), t) =
    maxWaitingTime(ncs, pidset,
                   min(maxTime(p, ncs, R(0, 1)), t))

  % find time p is no longer frustrated by (any phase in) ncs
  maxTime(p, emptyPendingSet, t) = t
  maxTime(p, add(p0, t0, ncs), t)   =
    if(conflict(p, p0),
      maxTime(p, ncs, max(t, t0)),
      maxTime(p, ncs, t))

% End Dynamic Accounting Functions Definitions


% Begin Process Specifications

act  detect                                 : Sid
     setGreen, setYellow, setRed            : Pid
     sendWeighedTime, receiveWeighedTime, commWeighedTime
                                            : Pid # Time # Int
     sendRealization, receiveRealization, commRealization
   : Pid # Time

comm sendWeighedTime   | receiveWeighedTime  = commWeighedTime
```

```
     sendRealization  | receiveRealization  = commRealization

proc PclearedRed(p: Pid, wt: WeighedTimeTable , w: Time) =
       integral( [and(leq(R(0,1), v), lt(v, w))],
         sum(s: Sid, detect(s)[v]
                   . PclearedRed(p, update(p, wt, s), sbtr(w, v))
       ))
     + integral( [and(leq(R(0,1), v), lt(v, w))],
         sendWeighedTime(p, weighedTime(wt), totalTp(wt))[v]
         . PclearedRed(p, wt, sbtr(w, v))
       )
     + integral( [and(leq(R(0,1), v), lt(v, w))],
       sum(t: Time, receiveRealization(p, t)[v]
           . setGreen(p) . Pgreen(p, t)
       ))
     + tau[w] . PclearedRed(p, calculateWeight(p, wt), cycleTime)

     PnonClearedRed(p: Pid, wt: WeighedTimeTable, w: Time) =
       integral( [and(leq(R(0,1), v), lt(v, w))],
         sum(s: Sid, detect(s)[v]
                 . PnonClearedRed(p, update(p, wt, s), sbtr(w, v))
       ))
     + tau[w] . PclearedRed(p, calculateWeight(p, wt), cycleTime)

     Pgreen(p: Pid, t: Time) = setYellow(p)[t] . Pyellow(p)
     Pyellow(p: Pid) = setRed(p)[yellowTime(p)]
       . PnonClearedRed(p, initStreams(p), clearanceDuration(p))

     % CP stands for Control Process

     CPreceive(add(p: Pid, cs: PidSet): PidSet,
               ncs: PendingSet, ws: WaitingSet) =
         sum(greenTime: Time, sum(trafficParticipants: Int,
           receiveWeighedTime(p, greenTime, trafficParticipants)
               [R(0,1)] )
           . CPreceive(cs, ncs,
               add(p, greenTime, trafficParticipants, ws) ) )

     CPreceive(emptyPidSet:PidSet, ncs:PendingSet,
                                             ws:WaitingSet) =
         CPsend(peel(ws), ncs, realizations(ncs, ws))

     CPsend(cs:PidSet, ncs:PendingSet,
             add(p:Pid, t:Time, ps:PendingSet) : PendingSet) =
         sendRealization(p, t)[R(0,1)] . CPsend(extract(p, cs),
          add(p, plus(plus(t, yellowTime(p)),
              clearanceDuration(p)), ncs), ps)

     CPsend(cs:PidSet, ncs:PendingSet,
                                emptyPendingSet:PendingSet) =
```

```
        tau[controlCycleTime]
        . CPreceive(newCs(cs, ncs), newNcs(ncs), emptyWaitingSet)

    % RTRS stands for Realtime Traffic Regulation System

    RTRS =
      hide
        (    {commRealization, commWeighedTime},
        encap
          (  {sendWeighedTime, receiveWeighedTime,
              sendRealization, receiveRealization},

           CPreceive(actualIntersectionPidSet, emptyPendingSet,
                                                  emptyWaitingSet)
           ||
                   merge(p : actualIntersectionPidSet,
                   PclearedRed(p, initStreams(p), cycleTime))
           )
         )
% End Process Specifications
```

An Experiment in Implementing Process Algebra Specifications in a Procedural Language

C. Groza
Department of Computer Science
Technical University of Timişoara
Bd. V. Pârvan 2, 1900 Timişoara, România
email: calin@utt.ro

Abstract

This paper presents a method for the implementation of process algebra specifications in a (structured) imperative language. The original specification is written in the Process Specification Formalism (PSF). The actual implementation is made in an extension of the C language for concurrent programming. First we make an informal presentation and then we consider a formal approach.

1 Introduction

The motivation to have a systematic approach for the implementation of process algebra specifications came during the development of an application which uses a communication protocol. On one side there was a formal specification of the protocol. On the other side, in order to use a protocol in the application it was required to have an implementation in a structured imperative language. The purpose of this paper is to present a method to bridge the gap between process algebra specifications and implementation in software development. For the specifications we used the Process Specification Formalism (PSF) and for the implementation the C language extended with functions for processes management and communication.

The implementation of process algebra specifications was also addressed in [10]. There, the process algebraic specifications were implemented in POOL-T – an object oriented language for concurrent programming.

The next section presents a brief description of the PSF formalism. Section 3 describes the extension of the C language used in the implementation. In Section 4 we present, informally, the implementation of the basic PSF constructs. A formal approach to the implementation is presented in Section 5 and in Appendix A.

2 The Process Specification Formalism (PSF)

The Process Specification Formalism (PSF) is an algebraic formalism used for the specification of concurrent systems. The formal definition of PSF can be found in [MV93]. A set of tools (such as compiler, simulator) are available to analyse a specification.

PSF is based on a formalism for algebraic specification of data: Algebraic Specification Formalism (ASF) [2] and a process algebra: Algebra for Communicating Processes (ACP) [1]. PSF supports the modular construction of specifications and parameterization of modules.

A PSF specification contains two kinds of modules:

Data modules contain the algebraic specifications of data. The major parts are: the definition of the *signature* and the description of the properties by means of *equations*.

Process modules contain the definition of PSF processes. They are defined as a series of atomic actions combined by operators. Atomic actions are the basic and indivisible elements of processes. The basic operators are: sequential composition, parallel composition, and alternative composition. Atomic actions and operators are used in the (recursive) definition of processes.

Parallel processes can communicate with each other. The atomic actions used in communications are declared in a separate section of the process modules.

PSF does not allow explicit representation of time. Nevertheless it is possible to introduce processes which represent timers. Timers are often used in communication protocols in order to deal with lost messages.

The following example is a specification of a *producer-consumer* system. There are two processes, the Producer and the Consumer which communicate with each other through a port:

```
data module Data
begin
  exports
    begin
    sorts
      DATA
    functions
      data-unit: -> DATA
    end
end Data

process module Prod-Cons
begin
  exports begin
    atoms
      send, recv, comm: DATA
      produce-data
      consume-data: DATA
    processes
      Producer, Consumer, Prod-Cons
    end
  imports
    Data
  sets of atoms
    H={send(d),recv(d)|d in DATA}
  communications
    send(d) | recv(d) =
        comm (d) for d in DATA
  definitions
    Producer = produce-data.
      send(data-unit).Producer
    Consumer = sum(d in DATA,recv(d).
      consume-data(d)). Consumer
    Prod-Cons =
      encaps(H, Producer || Consumer)
  end Prod-Cons
```

3 EC: an extension of C for concurrent programming

The language used in the implementation is C extended with functions for process management and interprocess communications. The extension (called EC) adds features usually offered by concurrent programming languages.

In order to differentiate the EC processes from PSF processes, the former will be called *tasks*. A *task* is defined by the its *code* and *data*. The code is executed sequentially, thus, it is not possible to have parallel execution of different parts of a task. The tasks communicate with each other by sending/receiving messages through *ports*. When one task issues a *send* to a port and another tasks issues a *recv* to the same port, the tasks synchronize and exchange data. The communication primitives are synchronous: when one task gets to a communication primitive it is blocked until another task executes the complementary primitive. Two boolean functions can be used to check if there is a message or a request pending at a port.

3.1 Tasks

A task is defined by both the code to be executed and the data. The code is defined as a C function.

ECtcreate - creates a task

```
#include <ec.h>
int ECtcreate( TASK (*func)(), int nargs, ... );
```

`ECtcreate` creates a new task which executes the code defined in function `func`. `nargs` is the number of simple-type (int) arguments given to the task. The function returns -1 on error and a non-negative integer on success. `TASK` is another name for the type `void`. It was introduced to increase the readability of the code. The functions which define a task code return the type `TASK`.

3.2 Communications

Communications in PSF are synchronous. We have implemented a similar mechanism in EC. Another solution could have been an event-driven mechanism with direct communication between tasks similar to that used in [9].

The following functions are used for communication:

ECpcreate - creates a port

```
#include <ec.h>
int ECpcreate(PORT *p, int data_size);
```

`PORT` is the data structure for ports. `ECpcreate` initializes the data of the port pointed to by p. The second argument is size of the data transmitted through the port.

ECsend, ECrecv - sends data to port, receives data from port

```
#include <ec.h>
void ECsend(PORT *port, void *pdata);
void ECrecv(PORT *port, void *pdata);
```

ECsend sends to the port indicated by p the data stored at the address pdata. ECrecv receives data from the port pointed to by p and stores it at the address pdata. The functions block the current task until another task executes the complementary communication function.

ECexist_data, ECexist_request - these functions check if there is data at a port or a request for data respectively.

```
#include <ec.h>
bool ECexist_data(PORT *p);
bool ECexist_request(PORT *p);
```

ECexist_data returns TRUE if another process sent (using ECsend) data to the port p and FALSE otherwise. ECexist_request checks if there is a request for data (another task issued a call to ECrecv) at the port p. These functions do not block the task.

3.3 Timers

We defined the data structure TIMER and the following functions for the implementation of timers in EC programs:

ECstart_timer - starts a timer

```
#include <ec.h>
void ECstart_timer(TIMER *t, long usec);
```

TIMER is a data structure used in the implementation of the timers. ECstart_timer updates the timer data. The time constant, in microseconds, is given as the second argument of the function.

ECtimeout - indicates whether a "time-out" occurred

```
#include <ec.h>
bool ECtimeout(TIMER *t);
```

ECtimeout returns TRUE if the time elapsed since the task issued ECstart_timer is greater than the time constant of the timer.

3.4 The run-time environment of EC programs

For the implementation of the run-time environment of EC programs we have used the lightweight processes library available in the UNIX environment. At the EC level, the application is a collection of tasks which run in parallel and communicate synchronously through ports. The EC level is implemented on the top of the lightweight processes environment. At the lightweight processes level an application consists of a set of threads which are running in parallel and communicate synchronously with each other.

The main primitives offered by the lightweight processes library are:

- create and destroy lightweight processes (also called threads).
- synchronous communication between threads. A thread can send a message to another thread and then it waits for a reply or, it can wait to get a message. When the *rendezvous* happens, the threads exchange data, the receiver is unblocked, and, it can resume the sender by sending a reply.

The EC tasks are implemented by threads. The task parameters are passed to the thread when it is created. The communication between EC tasks is indirect, through ports. In the lightweight processes environment, the communication between threads is direct. Thus, in order to implement the EC communications a thread is created for every port. The EC functions `ECsend` and `ECrecv` are implemented by communications between the threads for tasks and the threads for ports. Every port contains a queue in which the messages are stored until rendezvous.

In the implementation of the timers we use an internal clock. Function `ECstart_timer` stores the current time and the time constant in the timer data. Function `ECtimeout` compares the current time with the timer data and decides if a "timeout" occured.

The complete code of the EC run-time environment is presented in [3].

4 From PSF specifications to EC programs

In this section we show, informally, how an EC implementation can be derived from a PSF specification and which are the limitations.

The primary goal is to determine the control structures in the EC implementation based on the PSF process definitions. We have identified a subset of PSF often used in the specification of real systems. We show for this subset, by means of examples, how to make an EC implementation. In section 5 we present a formalization of this method. The focus is on process definitions, but, we present also how other parts of PSF specifications (data, communications) are implemented.

4.1 Data

There are two aspects which are related to data:

1. The definition of abstract data types. In PSF specifications the data modules contain the algebraic specification of data. In EC, abstract data types are defined using the C data structures and functions.
2. The use of data. In PSF, the processes and the atomic actions can have parameters. In EC, data is used as parameters of the tasks and functions and in variables, which form an environment for computation.

The abstract data types specified in the PSF data modules are implemented in EC by means of C data structures and functions. Although both C and PSF provide support for the definition of abstract data types, a number of differences makes difficult a direct conversion of PSF data definition into EC code. First, PSF uses an algebraic approach for data definition while C uses an imperative approach. Second, the syntax of the interface with the abstract data types is more flexible in PSF than in C. For example, in PSF the names of the functions can be overloaded but in C the names of the functions must be unique.

We have implemented several communication protocols from their PSF specifications. In these cases the data definition was simple and we used an ad-hoc implementation in C of the abstract data types specified in PSF. In general, the PSF sorts have corresponding C data types in the implementation and PSF functions have correspondent C functions. For example, let us consider the data structure used in the specification of the Positive Acknowledgement with Retransmission Protocol to describe the structure of the messages sent through a channel. On the left side is the PSF specification and on the right side is the EC implementation.

⟨Frames - psf⟩≡

```
data module Frames
begin
  exports begin
    sorts
      FRAME
    functions
      frame: DATA#BIT -> FRAME
      ce:             -> FRAME
      error: FRAME    -> BOOLEAN
      bit: FRAME      -> BIT
      datum: FRAME    -> DATA
    end
  imports
    Booleans, Bits, Data

  variables
    b: -> BIT
    d: -> DATA
  equations
    [01] error(frame(d,b)) = false
    [02] error(ce) = true
    [03] bit(frame(d,b)) = b
    [04] datum(frame (d,b)) = d
end Frames
```

⟨Frames - ec⟩≡

```
typedef struct {
        int error;
        BIT b;
        DATA d;
} FRAME;

DATA datum(FRAME f)
{
        return f.d;
}

BIT bit(FRAME f)
{
        return f.b;
}

int frame_error(FRAME f)
{
        return f.error;
}
```

In PSF, data occurs in process definitions as parameters of processes and atomic actions. One important difference between PSF and EC in using data is the way variables take values. In PSF this is done by matching terms while EC provides the assignment statement. We impose the following restriction on the PSF specification: matching is allowed only between a free variable and an expression. More general term matching can be converted to this simple case and conditional expressions.

4.2 Processes

If a PSF specification contains only regular process expressions it is possible to make an implementation based on the finite automaton model. The implementation of an entire system as one finite automaton has disadvantages. For example, a single finite automaton assumes that all the processes share a global memory which contains the state and a representation of the transitions. This assumption is not justified in applications in which the processes are physically distributed and do not share a global memory.

Instead of using a single finite automaton, our implementation uses a set of finite automata which communicates with each other through messages. The implementation is based on the following rules:

- the PSF specification must contain only regular process definitions;
- for every PSF process which occurs in parallel composition the implementation is an EC task.
- every task executes the code of a finite automaton which implements a collection of PSF processes.

Throughout the rest of this section we will detail these rules. First we present how the process definitions which use parallel composition determine the tasks in the implementations. Then, we present how the transitions in the finite automata can be determined from the rest of PSF definitions.

4.3 Parallel composition

The process definitions which are using parallel composition determine which are the tasks in the implementation. For example, let us consider the process P defined as the parallel composition of the processes Q and R:

⟨*PSF - parallel composition*⟩≡

```
P = Q || R
```

⟨*EC - task definitions*⟩≡

```
TASK Q ( )
{
        ⟨code for the finite automaton Q⟩
}

TASK R ( )
{
        ⟨code for the finite automaton R⟩
}

TASK P ( )
{
  ECtcreate(Q, 0);
  ECtcreate(R, 0);
}
```

The code for the finite automaton has the form:

⟨*code for the finite automaton Q*⟩≡

```
enum {Q, ⟨other state names⟩} state;
state=Q;
while ( TRUE ) {
        ⟨code for transitions in the finite automata Q⟩
}
```

The implementation contains the definitions of the tasks Q and R which correspond to the operands of the PSF definition. They execute the code of finite automata. The implementation contains also the definition of task P corresponding to the PSF process P. If the operands of the parallel composition have parameters the task definitions have parameters too.

Often, the form of process definitions with parallel composition is:

```
P = encaps ( H, Q || R || S ... )
```

where H is an encapsulation set which contains the components of communications. The encapsulation operator is used to enforce the communication between parallel processes. EC does not contain an explicit element to express the *encapsulation* operator, but, because the EC communication functions are synchronous, the execution of the tasks in parallel implements not only the parallel composition but also the encapsulation.

4.4 Sequential composition

The basic elements which occur in sequential compositions are atomic actions and processes. For every PSF communication the implementation contains a declaration of a port:

```
atoms
  s3, r3, c3 : FRAME                    PORT port3;
communications
  s3(f)|r3(f) = c3(f) for f in FRAME
```

In general, it is possible to identify a *sender* and a *receiver* in communications, so, the atomic actions are divided in *send* and *receive* atomic actions. The *send* atoms are implemented by calls to the EC function `ECsend`.

```
s3(f)                                   ECsend(&port3, &f)
```

In the subset of PSF that we use the argument of a *recv* atom must be a free variable. Therefore, *recv* atomic actions always occur in a generalized alternative composition. The implementation is a call to the function `ECrecv` which assigns the value sent by `ECsend` to the variable given as argument to `ECrecv`.

```
sum(f in FRAME, r3(f))                  ECrecv(&port3, &f)
```

Although, `sum(f in FRAME, r3(f))` is a process expression and not an atomic actions we will assimilate this type of process expressions to the atomic action *recv*.

For every PSF process which occurs in a sequential composition the implementation contains a state in the finite automaton. If the process has parameters, the EC state is attributed with data. The attributes are implemented as local variables within the scope of a task. We will use in the following examples the function `attrib` to express the setting of a state attribute:

`attrib ( S, Val )` - assigns the value `Val` to the attribute of state `S`.

The implementation of the sequential compositions contains a sequence of calls to EC communication functions and the transition to a new state.

```
⟨PSF - sequential composition⟩≡          ⟨EC - sequential composition⟩≡
  P(f1) = s2(f1).                          if ( state == P ) {
    sum(f2 in F, r3(f2). Q(f2))              ECsend(&port2, &f1);
                                             ECrecv(&port3, &f2);
                                             attrib(Q, f2);
                                             state = Q;
                                           }
```

4.5 Alternative composition

We present the implementation of three kinds of alternative composition:

1. the operands begin with communication atomic actions,
2. the operands are conditional expressions, and,

3. the non-deterministic alternative composition.

In the first case the implementation is a loop in which the alternative ports are *polled* to see if there is data available.

```
⟨PSF - alternative composition⟩≡
  P = sum( d1 in DATA1,
           r1(d1).Q1(d1))
      +
      sum( d2 in DATA2,
           r2(d2).Q2(d2))
```

```
⟨EC - alternative composition⟩≡
  if ( state == P ) {
    while ( TRUE ) {
      if ( ECexist_data (&port1)) {
        ECrecv(&port1, &d1);
        attrib( Q1, d1);
        state = Q1;
        break;
      }
      if ( ECexist_data (&port2 )) {
        ECrecv(&port2, &d2);
        attrib( Q2, d2);
        state = Q2;
        break;
      }
    }
  }
```

The implementation uses an asymmetric communication model. In every communication it is possible to identify an active part and a passive one. The active part "triggers" the communication. The passive part watches the port to see if the active part is present. The implementation assumes that the prefixes of the alternatives are passive atomic actions. Often, the process expressions which occur in alternative composition start with "read" atomic actions. This means that, in general, the *receive* atoms are the passive part. In these cases we use function `ECexist_data` to check if data is available at the port. But in some cases the prefix of the alternatives are *send* atoms. For example, in the Concurrent Alternating Bit Protocol (CABP) [8], the process Sender can alternatively send data or receive an acknowledgement. In this case *send* will be implemented as the passive part of the communication because it occurs in an alternative composition. In the implementation we will use the function `ECexist_request` to check if there is a "request" for data at the Sender.

One disadvantage of this implementation is that it uses a *busy wait* solution for synchronization between tasks. If one is mainly interested in the use of the EC code as a pseudo code description, the implementation should be seen as a *select ... accept* structure available in concurrent programming languages based on the *rendezvous* paradigm such as Ada. If one is interested in the speed of the implementation a solution is to use a new function, `ECtresched` (tasks reschedule):

ECtresched - selects another task to run

```
#include <ec.h>
void ECtresched(void);
```

the scheduler selects a new task to run based on a round-robin policy.

The call to `ECtresched` is inserted at the end of the loop body. Thus, after polling the ports once, if there is not data available, a new task will take the control. The example given above becomes:

```
P =   sum( d1 in DATA1,
             r1(d1).Q1(d1))
      +
      sum( d2 in DATA2,
             r2(d2).Q2(d2))
```

```
if ( state == P ) {
  while ( TRUE ) {
    if ( ECexist_data (&port1)) {
      ECrecv(&port1, &d1);
      attrib( Q1, d1);
      state = Q1;
      break;
    }
    if ( ECexist_data (&port2 )) {
      ECrecv(&port2, &d2);
      attrib( Q2, d2);
      state = Q2;
      break;
    }
    /* reschedule the tasks */
    ECtresched();
  }
}
```

Another case of alternative composition is when the operands are conditional expressions. The implementation uses conditional statements to choose one alternative.

⟨*PSF - conditional expressions*⟩≡

```
P(d) = [func(d)=true] -> Q
         +
       [func(d)=false] -> R
```

⟨*EC - conditional expressions*⟩≡

```
if ( state == P ) {
    if ( func(d) ) {
         state = Q;
    }
    if ( not(func(d)) ) {
        state = R;
    }
}
```

Let us consider the case of a process P which receives an acknowledgement. The acknowledgement can be correct or it can be damaged. This situation is specified in PSF using alternative composition:

```
P = r6(ack).Q + r6(ack-error).R
```

Because the arguments of *recv* atomic actions must be free variables, in order to make an implementation, the specification is rewritten using conditional expressions. The implementation combines the rules for sequential composition and for alternative composition of conditional expressions.

```
P = sum ( a in ACK, r6(a).(
      [error(a)=false] -> Q
      +
      [error(a)=true] -> R
   ))
```

```
if ( state == P ) {
    ECrecv(&port6, &a);
    if (  not(error(a)) ) {
        state = Q;
    }
    if ( error(a) ) {
        state = R;
    }
}
```

For the specification of internal steps the formalism provides the atomic action *skip*. It is used in specifications to express nondeterministic choices. For example, let us consider a channel which transmits a frame. The frame can be transmitted correctly, it can be lost or it can be damaged.

⟨*PSF - nondeterministic alternatives*⟩≡

```
K1(f) =
  skip.K2          -- lost
  +
  skip.s4(ce).K2   -- damaged
  +
  skip.s4(f).K     -- correct
```

⟨*EC - non-deterministic alternatives*⟩≡

```
if ( state == K1 ) {
  switch ( random() % 3 ) {
    case 0: state = K2;
      break;
    case 1: f.error = TRUE;
      ECsend(&port4, &f); state = K2;
      break;
    case 2: ECsend(&port4, &f);
      state = K;
      break;
  }
}
```

The implementation uses a function which generates random numbers. The result is used to select a branch in a *switch* statement.

We present the implementation for timers in the context of alternative composition. Timers occur, in general, in situations informally described by: "receive a message in maximum *n* microseconds" which can be transformed in "receive message OR receive time-out". A timer is specified in PSF as:

⟨*PSF - specification of a timer*⟩≡

```
T = r1(start-timer).T1
T1 = r1(start-timer).T1 + s3(time-out).T
```

In the case of the process `T` the implementation uses the rules given for sequential composition while in the case of `T1` the implementation is similar to the alternative composition when the operands start with communication atoms. The difference is that for the alternative starting with the atomic action `s3(time-out)` the implementation does not contain the test for "request" – `ECexit_request`, but the test for "timeout" – `ECtimeout`.

⟨*EC implementation of a timer*⟩≡

```
#define TIME_CONSTANT 10000 /* usec */

TIMER Timer;
TASK T ()
{
  enum { T, T1 } state;
  enum { TIME_OUT_SIGNAL,
    START_TIMER_SIGNAL }
     timer_signal;

  state = T;
  while ( TRUE ) {
    if (state == T) {
      ECrecv(&port1,
        &timer_signal);
      ECstart_timer(&Timer,
        TIME_CONSTANT);
      state = T1;
    }
    if (state == T1) {
      while ( TRUE ) {
        if (ECexist_data(&port1)) {
           ECrecv(&port1,&timer_signal);
           ECstart_timer(&Timer,
             TIME_CONSTANT);
           state=T1;
           break;
        }
        if (ECtimeout(&Timer)) {
          timer_signal=TIME_OUT_SIGNAL;
          ECsend(&port3,&timer_signal);
          state=T;
          break;
        }
        ECtresched();
      }
    }
  }
}
```

Based on the examples presented in this section we have implemented several communication protocols: Alternating Bit Protocol (ABP), Positive Acknowledgement with Retransmission (PAR) Protocol, Concurrent Alternating Bit Protocol (CABP), Simple Token Ring (STR) Protocol [8]. In [3] we presented the complete PSF specification and the EC implementation of the Positive Acknowledgement with Retransmission protocol based on the method presented in this section.

4.6 Performance

We have compared the speed of the EC implementation with the the speed of the PSF simulator, a tool available in the PSF-Toolkit [8]. The test consists in the execution of three protocols (ABP, CABP and PAR without real timers) for a fixed amount of time, counting the number of transitions. The following table contains the results (`tr/sec` means transitions per second).

	PSF-simulator	EC implementation
ABP	11 `tr/sec`	2967 `tr/sec`
CABP	10 `tr/sec`	2115 `tr/sec`
PAR	11 `tr/sec`	2238 `tr/sec`

The results show that the EC implementation is significantly faster than the simulation.

5 Towards a formal specification of the implementation

Section 4 presents, informally, how the PSF specification can be used to make an EC implementation. The complete specification and the implementation of the Positive Acknowledgement with Retransmission protocol is presented in [3]. The examples are instantiations of the "rules" to transform PSF structures into EC code. We made a formal specification of this rules. The specification is not complete, in the sense that it does not generate an EC "ready-to-run" program. However, it generates the main control structures of the EC implementation and suggests the operations on data.

We make a summary of the restrictions and assumptions on the PSF specification and EC implementation introduced in Section 4:

- the examples cover only a subset of PSF;
- the implementation of data is not generated from the PSF specification;
- we used the following name conventions: *send* atoms have the form `s Nat ( Data )`, *recv* atoms `r Nat ( Data )`, and communication atoms `c Nat ( Data )`. All communications have the the form:

```
communications
  s⟨Nat⟩(⟨Data⟩) | r⟨Nat⟩(⟨Data⟩) = c⟨Nat⟩(⟨Data⟩) for ⟨Data⟩ in DATA
```

- we made the implicit assumption that the specification defines one set which contains all the components of communications:

```
sets
  of atoms
    H = { r1(d1), s1(d1) | for d1 in DATA1}
        + { r2(d2), s2(d2) | for d2 in DATA2}
        + ...
```

- the parallel compositions are "encapsulated" with `H` as encapsulation set.
- the operators used in process expressions are:
    1. parallel composition,
    2. sequential composition, and,
    3. three kinds of alternative composition: in one the alternatives begins with atoms which are part of communications, in the second the alternatives begins with the atom *skip*, in the third case the alternatives are guarded with conditions.
- the PSF specification does not make explicit use of timers. Some "extra" information is required in order to make an implementation with real timers.

For the formal specification of the implementation we used the ASF+SDF Formalism. It consists of two formalisms: the Syntax Definition Formalism (SDF) [4, 5] for describing syntax of a context-free language and the Algebraic Specification Formalism (ASF) [2] for describing semantics by means of (conditional) equations. We have made a formal specification of a transformation from the PSF subset which satisfies the conditions presented above into EC. The major parts of the specification are:

- a simple definition of the EC syntax.
- a specification of the syntax of the PSF process definitions.
- a specification of the function `map` which translates PSF structures into EC code.

We developed the ASF+SDF specification in the ASF+SDF Meta-environment [6]. An important feature offered by the environment is the possibility to "execute" the specifications. In our case the input is a set of PSF process definitions and the output is a "primitive" EC implementation.

6 Conclusions

This paper presents the implementation of PSF specifications in an extension of C for concurrent programming. In order to use this method the specification must satisfy some conditions which are discussed in Sections 4 and 5. The conditions are not too restrictive for the specifications of real systems. We used the method for the implementation of process algebra specification presented above for the implementation of four protocols: Alternating Bit Protocol (ABP), Positive Acknowledgement with Retransmission Protocol (PAR), Concurrent Alternating Bit Protocol (CABP), and Simple Token Ring Protocol (STR). The PSF specifications for these protocols are presented in [8].

In Section 5 we consider a formalization of the transformation from PSF to EC. Future work will be oriented in this direction. Based on the rules presented in this paper and a compiler for data, it would be possible to make a compiler of the PSF specifications into EC code.

The present implementation is based on a synchronous communication mechanism. Another direction for future work is to study the possibility to make the implementation using other mechanisms for communication such as asynchronous communication or event-driven communication.

References

[1] J.C.M. Baeten and W.P. Weijland. *Process Algebra.* Cambridge Tracts in Theoretical Computer Science 18. Cambridge University Press, 1990.

[2] J.A. Bergstra, J. Heering, and P. Klint. The algebraic specification formalism ASF. In J.A. Bergstra, J. Heering, and P. Klint, editors, *Algebraic Specification*, ACM Press Frontier Series, pages 1–66. The ACM Press in co-operation with Addison-Wesley, 1989. Chapter 1.

[3] C. Groza. An experiment in implementing process algebra specifications in an imperative language. Report P9412, University of Amsterdam, !994.

[4] J. Heering, P.R.H. Hendriks, P. Klint, and J. Rekers. The syntax definition formalism SDF - reference manual. *SIGPLAN Notices*, 24(11):43–75, 1989.

[5] J. Heering and P. Klint. The syntax definition formalism SDF. In J.A. Bergstra, J. Heering, and P. Klint, editors, *Algebraic Specification*, ACM Press Frontier Series, pages 283–297. The ACM Press in co-operation with Addison-Wesley, 1989. Chapter 6.

[6] P. Klint. A meta-environment for generating programming environments. *ACM Transactions on Software Engineering Methodology*, 2(2):176–201, 1993.

[7] D.E. Knuth. Literate programming. *The Computer Journal*, 27(2):97–111, 1984.

[8] S. Mauw and G.J. Veltink, editors. *Algebraic Specification of Communication Protocols.* Cambridge Tracts in Theoretical Computer Science 36. Cambridge University Press, 1993.

[9] A.S. Tanenbaum. *Computer networks.* Prentice-Hall International, Englewood Cliffs, 1989.

[10] J.L.M. Vrancken. The implementation of process algebra specifications in POOL–T. Report P8807, Programming Research Group, University of Amsterdam, 1988.

A A formal specification of the implementation of PSF process definitions

This section presents a formal specification of the implementation of PSF process definitions. The formalism used for the specification is ASF+SDF . An ASF+SDF specification consist of a sequence of named modules. Each module may contain:

Imports of other modules.

Sorts declarations defining the sorts of a signature.

Lexical syntax defining the lexical tokens.

Context-free syntax defining the concrete syntax of the functions in the signature.

Variables to be used in equations.

Equations define the meaning of the function defined in the context-free syntax.

The ASF+SDF specification contains the definition of syntax and the equations. It is presented in a "literate programming" layout [7]. Informal description and "chunks" of specification are mixed. The whole ASF+SDF specification can be obtained by "assembling" the chunks.

```
⟨PSF2EC⟩≡
  ⟨PSF2EC syntax⟩
  ⟨PSF2EC equations⟩
```

```
⟨PSF2EC syntax⟩≡
    imports Layout Identifiers  Integers
    exports
       variables
         Id[0-9']*          -> ID
         Int[0-9']*         -> INT
      ⟨Data⟩
      ⟨EC syntax⟩
      ⟨Process expressions⟩
      ⟨Atomic actions⟩
      ⟨Simple processes⟩
      ⟨Parallel composition⟩
      ⟨Sequential composition⟩
      ⟨Alternative composition - 1⟩
      ⟨Alternative composition - 2⟩
      ⟨Alternative composition - 3⟩
      ⟨Process definitions⟩
```

```
⟨PSF2EC equations⟩≡
    equations
       ⟨Equations⟩
```

The syntax of data is the same in PSF definitions as in EC programs. This specification of data is very simple. It allows only prefix functions and variables.

```
⟨Data⟩≡
     sorts DATA
     context-free syntax
        ID                                  -> DATA
        INT                                 -> DATA
        ID "(" {DATA ","}+ ")"              -> DATA
     variables
        Data[0-9']*                         -> DATA
```

For the syntax of EC programs we use a simplified form of the syntax of a C-like imperative language.

```
⟨EC syntax⟩≡
     sorts   EC-CODE EC-DNAME EC-DEF
```

Task definitions:

```
⟨EC syntax⟩+≡
   context-free syntax
       "TASK" ID "(" {DATA","}* ")"  EC-CODE -> EC-CODE
```

EC statements:

```
⟨EC syntax⟩+≡
       while "(" DATA ")"  EC-CODE        -> EC-CODE
       switch "(" DATA ")"  EC-CODE       -> EC-CODE
       case DATA ":" EC-CODE              -> EC-CODE
       break ";"                          -> EC-CODE
       if "(" DATA ")"   EC-CODE          -> EC-CODE
       "{"  EC-CODE*  "}"                 -> EC-CODE
       DATA ";"                           -> EC-CODE
```

The code of the EC program is organized in EC definitions. The EC definition has a name and a body. The body is a piece of EC code which can contain references to other EC definitions. The EC program is obtained from the EC definitions by unfolding the references. This approach, called "literate programming" is presented in [7].

```
⟨EC syntax⟩+≡
       "<-<" ID* ">->"                    -> EC-DNAME
       EC-DNAME "=" EC-CODE               -> EC-DEF
       EC-DNAME                           -> EC-CODE
       "[" EC-DEF* "]"                    -> EC-DEF
   variables
       EcCodes[0-9']*                     -> EC-CODE*
       EcDefs[0-9']*                      -> EC-DEF*
```

```
⟨Equations⟩≡
     [ec-1] {EcCodes{EcCodes'}EcCodes''} = {EcCodes EcCodes' EcCodes''}
     [ec-2] [EcDefs [EcDefs'] EcDefs''] = [EcDefs EcDefs' EcDefs'']
```

The declaration of the sorts for process expressions, bracketed process expressions and the function map which transforms a process expression into EC code is:

```
⟨Process expressions⟩≡
   sorts PROC-EXPR
   context-free syntax
     "(" PROC-EXPR ")"                    -> PROC-EXPR
     map "(" PROC-EXPR ")"                -> EC-CODE
   variables
     ProcExpr[0-9']*                      -> PROC-EXPR
```

```
⟨Equations⟩+≡
     [map-01] map ( ( ProcExpr ) ) = map ( ProcExpr )
```

The specification of atomic actions uses several conventions often used in PSF specifications. For example, a typical *send* atomic action is `s4(d)` where `s` indicates a *send*, 4 is the number of the port and `d` is the sent data. A similar convention holds for the *recv* atomic action. In this case a typical example is `r4(d)`. *Recv* atoms occur only in generalized alternative composition.

```
⟨Atomic actions⟩≡
   sorts ATOM
   context-free syntax
```

```
      s NAT "(" DATA ")"                                     -> ATOM
      sum "(" ID in ID "," r NAT "(" DATA ")"  ")"           -> ATOM
   variables
      Atom[0-9']*                                            -> ATOM
```

We use an internal representation for the atomic actions:

⟨*Atomic actions*⟩+≡

```
    context-free syntax
      atom-send ( NAT , DATA )                               -> ATOM
      atom-recv ( NAT , ID )                                 -> ATOM
```

⟨*Equations*⟩+≡

```
      [atom-01] s Nat ( Data ) = atom-send ( Nat, Data )
      [atom-03] sum ( Id in Id', r Nat ( Id ) ) = atom-recv ( Nat , Id )
```

The processes which have a name are called "simple processes":

⟨*Simple processes*⟩≡

```
    sorts SIMPLE-PROC
    context-free syntax
      ID                                          -> SIMPLE-PROC
      ID "(" DATA ")"                             -> SIMPLE-PROC
      SIMPLE-PROC                                 -> PROC-EXPR
      name ( SIMPLE-PROC )                        -> ID
    variables
      SProc[0-9']*                                -> SIMPLE-PROC
```

⟨*Equations*⟩+≡

```
      [proc-1] name ( Id )            = Id
      [proc-2] name ( Id ( Data ) ) = Id
      [map-02] map ( Id )             = assign(state, Id );
      [msp-03] map ( Id ( Data ) )  = { attrib ( Id   , Data );
         assign(state, Id ); }
```

The specification of the parallel composition is:

⟨*Parallel composition*⟩≡

```
    sorts PAR-EXPR
    context-free syntax
      SIMPLE-PROC "||" { SIMPLE-PROC "||"}+       -> PAR-EXPR
      encaps ( ID , PAR-EXPR )                    -> PAR-EXPR
      PAR-EXPR                                    -> PROC-EXPR
      map-par-tcreate (  { SIMPLE-PROC "||"}+ )  -> EC-CODE
      map-par-tdef (  { SIMPLE-PROC "||"}+)      -> EC-DEF
    variables
      ParExpr                                     -> PAR-EXPR
      SProcs                                      -> { SIMPLE-PROC "||"}+
```

⟨*Equations*⟩+≡

```
    [par-01] encaps ( Id ,  ParExpr )             = ParExpr
    [par-02] map-par-tcreate ( SProc )            = ECtcreate ( name(SProc));
    [par-03] map-par-tcreate ( SProc || SProcs ) =
                 { map-par-tcreate ( SProc )
                 map-par-tcreate ( SProcs)}
    [par-04] map-par-tdef ( SProc )               =
                 <-< Code for task name(SProc) >-> =
                 TASK name ( SProc )  ( ) {
                   while ( TRUE ) { assign ( state, name ( SProc));
                     <-< code for transitions name ( SProc) >->
                     }
                 }
    [par-05] map-par-tdef ( SProc || SProcs )     =
                 [ map-par-tdef(SProc) map-par-tdef(SProcs) ]
```

The syntax and the equations for translating the sequential composition:

⟨*Sequential composition*⟩≡

```
    sorts COMM-EXPR
    context-free syntax
      ATOM "." PROC-EXPR                                              -> COMM-EXPR
      sum "(" ID in ID "," r NAT "(" DATA ")" "." PROC-EXPR ")"  -> COMM-EXPR
      COMM-EXPR                                                       -> PROC-EXPR
```

⟨*Equations*⟩+≡
```
    [co-01] sum ( Id in Id', r Nat ( Id ) . ProcExpr ) =
                atom-recv ( Nat , Id ) . ProcExpr
    [co-02] map (   atom-send ( Nat , Data ) . ProcExpr ) =
                { ECsend ( refport ( Nat ) , refdata (Data) ) ; map ( ProcExpr) }
    [co-03] map ( atom-recv ( Nat , Id ) . ProcExpr ) =
                { ECrecv ( refport( Nat ) , refvar(Id) ) ; map ( ProcExpr )}
```

The alternative composition when all the alternatives are prefixed by atomic actions:

⟨*Alternative composition - 1*⟩≡
```
    context-free syntax
      COMM-EXPR "+" { COMM-EXPR "+"}+                  -> PROC-EXPR
      map-alt1 "(" { COMM-EXPR "+" }+ ")"              -> EC-CODE
      exist-func ( ATOM )                              -> DATA
    variables
      CommExpr                                         -> COMM-EXPR
      CommExprs                                        -> { COMM-EXPR "+"}+
```

⟨*Equations*⟩+≡
```
    [alt1-01] map ( CommExpr + CommExprs) =
                  while ( TRUE )  map-alt1 ( CommExpr + CommExprs)
    [alt1-02] map-alt1 ( Atom. ProcExpr ) =
                  if (exist-func(Atom)) {map(Atom.ProcExpr) break;}
    [alt1-03] map-alt1 ( CommExpr + CommExprs) =
                  { map-alt1 ( CommExpr) map-alt1 (CommExprs) }
    [alt1-04] exist-func ( atom-send ( Nat , Data )  ) =
                  ECexist-request ( refport( Nat ) )
    [alt1-05] exist-func ( atom-recv ( Nat , Id )  ) =
                  ECexist-data ( refport( Nat ) )
```

The alternative composition when the operands are conditional expressions:

⟨*Alternative composition - 2*⟩≡
```
    sorts BCOND COND-EXPR
    context-free syntax
      DATA "=" BOOL                                  -> BCOND
      "[" BCOND "]" "->" PROC-EXPR                   -> COND-EXPR
      COND-EXPR "+" { COND-EXPR "+" }+               -> PROC-EXPR
      map-alt2 "(" { COND-EXPR "+" }+ ")"            -> EC-CODE
      map-bcond ( BCOND )                            -> DATA
    variables
      CDExpr                                         -> COND-EXPR
      CDExprs                                        -> { COND-EXPR "+" }+
      BCond[0-9']*                                   -> BCOND
```

⟨*Equations*⟩+≡
```
    [alt2-01] map ( CDExpr  + CDExprs ) = { map-alt2 ( CDExpr) map-alt2 ( CDExprs) }
    [alt2-02] map-alt2 (   [ BCond ] -> ProcExpr     ) =
                           if ( map-bcond ( BCond ) ) map ( ProcExpr )
    [alt2-03] map-alt2 ( CDExpr + CDExprs ) =
                           { map-alt2 ( CDExpr ) map-alt2 ( CDExprs ) }
    [alt2-04] map-bcond  ( Data = true ) = Data
    [alt2-05] map-bcond ( Data = false ) = neg ( Data )
```

In the third kind of alternative composition the operands are non-deterministic alternatives:

⟨*Alternative composition - 3*⟩≡
```
    sorts NONDET-EXPR
    context-free syntax
      skip "." PROC-EXPR                                      -> NONDET-EXPR
      NONDET-EXPR "+" { NONDET-EXPR "+"}+                     -> PROC-EXPR
      map-alt3 "(" INT "," { NONDET-EXPR "+"}+ ")"            -> EC-CODE
      no "(" { NONDET-EXPR "+"}+ ")"                          -> INT
    variables
      NDExpr                                                  -> NONDET-EXPR
      NDExprs                                                 -> {NONDET-EXPR "+"}+
```

⟨*Equations*⟩+≡

```
    [alt3-01] map ( NDExpr + NDExprs ) = switch (xrand(no(NDExpr + NDExprs)))
                   map-alt3(0, NDExpr + NDExprs)
    [alt3-02] map-alt3 ( Int , skip. ProcExpr ) = case Int:{map(ProcExpr)break;}
    [alt3-03] map-alt3 ( Int, NDExpr + NDExprs ) =
                   { map-alt3 ( Int , NDExpr) map-alt3 ( Int+1, NDExprs) }
    [no-01] no ( NDExpr ) = 1
    [no-02] no ( NDExpr + NDExprs ) = 1 + no ( NDExprs )
```

The translation of process definitions results in a collection of EC definitions:

⟨*Process definitions*⟩≡

```
    sorts PSF-DEF
    context-free syntax
      SIMPLE-PROC "=" PROC-EXPR                        -> PSF-DEF
      map ( PSF-DEF* )                                 -> EC-DEF
      map-def ( PSF-DEF )                              -> EC-DEF
    variables
      PsfDefs                                          -> PSF-DEF*
      PsfDef                                           -> PSF-DEF
```

⟨*Equations*⟩+≡

```
    [map-1] map ( PsfDef PsfDefs ) = [ map-def ( PsfDef )  map ( PsfDefs ) ]
    [map-2] map ( )  = [ ]
    [def-01] map-def ( SProc = SProc' || SProcs) =
                   [   <-< Code for task name(SProc) >-> =
                           map-par-tcreate ( SProc' || SProcs )
                       map-par-tdef (SProc' || SProcs )
                   ]
    [def-03] map-def ( SProc = SProc' ) = <-<Code for state name(SProc)>->=
                           if ( eq ( state, name(SProc))) map ( SProc' )
    [def-03] map-def ( SProc = CommExpr ) = <-<Code for state name(SProc)>->=
                           if ( eq ( state, name(SProc))) map ( CommExpr )
    [def-03] map-def ( SProc = CommExpr + CommExprs ) =
                   <-<Code for state name(SProc)>-> =
                   if ( eq ( state, name(SProc))) map ( CommExpr + CommExprs )
    [def-03] map-def ( SProc = NDExpr + NDExprs ) =
                   <-<Code for state name(SProc)>-> =
                   if (eq(state, name(SProc))) map(NDExpr + NDExprs)
    [def-03] map-def ( SProc = CDExpr + CDExprs ) =
                   <-<Code for state name(SProc)>-> =
                   if ( eq ( state, name(SProc))) map ( CDExpr + CDExprs )
```

Graph Isomorphism Models for Non Interleaving Process Algebra

J.C.M. Baeten
Department of Computer Science, Eindhoven University of Technology, Eindhoven, The Netherlands

J.A. Bergstra
Programming Research Group, University of Amsterdam, Amsterdam, The Netherlands; and Department of Philosophy, Utrecht University, Utrecht, The Netherlands

We present a simple and intuitive model for the syntax of ACP based on graph isomorphism. We prove an expressivity result, and use the model to determine the number of states of a process.
Note: This research was supported in part by ESPRIT basic research action 7166, CONCUR2. The second author is also partially supported by ESPRIT basic research action 6454, CONFER.

1. Introduction.

The purpose of this paper is to provide a very simple model of the syntax of ACP [7]. This model, based on graph isomorphism, provides a clear explanation of the meaning of the primitives of ACP but it satisfies fewer axioms. In particular, it is non-interleaving in the sense of [3].

We feel that the graph isomorphism model is the simplest and most convincing one presented thus far in the literature on ACP. Of course it has various drawbacks: because it is non-interleaving and because it is concrete (in the sense of [1]) it is not very well suited for equational protocol verification. The practical merit of the graph isomorphism model is that it allows a precise state count of systems. We propose to use this model if the number of states of a process description is referred to. To this end we provide some examples.

We are unaware of a similar model in the literature. Of course most constructions have been given already in [8] but that paper failed to identify the graph isomorphism structure as a model for the syntax of ACP in its own right.

Acknowledgement: The authors thank S. Mauw and P. Rambags (both Eindhoven University of Technology) for useful comments.

2. Process graphs modulo isomorphism.

2.1 Definition. We introduce a set of process graphs as follows. Let A be a given set, and let κ, λ be two infinite cardinal numbers with $\kappa \geq \lambda$. A process graph g of cardinality $< \kappa$, with out-degree (branching degree) $< \lambda$, over a set of labels A is a quadruple $\langle S, \rightarrow, begin, end\rangle$ where

- S is a set, the set of states,
- begin $\in$ S, the start state
- end $\in$ S, the end state
- $\rightarrow \subseteq S \times A \times S$ is the transition relation,

and we have the following conditions:

- $1 < |S| < \kappa$
- begin $\neq$ end
- $\forall s \in S \; |\{t \in S : \exists a \in A \langle s,a,t\rangle \in \rightarrow\}| < \lambda$, the out-degree is $< \lambda$
- $\{s \in S : \exists a \in A \langle s, a, begin\rangle \in \rightarrow\} = \varnothing$, the start state has no incoming edges
- $\{s \in S : \exists a \in A \langle end, a, s\rangle \in \rightarrow\} = \varnothing$, the end state has no outgoing edges.

We call any state different from the start state or the end state an *interior state.* We write $s \xrightarrow{a} t$ for $\langle s,a,t\rangle \in \rightarrow$. We refer to the four components of a process graph g by S(g), $\rightarrow$(g), begin(g) and end(g), respectively. $\mathcal{G}(A, \kappa, \lambda)$ is the set of all process graphs over a set of labels A of cardinality $< \kappa$ with out-degree $< \lambda$. $\mathcal{G}(A, \aleph_0, \aleph_0)$ is the set of finite process graphs, $\mathcal{G}(A, \kappa, \aleph_0)$ contains only finitely branching process graphs.

Note that we require that start and end state are always different. This allows us to give an intuitive definition for alternative composition (without root unwinding). An alternative to the present definition is to allow several end states (and several start states). We restrict ourselves to the simplest definition here.

2.2 Definition. Let g,h $\in \mathcal{G}(A, \kappa, \lambda)$. A bijection ϕ between states of g and states of h is called an *isomorphism* if:

1. ϕ(begin(g)) = begin(h), ϕ(end(g)) = end(h)
2. $s \xrightarrow{a} t \Leftrightarrow \phi(s) \xrightarrow{a} \phi(t)$.

We say g,h are *isomorphic*, $g \approx h$, if there is an isomorphism between g and h. Obviously, isomorphism is an equivalence relation on process graphs. We can divide out this equivalence, and obtain the algebras $\mathcal{G}(A, \kappa, \lambda)/\approx$. Basically, this means that in these algebras the names of the states do not matter. This allows us to take disjoint unions of state spaces in the following definitions of operators on process graphs modulo isomorphism.

2.3 Definition. We define several operators on process graphs modulo isomorphism, i.e. on the algebras $\mathcal{G}(A, \kappa, \lambda)/\approx$. First, constants.

1. *Atomic action.* Let a $\in$ A. **a** = $\langle\{b, e\}, \{\langle b, a, e\rangle\}, b, e\rangle$.
2. *Deadlock.* $\delta = \langle\{b, e\}, \varnothing, b, e\rangle$

Abusing notation, we usually write a for **a**.

Next, operators.

3. Alternative composition. Let g,h ∈ $\mathcal{G}$(A, κ, λ) be given. Assume that the set of states of g is disjoint from the set of states of h (since we consider process graphs modulo isomorphism, we can always ensure that this is the case). The set of states of g+h consists of the interior states of g, the interior states of h, and two new states begin, end. The transition relation is given by:

a. all transitions between interior states of g or h

b. a transition begin $\xrightarrow{a}$ end whenever there is a transition begin(g) $\xrightarrow{a}$ end(g) or begin(h) $\xrightarrow{a}$ end(h)

c. for interior s in g, a transition begin $\xrightarrow{a}$ s if begin(g) $\xrightarrow{a}$ s, a transition s $\xrightarrow{a}$ end if s $\xrightarrow{a}$ end(g)

d. for interior s in h, a transition begin $\xrightarrow{a}$ s if begin(h) $\xrightarrow{a}$ s, a transition s $\xrightarrow{a}$ end if s $\xrightarrow{a}$ end(h).

Note that a + a ≈ a, and g + h ≈ h + g, g + (h + k) ≈ (g + h) + k, g + δ ≈ g for all g,h,k.

4. Sequential composition. Let g,h ∈ $\mathcal{G}$(A, κ, λ) be given with disjoint state sets. The set of states of g·h consists of the interior states of g, the interior states of h, the states begin(g), end(h) and a new state link. begin(g·h) = begin(g), end(g·h) = end(h). The transition relation is given by:

a. all transitions between states of g or h that are still in the set of states of g·h

b. a transition s $\xrightarrow{a}$ link whenever there is a transition s $\xrightarrow{a}$ end(g)

c. a transition link $\xrightarrow{a}$ s whenever there is a transition begin(h) $\xrightarrow{a}$ s.

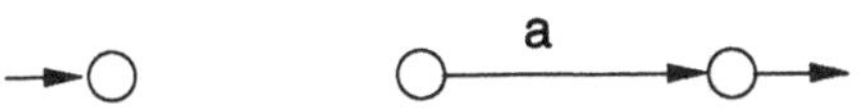

Fig. 1a. δ·a.

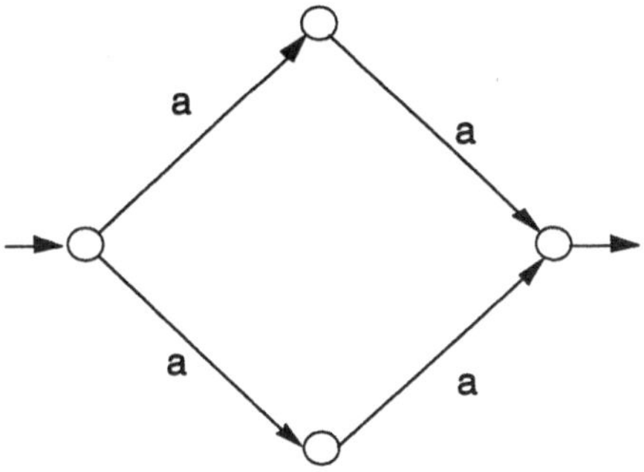

Fig. 1b. a·a + a·a.

Examples: fig. 1a shows a process graph (modulo isomorphism) for δ·a. The start state is denoted by a small unlabeled incoming arrow, the end state by a small unlabeled outgoing arrow. Note that this graph is not isomorphic to the graph of δ. Fig. 1b shows a process graph for $a{\cdot}a + a{\cdot}a$. Note that this graph is not isomorphic to the graph of $a{\cdot}a$, so $g + g \approx g$ does not hold for all process graphs. Similarly, $(a + b){\cdot}a$ is not isomorphic to $a{\cdot}a + b{\cdot}a$ (if $b \neq a$), so $(g + h){\cdot}k \approx g{\cdot}k + h{\cdot}k$ does not hold in general. We do have the identity $(g{\cdot}h){\cdot}k \approx g{\cdot}(h{\cdot}k)$ for all graphs.

5. *Parallel composition.* Let $g,h \in \mathcal{G}(A, \kappa, \lambda)$ be given. Let a partial, commutative and associative function $\gamma: A \times A \to A$ be given, the communication function. The set of states of $g \| h$ is the cartesian product of the states of g and the states of h. $begin(g \| h) = \langle begin(g), begin(h) \rangle$, $end(g \| h) = \langle end(g), end(h) \rangle$.

The transition relation is given by:

a. for each state s in g, and each transition $t \xrightarrow{a} t'$ in h, there is a transition $\langle s,t \rangle \xrightarrow{a} \langle s,t' \rangle$

b. for each state t in h, and each transition $s \xrightarrow{a} s'$ in h, there is a transition $\langle s,t \rangle \xrightarrow{a} \langle s',t \rangle$

c. for each pair of transitions $\langle s,t \rangle \xrightarrow{a} \langle s',t \rangle$, $\langle s,t \rangle \xrightarrow{b} \langle s,t' \rangle$ such that $\gamma(a,b)$ is defined, say $\gamma(a,b) = c$, there is a transition $\langle s,t \rangle \xrightarrow{c} \langle s',t' \rangle$.

Note that $a \| b \approx a{\cdot}b + b{\cdot}a$ (if $\gamma(a,b)$ is undefined). However, $a{\cdot}a \| b$ is not isomorphic to $a{\cdot}(a{\cdot}b + b{\cdot}a) + b{\cdot}a{\cdot}a$ (the former graph has 6 states, the latter 7).

6. *Left merge.* The graph of $g \mathbin{⌊\!⌊} h$ has the same states as the graph of $g \| h$, and the same transitions except that the transitions $\langle begin(g), begin(h) \rangle \xrightarrow{a} \langle s, t \rangle$ with $t \neq begin(h)$ are omitted.

7. *Communication merge.* The graph of $g \mid h$ has the same states as the graph of $g \| h$, and the same transitions except that the transitions $\langle begin(g), begin(h) \rangle \xrightarrow{a} \langle s, t \rangle$ with $t = begin(h)$ or $s = begin(g)$ are omitted.

8. *Encapsulation.* Let $g \in \mathcal{G}(A, \kappa, \lambda)$ be given, and let $H \subseteq A$. The graph of $\partial_H(g)$ has the same states as the graph of g, and the same transitions except that all transitions $s \xrightarrow{a} s'$ with $a \in H$ are omitted.

9. *Renaming.* Let $g \in \mathcal{G}(A, \kappa, \lambda)$ be given, and let $f: A \to A$ be a given function. The graph of $\rho_f(g)$ has the same states as the graph of g, the same $begin$ and end, and each transition $s \xrightarrow{a} s'$ is replaced by a transition $s \xrightarrow{f(a)} s'$.

10. *Conditional operator.* Let $g \in \mathcal{G}(A, \kappa, \lambda)$ be given. The graph of $true :\to g$ is the same as the graph of g, and the graph of $false :\to g$ is obtained from the graph of g by removing all edges starting in the begin state. The ternary *if...then...else...* operator is defined by (b is a boolean):

$$g \lhd b \rhd h = (b :\to g) + (\neg g :\to h).$$

It is more involved to define a conditional operator over a general boolean algebra (other than {$true$, $false$}), as we did in [2]. We sketch part of this in section 6.

11. *Finite state operator.* Let $g \in \mathcal{G}(A, \kappa, \lambda)$ be given, let St be a finite set, let $act: A \times St \to A$ be a partial function, let $eff: A \times St \to St$ be a total function and let $T \in St$. The

set of states of $\lambda_T(g)$ is the the cartesian product $(S(g) - \{end(g)\}) \times St$ together with the singleton $\{end(g)\}$. $begin(\lambda_T(g)) = \langle begin(g), T\rangle$, $end(\lambda_T(g)) = end(g)$.
The transition relation is given by:

a. Suppose $s \xrightarrow{a} s'$ is a transition in g, $t \in St$ and $act(a, t)$ is undefined. In this case, we do not have a transition.

b. Suppose $s \xrightarrow{a} s'$ is a transition in g, $s' \neq end(g)$, $t \in St$ and $act(a, t)$ is defined. In this case, we have a transition $\langle s,t\rangle \xrightarrow{b} \langle s',u\rangle$, where $b = act(a,t)$ and $u = eff(a,t)$.

c. Suppose $s \xrightarrow{a} end(g)$ is a transition in g, $t \in St$ and $act(a, t)$ is defined. In this case, we have a transition $\langle s,t\rangle \xrightarrow{b} end(g)$, where $b = act(a,t)$.

12. Priority operator. Let $g \in \mathcal{G}(A, \kappa, \lambda)$ be given, and let $<$ be a given partial ordering on A. The set of states of $\theta_<(g)$ is the set of states of g, and the set of transitions is a subset of the set of transitions of g, given by:
$s \xrightarrow{a} s'$ is a transition in $\theta_<(g)$ if for all $b > a$ we do not have a transition $s \xrightarrow{b} s''$ in g.

13. Ternary Kleene star operator. Let $g,h,k \in \mathcal{G}(A, \kappa, \lambda)$ be given with disjoint state sets. The set of states of $tks(g,h,k)$ is the set of states of $g{\cdot}k$ together with the set of the interior states of h. The begin state is $begin(g)$, the end state is $end(k)$. The transitions are those of $g{\cdot}k$, as given in 4, the transitions between interior states of h, and moreover:

a. a transition $s \xrightarrow{a} link$ whenever there is a transition $s \xrightarrow{a} end(h)$ in h (s interior)

b. a transition $link \xrightarrow{a} s$ whenever there is a transition $begin(h) \xrightarrow{a} s$ in h (s interior)

c. a transition $link \xrightarrow{a} link$ whenever there is a transition $begin(h) \xrightarrow{a} end(h)$ in h.

Note that $g{\cdot}h \approx tks(g, \delta, h)$.

14. Proper iteration and binary Kleene star. With the help of the ternary Kleene star operator as defined above, we can easily define the binary Kleene star operator of [6] and the proper iteration operator of [9].

Proper iteration: $g^{\oplus}h = tks(g, g, h)$

Binary Kleene star: $g^*h = h + g^{\oplus}h$.

2.4 Definition. Our model makes it possible to define a cardinality function on process graphs modulo isomorphism. Further, we can define the reverse of a process. Finally, we define an extra operator ξ, that limits a process to its set of reachable states.

1. The *cardinality* of a process graph, $|g|$, is the cardinality of its set of states. We can compute:

a. $|g \cdot h| = |g| + |h| - 1$

b. $|g + h| = |g| + |h| - 2$

c. $|g \parallel h| = |g| \times |h|$.

2. *Reverse operator.* If $g \in \mathcal{G}(A, \kappa, \lambda)$, then g^{-1} has the same set of states as g, $begin(g^{-1}) = end(g)$, $end(g^{-1}) = begin(g)$ and $s \xrightarrow{a} t$ is a transition in g^{-1} whenever $t \xrightarrow{a} s$ is a transition in g. Note that this operator commutes with all operators defined so far. It does not, however, commute with the following operator.

3. *Reachability operator.* Let a process graph $g \in \mathcal{G}(A, \kappa, \lambda)$ be given. We define its set of reachable states, $reach(g) \subseteq S(g)$ inductively:

   a. $begin(g) \in reach(g)$

   b. if $s \in reach(g)$, and $s \xrightarrow{a} s'$ is a transition in g, then $s' \in reach(g)$.

Now we define $\xi(g)$ as follows. The set of states of $\xi(g)$ is $reach(g) \cup \{end(g)\}$, with same begin state and end state, and only the transitions between reachable states. With the help of this reachability operator, we can formulate new versions of well-known identities, for instance we have $\xi(\delta \cdot g) \approx \delta$ for all process graphs g.

2.5 Theorem. The models $\mathcal{G}(A, \kappa, \lambda)$ are non-interleaving (in the sense of [3]).
Proof: Consider the process $a \cdot a \parallel b$ ($a \neq b$, and $\gamma(a,b)$ is undefined). If we have an interleaving model, then this process should equal $a \cdot (a \cdot b + b \cdot a) + b \cdot a \cdot a$. However, we found in 2.3.5 that this is not the case.

We conclude that the expansion theorem does not hold in the models $\mathcal{G}(A, \kappa, \lambda)$, and thus they are non-interleaving models.

3. Bisimulation.

We look at the familiar notion of bisimulation in the present setting. To this end, consider the following definition.

3.1 Definition.

We define the familiar notion of bisimulation on process graphs. Let $g,h \in \mathcal{G}(A, \kappa, \lambda)$. A relation R between states of g and states of h is called a *bisimulation* if:

1. $R(begin(g), begin(h))$, $R(end(g), end(h))$ and a begin or end state is not related to another state;
2. if $R(s, t)$ and $s \xrightarrow{a} s'$, then there is a t' such that $t \xrightarrow{a} t'$ and $R(s', t')$;
3. if $R(s, t)$ and $t \xrightarrow{a} t'$, then there is a s' such that $s \xrightarrow{a} s'$ and $R(s', t')$.

We say g,h are *bisimilar*, $g \leftrightarrows h$, if there is an bisimulation between g and h.

It is well-known that bisimulation is an equivalence relation on process graphs. We can divide out this equivalence, and obtain the algebras $\mathcal{G}(A, \kappa, \lambda)/\leftrightarrows$. Since bisimulation is also a congruence for all operators defined in section 2.3, we can define these operators on these algebras. We cannot, however, define the cardinality operator or the reverse operator any more. For the reachability operator, we have $\xi(g) \leftrightarrow g$, so this operator becomes the identity on process graphs modulo bisimulation.

3.2 Definition.

Let a process graph $g \in \mathcal{G}(A, \kappa, \lambda)$ be given. We say states s,t of g are *bisimulation equivalent*, $s \leftrightarrow t$, iff there is an bisimulation R between g and g such that R(s, t).

It is obvious that this defines an equivalence relation on states of g. We can divide out this equivalence relation, and obtain the reduced graph of g.

3.3 Definition. Let a process graph $g \in \mathcal{G}(A, \kappa, \lambda)$ be given. The *reduced graph* of g, $g/\leftrightarrow$ has as states the set of equivalence classes of bisimulation equivalent states of g, $begin(g/\leftrightarrow) = \{begin(g)\}$, $end(g/\leftrightarrow) = \{end(g)\}$ and $s/\leftrightarrow \xrightarrow{a} s'/\leftrightarrow$ iff $s \xrightarrow{a} s'$.

3.4 Remark. The graph isomorphism model is non-interleaving. The bisimulation model can be obtained from this model as a homomorphic image, and is interleaving. The bisimulation model, however, is not the *least* identifying model that is interleaving; in other words, there are models in between the graph isomorphism model and the bisimulation model that are still interleaving. Whether such model are useful, we do not know. Therefore, we choose not to present such a model here.

4. Expressivity.

In this section, we prove an expressivity result for the algebras $\mathcal{G}(A, \aleph_0, \aleph_0)/\approx$. We show that every finite process graph modulo isomorphism can be obtained from a single graph by using alternative, sequential, parallel composition and iteration, renaming and encapsulation operators. We do need an infinite set of atomic actions and an appropriate choice of the communication function in order to obtain this result.

4.1 Atomic actions.

Suppose we have a countable set of atomic actions A. Divide A into a countable set B and a disjoint countable set C. Suppose we have a bijection i from the set of finite subsets of B to C. Define a communication function γ as follows:

1. $\gamma(a,b) = i(\{a,b\})$ if $a,b \in B$
2. $\gamma(a, i(s)) = \gamma(i(s), a) = i(\{a\} \cup s)$ if $a \in B$, s a finite subset of B
3. $\gamma(i(s), i(s')) = i(s \cup s')$ if s,s' are finite subsets of B.

Note that this definition makes γ commutative and associative. Notice that this definition amounts to a *free* communication function on the set B. This is similar to the approach we used in [3].

4.2 The seed process.

Let a,b,c,d,e,f,g,h,k be distinct atoms from B. The seed process P is given in fig. 2. This process has 4 states, and is maximally connected: every state except end has an outgoing edge to every state except begin and every state except begin has an incoming edge from every state except end.

If Q is the same as process P except that the k-edge is omitted, then we have $P \approx Q + k$. Note that a further decomposition of P using the operators of section 2 is not possible.

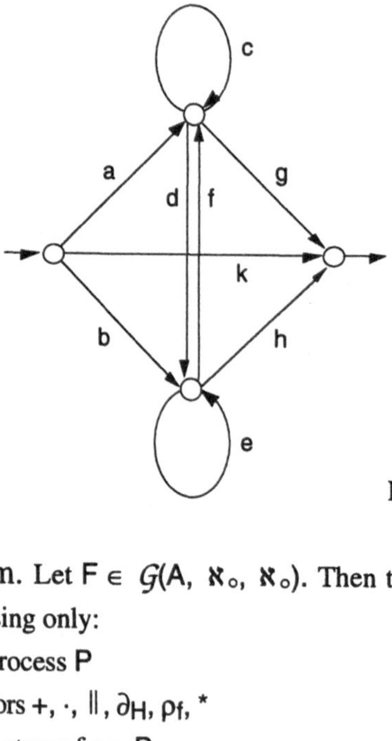

Figure 2. Seed process.

4.3 Theorem. Let $F \in \mathcal{G}(A, \aleph_0, \aleph_0)$. Then there is a graph $G \approx F$ and G can be constructed using only:

1. the seed process P
2. the operators $+, \cdot, \|, \partial_H, \rho_f, {}^*$
3. additional atoms from B.

We will construct the graph G in several stages.

4.4 Obtaining enough nodes.

First of all, we construct a graph G_1 that has at least as many nodes as F, and is maximally connected. Moreover, all edges of G_1 have distinct labels. Take a number $N \geq 1$ such that $2^N \geq |F| - 2$ (the number of internal nodes of F). Choose a set of distinct atoms $\{a_{k,j} : 1 \leq k \leq 9, 1 \leq j \leq N\} \subseteq B$. Define for each j the renaming f_j by:

$f_j(a)=a_{1,j}$, $f_j(b)=a_{2,j}$, $f_j(c)=a_{3,j}$, $f_j(d)=a_{4,j}$, $f_j(e)=a_{5,j}$, $f_j(f)=a_{6,j}$, $f_j(g)=a_{7,j}$, $f_j(h)=a_{8,j}$, $f_j(k)=a_{9,j}$

Define $P_j = \rho_{f_j}(P)$. This gives N copies of P, with all edge-labels distinct.

Now put $G_1 = \partial_H(P_1 \parallel P_2 \parallel \ldots \parallel P_N)$, with

$H = H_0 \cup H_1 \cup H_2$

$H_0 = \{a_{k,j} : 1 \le k \le 9,\ 1 \le j \le N\}$

$H_1 = \{i(s) \in C : |s| < N\}$

$H_2 = \{i(s) \in C : \exists k \in \{7,8,9\}, j \in \{1,\ldots,N\}\ a_{k,j} \in s \ \wedge\ \exists k \in \{1,\ldots,6\}, j \in \{1,\ldots,N\}\ a_{k,j} \in s\}$.

H_0 and H_1 together ensure that the only steps possible in G_1 are communications involving a step from each of the components, H_2 ensures that if one component does a termination step (a step to end) then all components do so simultaneously.

Now put $G_2 = \xi(G_1)$. Note that G_2 has exactly 2^N internal nodes and

- exactly one transition from begin to each internal node and to end
- exactly one transition from each internal node to each internal node (including itself)
- exactly one transition from each internal node to end.

Moreover, all transitions have distinct labels.

4.5 Exact number of nodes.

The second step is to reduce G_2 so that we obtain exactly the right number of nodes. Let ϕ be an injection from S(F) into S(G_2) that respects begin and end. Such an injection exists by choice of N. Let H_3 contain all labels of edges of h_2 that start from or end in a node outside the range of ϕ.

Put $G_3 = \xi \circ \partial_{H_3}(G_2)$. Then G_3 has exactly |F| nodes and still has the further properties of G_2 above.

4.6 Multiple steps.

Let M be the maximum number of distinct edges between any pair of nodes in F. We will modify G_3, so that each transition is replaced by M distinct transitions. To this end, let $b_1, \ldots, b_M, c_1, \ldots, c_M$ be fresh atoms in B (distinct from all $a_{i,j}$ and pairwise distinct). Put $G_4 = b_1 + \ldots + b_M$, $G_5 = c_1 + \ldots + c_M$. Next,

$G_6 = \partial_{H'}(G_3 \parallel G_4{}^*G_5)$, with

$H' = H_4 \cup H_5 \cup H_6 \cup H_7$

$H_4 = \{b_m, c_m : 1 \le m \le M\}$

$H_5 = \{i(s) \in C : |s| = N\}$

$H_6 = \{i(s \cup \{c_m\}) \in C : i(s)$ labels a non-terminating step in G_3 and $1 \le m \le M\}$

$H_7 = \{i(s \cup \{b_m\}) \in C : i(s)$ labels a terminating step in G_3 and $1 \leq m \leq M\}$.
The encapsulation here ensures that the only steps possible in G_3 are communications between a step of G_3 and a step of the other component, and moreover that both components only terminate together.

Now put $G_7 = \xi(G_6)$. G_7 has exactly $|F|$ nodes and

- exactly M transitions from begin to each internal node and to end
- exactly M transitions from each internal node to each internal node (including itself)
- exactly M transitions from each internal node to end.

Moreover, all transitions have distinct labels.
We have now constructed a graph into which F can be embedded, after a suitable relabeling of edges. What remains now is to define this renaming, and trim away all superfluous edges.

4.7 Number of edges, labels of edges.

Now we define a renaming function f and an encapsulation set H" as follows:
Take a pair of nodes $\langle n,m \rangle$ in G_7. If either $n = \text{end}(G_7)$ or $m = \text{begin}(G_7)$, do nothing. Otherwise, there are M edges from n to m, say with labels $d_1, \ldots, d_M$.

Since ϕ is a bijection between $S(F)$ and $S(G_7)$, $n = \phi(s)$, $m = \phi(t)$ for certain nodes s,t in F. Suppose there are K edges between s and t in F, with labels $e_1, \ldots, e_K$, $0 \leq K \leq M$. Put $f(d_1) = e_1, \ldots, f(d_K) = e_K$, and put $d_{K+1}, \ldots, d_M$ into H".
Do the same for every node-pair in G_7. Since all labels in G_7 are distinct, f is well-defined, and dom(f) and H" are disjoint. Define

$G = \partial_{H''} \circ \rho_f(G4)$. By construction we have $G \approx F$.

5. Application: counting states.

The graph isomorphism model allows us to determine the number of states of a process. As an example, we consider the familiar Alternating Bit Protocol. We take the description of [4], and recast this in order to use iteration operators instead of recursive equations. We remark that [5] contains a description and verification of a simplified ABP (due to [10]), also using iteration instead of recursion.

We assume that we have a data set D, with $|D| = n$, that are to be transmitted from sender S to receiver R using unreliable channels K,L. $B = \{0,1\}$. The communication links are as shown in fig. 3. We use the standard communication function given by $\gamma(r_k(x), s_k(x)) = c_k(x)$ (see [4]). We have the following specifications.

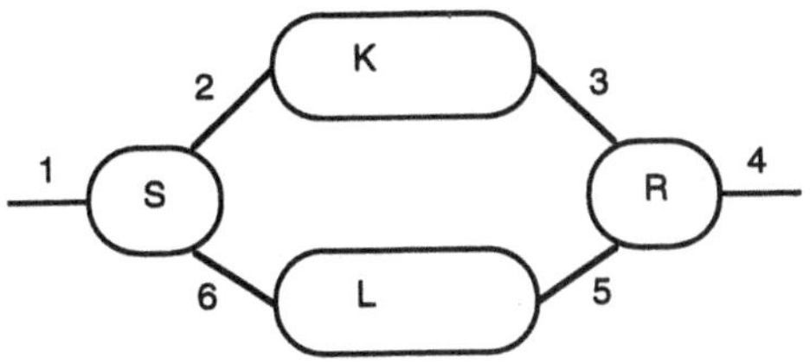

Figure 3. ABP.

5.1 Channels.

$$K = \Big(\sum_{d \in D, b \in B} r_2(db) \cdot (i \cdot s_3(db) + i \cdot s_3(\bot)) \Big) * \delta$$

$$L = \Big(\sum_{b \in B} r_5(b) \cdot (i \cdot s_6(b) + i \cdot s_6(\bot)) \Big) * \delta$$

Following the definitions in section 2, we find that K has $12n + 3$ states, and L has 15 states. The graph of L is shown in fig. 4. The labels in the second part are the same as those in the first part, and are omitted. The graph of K is similar, except that the branching in the begin state and the middle state is of size $2n$.

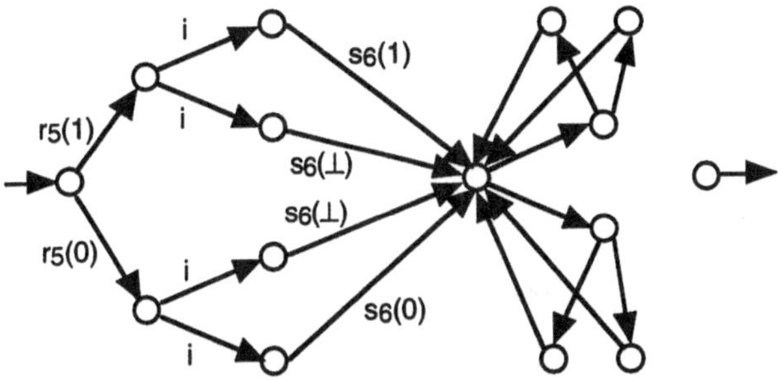

Figure 4. Acknowledgement channel L.

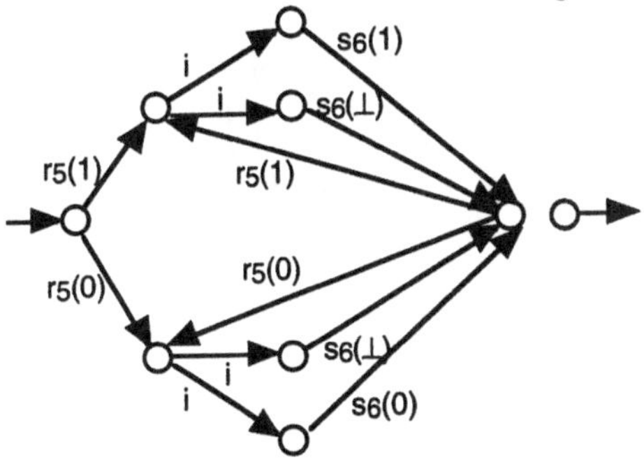

Figure 5. Reduced graph of L.

We can divide out bisimulation equivalence, and then the states in the second part of the graph are cancelled. We show the reduced graph of L in fig. 5. We find in this case that

K has $6n + 3$ states, and L 9. If we want to define these reduced processes in our syntax, it is not sufficient to use the operators of section 2, but have to follow in essence the construction of section 4.

To sketch this in case of L, we take three copies of the seed process P, with sets of atoms $a_1,...,a_9$ resp. $b_1,...,b_9$ resp. $c_1,...,c_9$, take the parallel composition, encapsulate all actions except the following 12 ternary communications, apply the following renaming and lastly apply the reachability operator.

- rename the communication of a_1, b_2 and c_2 and of a_6, b_4 and c_5 into $r_5(1)$
- rename the communication of a_2, b_2 and c_1 and of a_5, b_4 and c_6 into $r_5(0)$
- rename the communication of a_3, b_6 and c_5, of a_3, b_5 and c_6, of a_5, b_6 and c_3 and of a_6, b_6 and c_3 into i
- rename the communication of a_4, b_3 and c_5 into $s_6(1)$
- rename the communication of a_4, b_3 and c_4 into $s_6(0)$
- rename the communication of a_4, b_6 and c_4 and of a_5, b_3 and c_4 into $s_6(\perp)$.

It is obvious that this definition of L does not add to our understanding of the process. We will omit such descriptions with a minimal number of states in the sequel.

We conclude that the number of states of a process depends very much on the specification of the process in the syntax, and that the simplest and most intuitive notation usually does not have a minimal number of states.

5.2 Sender.

$$S = (S0 \cdot S1) * \delta$$

$$Sb = \sum_{d \in D} r_1(d) \cdot tks(s_2(db), (r_6(1\text{-}b)+r_6(\perp)) \cdot s_2(db), r_6(b)) \qquad \text{for } b = 0,1.$$

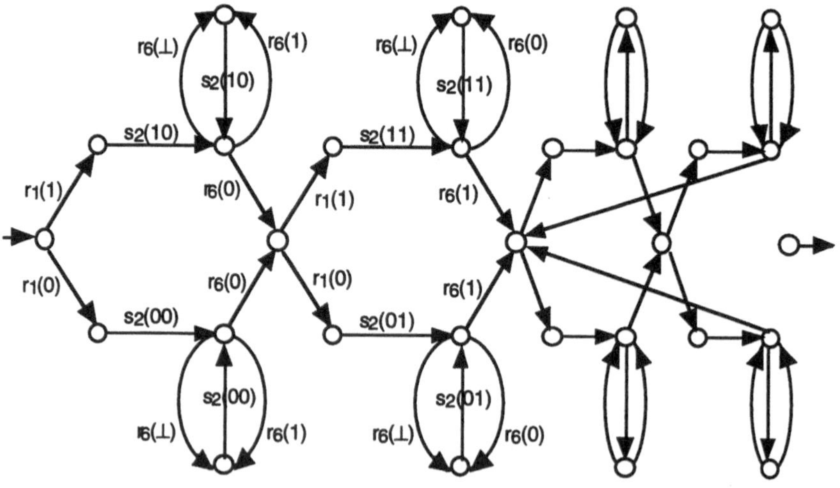

Figure 6. Graph of S, for D = {0, 1}.

We find that each Sb has 3n + 2 states, and S has 12n + 5 states. The sender will have even more states if we do not use the ternary tks operator, but instead the binary iteration operators. The graph of S is shown in fig. 6, in case D = {0, 1}. Again, the labels in the second part are the same as in the first part, and are omitted. Dividing out bisimulation equivalence, we get that each Sb has 2n + 2 states, and S has 4n + 4. We show the reduced graph of S in fig. 7.

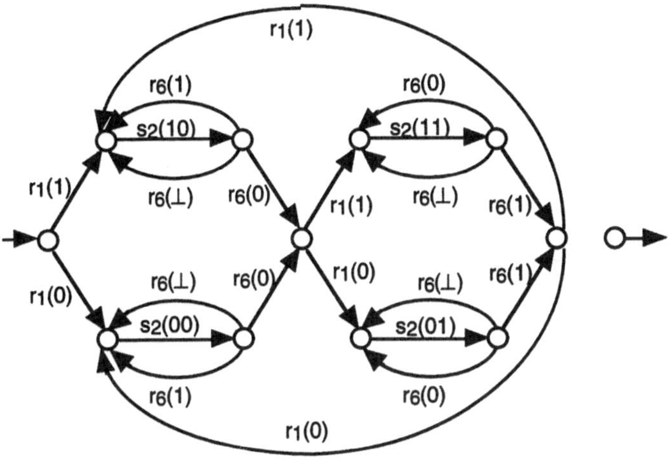

Figure 7. Reduced graph of S, for D = {0, 1}.

5.3 Receiver.

$R = (R1 \cdot R0) * \delta$

$$Rb = \Bigg(\Big(\big(\sum_{d \in D} r_3(db) + r_3(\perp)\big) \cdot s5(b)\Big) * \Big(\sum_{d \in D} r_3(d(1\text{-}b)) \cdot s_4(d)\Big)\Bigg) \cdot s_5(1\text{-}b)$$

We find that each Rb has 2n + 6 states, and R has 8n + 20 states. Dividing out bisimulation equivalence, we get that each Rb has n + 5 states, and R has 2n + 9. We show the graph of R in fig. 8, in case D = {0, 1}, with the same conventions as above, and the reduced graph in fig. 9.

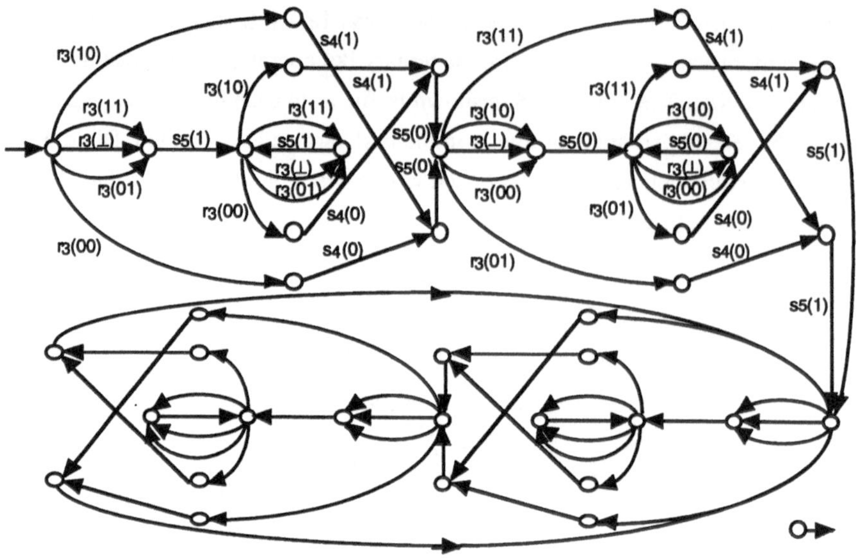

Figure 8. Graph of R, for D = {0, 1}.

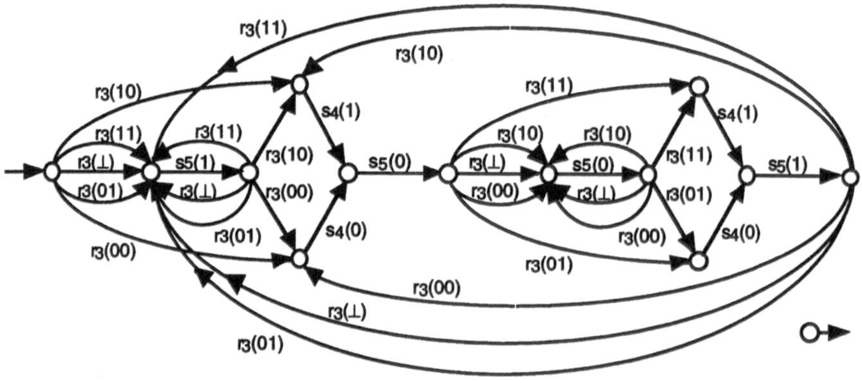

Figure 9. Reduced graph of R, for D = {0, 1}.

5.4 Protocol.

The protocol is now given by $ABP = \partial_H(S \parallel K \parallel L \parallel R)$, with

$$H = \{r_k(x), s_k(x) : k \in \{2,3,5,6\}, x \in (D \times B) \cup B \cup \{\perp\}\}.$$

By section 2, the process ABP has $(12n + 3)(12n + 5)(8n + 20)15 = 17280n^3 + 54720n^2 + 30600n + 4500$ states. Of course, most of these states are not reachable. It is much more interesting to determine the number of states of $\xi(ABP)$, i.e. the number of reachable states of the protocol.

Although the expansion theorem does not hold in our graph isomorphism model, a restricted form of it does hold, and we can use this to calculate the number of reachable states. This calculation is based on the following two identities:

- $\xi(ax \mathbin{\|\!\_} y) \approx a \cdot \xi(x \parallel y)$
- $\xi((ax \mid by) \mathbin{\|\!\_} z) \approx c \cdot \xi(x \parallel y \parallel z)$ if $\gamma(a, b) = c$.

Thus, if a parallel composition of processes yields a process that is really sequential (i.e. in all states there is only one possibility to proceed, either an autonomous step of one component, or a synchronisation between two components), we can use these identities to calculate the state graph. In the following, we present the results for the Alternating Bit Protocol.

5.5 Number of reachable states.

The ABP starts with an initialisation phase, where the components have not all reached the iteration parts of their behaviour. We show this initialisation phase in fig. 10. Open circles denote n states, one for every element of D, closed circles denote just one state. The initialisation phase ends in one of the two closed circles on the top left. These two are states the protocol can return to later. We show the iteration behaviour in fig. 11. We count $67n + 3$ states in the initialisation phase, plus 1 end state.

The rest of the reachable states of the ABP are shown in fig. 11. The two closed circles on the top left are the same as those shown in fig. 10. We count $58n + 4$ states in this part. Thus, in total we have $125n + 8$ states. Dividing out bisimulation equivalence, this reduces to $34n + 4$ states. Without separate begin and end states, we have the traditional $34n + 2$ states, as shown e.g. in [4].

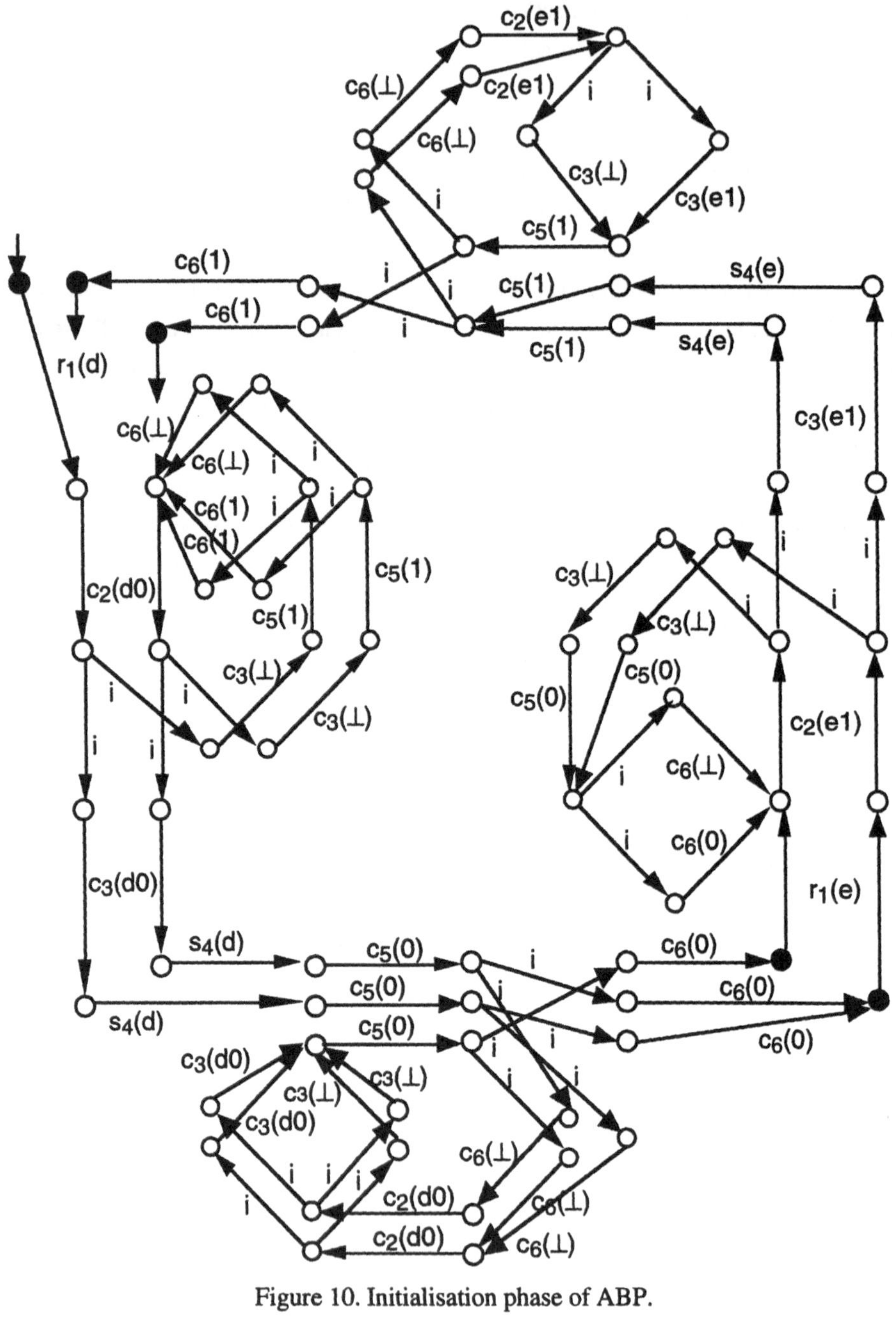

Figure 10. Initialisation phase of ABP.

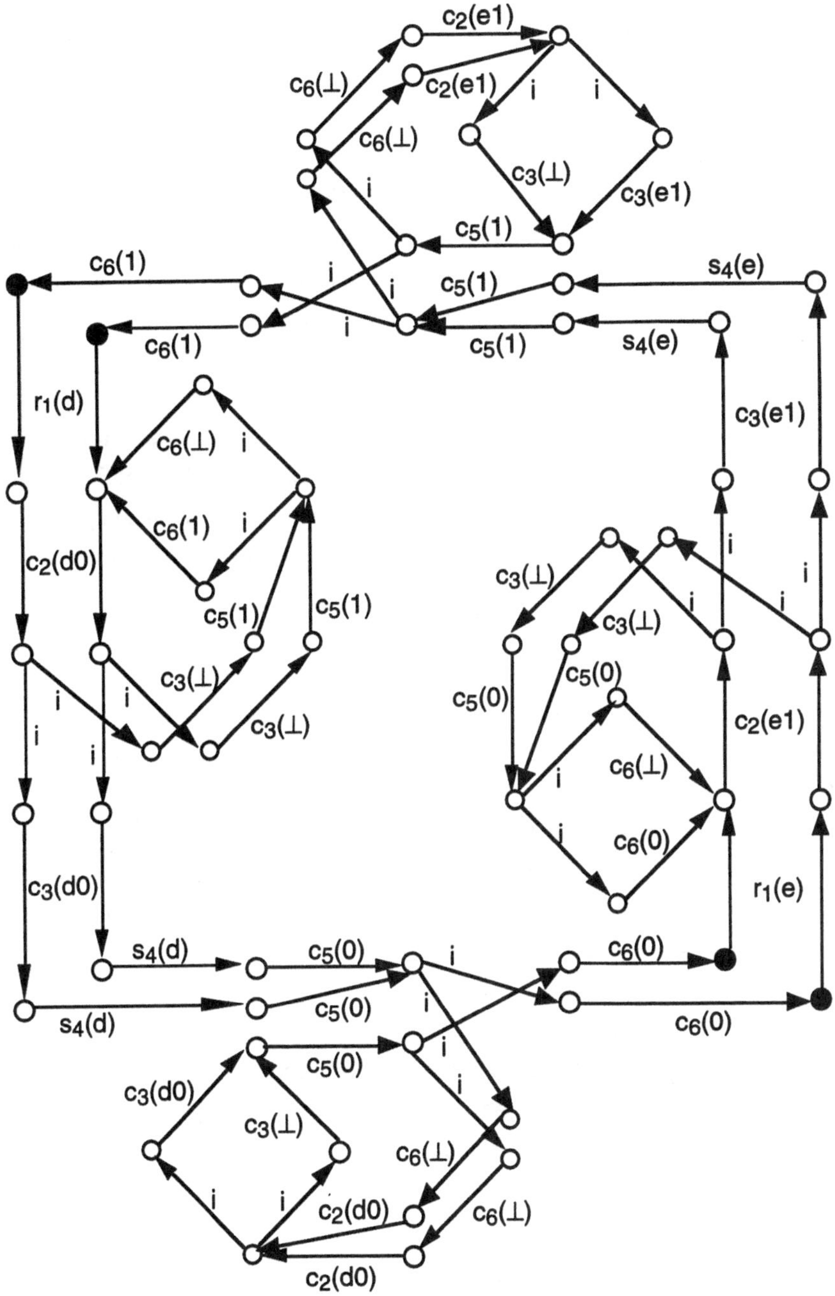

Figure 11. Iteration phase of ABP.

6. Conditions.

We can redo the theory of the previous sections in case we have conditions over a general boolean algebra. We use the theory developed in [2], and just mention a few key items, glossing over many details.

6.1 Boolean algebra.

Let $\mathbb{B}$ be a boolean algebra, with constants true, false and operators $\vee$, $\wedge$, $\neg$ We use letters ϕ, ψ to range over $\mathbb{B}$. A *valuation* is a homomorphism from $\mathbb{B}$ into {true, false}. For each process x, there is a process $\phi :\rightarrow x$ (if ϕ, then x). We redefine the model of process graphs, in order to take conditions on edges into account.

6.2 Process graphs over a boolean algebra.

Before, we had that a process graph is a quadruple $\langle S, \rightarrow, begin, end\rangle$ with $\rightarrow \subseteq S \times A \times S$, or, equivalently, the transition relation is a mapping from $S \times A \times S$ into {true, false}. Over a general boolean algebra $\mathbb{B}$, the transition relation is a mapping from $S \times A \times S$ into $\mathbb{B}$, and thus $(s \xrightarrow{a} t) \in \mathbb{B}$. The conditions on the transition relation are reformulated as follows:

- for all valuations v, states s $|\{t \in S : \exists a \in A\ v(s \xrightarrow{a} t) = true\}| < \lambda$,
- $\forall s \in S\ \forall a \in A\ (s \xrightarrow{a} begin(g)) = false$,
- $\forall s \in S\ \forall a \in A\ (end(g) \xrightarrow{a} s) = false$.

Isomorphism between two graphs g,h is now defined as expected. A bijection F between states of g and states of h is called an *isomorphism* if:

1. $F(begin(g)) = begin(h)$, $F(end(g)) = end(h)$
2. for all valuations v, $v(s \xrightarrow{a} t) = v(F(s) \xrightarrow{a} F(t))$.

Again, g,h are *isomorphic*, $g \approx h$, if there is an isomorphism between g and h.

6.3 Operators.

Definition of the operators in 2.3 is fairly straightforward. Writing down four interesting cases, condition b. in the definition of g+h becomes:

b'. $begin \xrightarrow{a} end = (begin(g) \xrightarrow{a} end(g)) \vee (begin(h) \xrightarrow{a} end(h))$.

In case of parallel composition, we calculate $\langle s,t\rangle$ \o\al(a,→) $\langle s',t'\rangle$ as the disjunction of all $(s \xrightarrow{b} s') \wedge (t \xrightarrow{c} t')$ for pairs b,c with $\gamma(b,c) = a$, plus $(s \xrightarrow{a} s')$ in case t=t', plus $(t \xrightarrow{a} t')$ in case s=s'.

In case of the conditional operator, $(begin \xrightarrow{a} s)$ in $\phi{:}\rightarrow g$ equals $\phi \wedge (begin \xrightarrow{a} s)$ in g.

Finally, in case of the priority operator, the value of a transition $s \xrightarrow{a} s'$ in $\theta_<(g)$ becomes the conjunction of its value in g and all $\neg(s \xrightarrow{b} s'')$ for $b > a$.

6.4 Reachability.

Let a process graph $g \in \mathcal{G}(A, \kappa, \lambda)$ be given. As in 2.4.3, we define its set of reachable states, $reach(g) \subseteq S(g)$ inductively:

a. $begin(g) \in reach(g)$

b. if $s \in reach(g)$, and there is a valuation v and an action a such that $v(s \xrightarrow{a} s') =$ true, then $s' \in reach(g)$.

Again, $\xi(g)$ is obtained from g by restriction to the set of reachable states.

6.5 Bisimulation.

Conditions 2 and 3 of the definition of bisimulation in 3.1 must be reformulated as follows (cf. [BAB92], section 8.2):

2'. if $R(s, t)$ and v is a valuation such that $v(s \xrightarrow{a} s') =$ true, then there is a t' such that $v(t \xrightarrow{a} t') =$ true and $R(s', t')$;

3'. if $R(s, t)$ and v is a valuation such that $v(t \xrightarrow{a} t') =$ true, then there is a s' such that $v(s \xrightarrow{a} s') =$ true and $R(s', t')$.

6.6 Expressivity.

We can still obtain an analogue of the expressivity result of section 4 in the present setting. Basically, the steps in 4.4, 4.5 remain the same, giving the required number of nodes, with exactly one transition between each pair (with first component not end, second component not begin). Next, we do not proceed as in 4.6, but instead, handle each node pair separately. Let $\langle n,m \rangle$ be a node pair. Since ϕ is a bijection, we can take certain nodes s,t in F with $n = \phi(s)$, $m = \phi(t)$. In case all transitions between s and t in F have condition false, do nothing. Otherwise there are $K \geq 1$ edges between s and t in g, with conditions $\phi_1, ..., \phi_K$ different from false. Take fresh atomic actions $b_0, b_1, ..., b_K$ in B.

We now consider two different cases. First, the case where $n = s =$ begin or $m = t =$ end. In this case, we know that a transition under consideration can be executed at most once. In this case, we can put $G_6 = \xi \circ \partial_H(G_3 \parallel (\phi_1{:}\rightarrow b_1 + ... + \phi_K{:}\rightarrow b_K))$, where the encapsulation enforces a communication between the edge between n and m in G_3 and one of the b_i. This replaces the step in 4.6. Next, we proceed by a renaming as in 4.7. Otherwise, the transition under consideration can be executed several times, and we have to use an iteration construct. We also need a termination clause. We take in this case $G_6 = \xi \circ \partial_H(G_3 \parallel (\phi_1{:}\rightarrow b_1 + ... + \phi_K{:}\rightarrow b_K)^* b_0)$,
where b_0 will synchronize with each termination step (again enforced by encapsulation).

7. Conclusion.

We have presented a simple and intuitive model for the syntax of ACP. This model is non-interleaving. We can use this model to calculate the number of states of a process. We found that this number of states depends very much on the representation of a process in the syntax. The most intuitive representation usually does not yield the minimal number of states. We have presented an expressivity result, that shows that every finite state graph in the model can be expressed in our syntax, starting from one seed process.

References.

1. J.C.M. Baeten & J.A. Bergstra. Global renaming operators in concrete process algebra. Inf. & Comp. 1988; 78:205-245.

2. J.C.M. Baeten & J.A. Bergstra. Process algebra with signals and conditions. In: M. Broy (ed.), Programming and mathematical method, Proc. Summer School, Marktoberdorf 1990, Springer Verlag 1992, pp. 273-323 (NATO ASI Series F 88).

3. J.C.M. Baeten & J.A. Bergstra. Non interleaving process algebra. In: E. Best (ed.), Proc. CONCUR'93, Hildesheim, Springer Verlag 1993, pp. 308-323 (Lecture Notes in Computer Science 715).

4. J.C.M. Baeten & W.P. Weijland. Process algebra. Cambridge Tracts in Theoretical Computer Science 18, Cambridge University Press 1990.

5. J.A. Bergstra, I. Bethke & A. Ponse. Process algebra with combinators. Technical Report P9319, Programming Research Group, University of Amsterdam 1993.

6. J.A. Bergstra, I. Bethke & A. Ponse. Process algebra with iteration and nesting. The Computer Journal 1994; 37 (to appear).

7. J.A. Bergstra & J.W. Klop. Process algebra for synchronous communication. Inf. & Control 1984; 60:109-137.

8. J.A. Bergstra & J.W. Klop. Algebra of communicating processes with abstraction. Theor. Comp. Sci. 1985; 37:77-121.

9. W.J. Fokkink & H. Zantema. Basic process algebra with iteration: completeness of its equational axioms. The Computer Journal 1994; 37 (to appear).

10. J. Parrow. Fairness properties in process algebra – with applications in communication protocol verification. Ph.D. Thesis, DoCS 85/03, Dept. of Computer Systems, Uppsala University 1985.

Process Specification in a UNITY Format

Jacob Brunekreef
Programming Research Group, University of Amsterdam
Amsterdam, The Netherlands

Abstract

In this paper a "UNITY Format" for process specifications is introduced. The format is based on conditions on process states and process data. Several aspects of this format are discussed: a straightforward normalisation of the parallel composition of processes, the relation between the ACP priority operator θ and conditions and, finally, the correspondence with a term rewriting system, which opens certain perspectives with respect to the validation and verification of a specification. Throughout the paper the simple and well-known PAR protocol serves as a running example.

1 Introduction

Different "specification styles" may be chosen for the specification of processes. In Vissers et al. ([20]) four specification styles are distinguished:

1. *Monolithic.* In the monolithic style only observable actions are presented and ordered as an alternative composition of sequences of actions. The only constructs applied are alternative composition, sequential composition and tail-recursion.

2. *Constraint–oriented.* In the constraint–oriented style also only observable actions are presented, but their temporal ordering is defined by a conjunction of different constraints. Parallel composition and encapsulation are now also applied as constructs. Hiding is forbidden, as all actions (or at least the resulting communications) have to be observable.

3. *State–oriented.* With the state–oriented style the system is regarded as a single resource whose state space is explicitly defined. Observable actions are presented in an alternative composition of conditional process terms, in which also changes in the state space are reflected. The same language constructs as in the monolithic style are applied.

4. *Resource–oriented.* In the resource–oriented style the external behaviour of a system is defined in terms of the parallel composition of separate resources. Internal actions (between the resources) are hidden. Each resource may be specified using any style. As with the constraint–oriented style, parallel composition and encapsulation are central constructs. Additionally, in the resource–oriented style hiding (abstraction) is used to make internal actions invisible.

In the field of ACP traditionally two styles are practised: for the specification of a single process and for the specification of overall system requirements the monolithic style is applied. On the other hand, systems composed of more than one process are specified in a resource–oriented style, with the separate processes identified as resources. Usually, these constituent processes themselves are specified in a monolithic style. Generally, in an ACP *verification* the external behaviour of a resource–oriented specification of a distributed system (e.g. a protocol) is proved to be equal to the monolithic requirement specification. This implies the transformation of the resource–oriented specification into a monolithic specification. This *normalisation* usually requires a lot of non–trivial calculations.

In this paper a state–oriented specification style for ACP is introduced and discussed. Because of its syntactical resemblance with the UNITY formalism of [10] we will call the proposed specification format the UNITY Format. In section 3 the format is introduced. It is based on conditions on process states and functions on data parameters. Several aspects of the format are discussed:

- In section 4 it is shown that normalisation of the parallel composition of two or more processes specified in the UNITY Format is rather simple.
- In section 5 it is shown that in this format a priority relation between actions can be translated to extra conditions on one or more summands.
- In section 6 a function on data parameters of a process is introduced. A Term Rewriting System (TRS) on this function is considered as an abstract representation of the process. It is discussed to what extent the TRS can be used for validation and verification purposes.

Preceding the discussion of these aspects, in section 2 some ACP axioms are repeated from the literature. These axioms are used in the proofs in the sections 4 and 5. No new axioms are introduced in this paper. Section 7 contains some concluding remarks. Throughout the paper the well-known PAR protocol is used as a running example.

Related work. The UNITY Format is close to the μCRL *linear process operator*, introduced in [7]. In [12] the relation between process algebras and finite state machines is studied. In [16] it is proved that the UNITY Format is universally expressive. Related work on the application of Term Rewriting Systems within the context of the verification of concurrent processes can be found in [13, 17].

2 Needed ACP axioms

The axiom system for "standard ACP" can be found in [5], published elsewhere in the ACP94 proceedings. In this section some additional axioms, concerning conditional process expressions and the priority operator θ, are repeated from the literature. The axioms in Table 1 (from [2]) deal with conditional process expressions. In this table two conditional operators are defined. The ternary operator $x \lhd c \rhd \delta y$ should be read as: "if c is true, then x else y", the binary operator $c :\rightarrow x$ should be read as "if c then x. The boolean condition c always evaluates to *true* or *false*. It is supposed that $\neg true = false$ and $\neg false = true$.

$x \lhd true \rhd y = x$	CO1
$x \lhd false \rhd y = y$	CO2
$true :\rightarrow x = x$	GC1
$false :\rightarrow x = \delta$	GC2
$c :\rightarrow x = x \lhd c \rhd \delta$	CG1
$x \lhd c \rhd y = c :\rightarrow x + (\neg c) :\rightarrow y$	CG2
$c :\rightarrow \delta = \delta$	GC9
$c :\rightarrow (x + y) = (c :\rightarrow x) + (c :\rightarrow y)$	GC10
$(c_1 \vee c_2) :\rightarrow x = (c_1 :\rightarrow x) + (c_2 :\rightarrow y)$	GC11
$c_1 :\rightarrow (c_2 :\rightarrow x) = (c_1 \wedge c_2) :\rightarrow x$	GC12
$(x \lhd c \rhd y) \mathbin{\|\!\!\lfloor} z = (x \mathbin{\|\!\!\lfloor} z) \lhd c \rhd (y \mathbin{\|\!\!\lfloor} z)$	CC1
$(x \lhd c \rhd y) \| z = (x \| z) \lhd c \rhd (y \| z)$	CC2
$x \| (y \lhd c \rhd z) = (x \| y) \lhd c \rhd (x \| z)$	CC3
$\partial_H(x \lhd c \rhd y) = \partial_H(x) \lhd c \rhd \partial_H(y)$	CC4

Table 1: Conditional ACP axioms.

In section 5 a relation between the ACP operator θ (the priority operator) and conditional process expressions is derived. The axioms in Table 2 (from [3], also given in [4]) define this operator. In this table an auxiliary operator is used: the binary *unless* operator, in [3] denoted by the symbol $\lhd$. In order to avoid any confusion between this operator and the left triangle of a conditional process expression, in this paper we will denote the unless operator with the symbol $\blacktriangleleft$. The last three axioms from Table 2 (from [2]) deal with the distributivity of the priority operator and the unless operator over a conditional process expression.

From these axioms the following identities can easily be derived. They will be used in the sections to come.

Lemma 2.1

1. $x \lhd c \rhd x = x$

2. $x \lhd c \rhd y = x \lhd c \rhd \delta + y \lhd \neg c \rhd \delta$

3. $(x \lhd c_1 \rhd \delta) \lhd c_2 \rhd \delta = x \lhd (c_1 \wedge c_2) \rhd \delta$

Proof: simple, see [8].

3 A UNITY Format for process specifications

In many ACP specifications it is common practise to specify a process with (a set of) linear equations, each in the following format:

$$X = a_1 \cdot X_1 + \ldots + a_i \cdot X_i + a_{i+1} + \ldots + a_j$$

See e.g. [1] for examples of specifications in this format. In this section we

$a \lhd\!\!\lhd b = a$	if $\neg(a < b)$	P1
$a \lhd\!\!\lhd b = \delta$	if $a < b$	P2
$x \lhd\!\!\lhd y \cdot z = x \lhd\!\!\lhd y$		P3
$x \lhd\!\!\lhd (y + z) = (x \lhd\!\!\lhd y) \lhd\!\!\lhd z$		P4
$x \cdot y \lhd\!\!\lhd z = (x \lhd\!\!\lhd z) \cdot y$		P5
$(x + y) \lhd\!\!\lhd z = x \lhd\!\!\lhd z + y \lhd\!\!\lhd z$		P6
$\theta(a) = a$		TH1
$\theta(x \cdot y) = \theta(x) \cdot \theta(y)$		TH2
$\theta(x + y) = \theta(x) \lhd\!\!\lhd y + \theta(y) \lhd\!\!\lhd x$		TH3
$x \lhd\!\!\lhd (y \lhd c \rhd z) = (x \lhd\!\!\lhd y) \lhd c \rhd (x \lhd\!\!\lhd z)$		PC1
$(x \lhd c \rhd y) \lhd\!\!\lhd z = (x \lhd\!\!\lhd z) \lhd c \rhd (y \lhd\!\!\lhd z)$		PC2
$\theta(x \lhd c \rhd y) = \theta(x) \lhd c \rhd \theta(y)$		THC

Table 2: Axioms for ACP_θ with conditions.

present a normal form for process specifications that is based on a single (recursive) equation, in which a process is specified as the alternative composition of a number of conditional process terms. The process name is parameterised with one or more data variables that contain state information about the process. Each conditional process term is built from an atomic action, followed by a condition, or the sequential composition of an atomic action and a recursive process call (usually with some substitutions in its parameter list), followed by a condition. The condition determines in which state the atomic action is enabled. Because of the resemblance with the UNITY formalism of [10] we will call this format the UNITY Format for process specifications. In this format a process X is specified as follows:

$$
\begin{aligned}
X(D) = \quad & a_1 \cdot X(D_1) \lhd c_1 \rhd \delta \\
+ \; & \ldots \\
+ \; & a_m \cdot X(D_m) \lhd c_m \rhd \delta \\
+ \; & a_{m+1} \lhd c_{m+1} \rhd \delta \\
+ \; & \ldots \\
+ \; & a_n \lhd c_n \rhd \delta
\end{aligned}
$$

D denotes a parameter list, $D_1 \ldots D_m$ denote a parameter list with substitutions for some of the elements of D. $a_1 \ldots a_n$ denote atomic actions. $c_1 \ldots c_n$ denote boolean conditions, based on elements of D. If a condition c_i is invariantly *true*, a summand can be written as $a_i \cdot X(D_i)$ or just a_i.

In the sequel we will only look at non–terminating processes, the non–recursive summands are left out. All results obtained in the coming sections can also easily be derived for terminating processes.

We will give an example in this format by giving a specification of the components of the well-known PAR protocol. PAR stands for *Positive Acknowledgement with Retransmission.* In this protocol, after the transmission of a frame, the sender waits for an acknowledgement from the receiver before a new frame is transmitted. The channels between sender and receiver are faulty: frames may be corrupted (resulting in a *checksum error* at the receiving side) or may get lost. If the receiver receives a frame with a checksum error no acknowledgement is sent back, the receiver just waits for a new frame to arrive. If the sender does not receive an acknowledgement within a certain time interval, a timeout is generated, which leads to a retransmission of the last frame sent. In order to distinguish retransmitted frames from new ones, a frame from the sender to the receiver contains an identification bit, which is flipped each time a new frame is transmitted. The specification of the protocol contains five components: the sender, the receiver, a timer process and two channels, K and L, between the sender and the receiver. Figure 1 shows the various components with their connections.

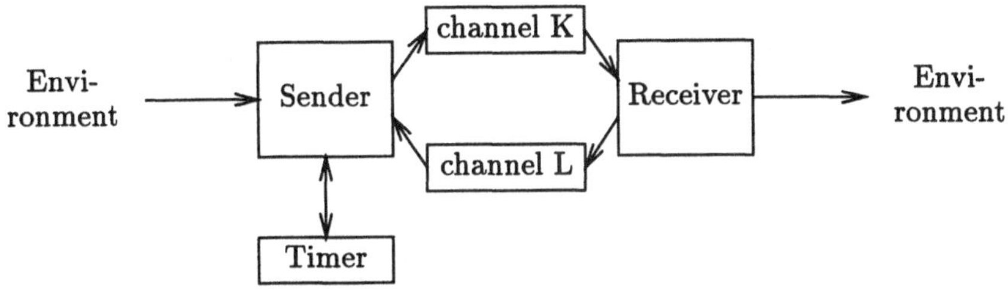

Figure 1: Components of the PAR protocol.

In the specification of the components below, the atomic read and send actions are labelled with two capitals. The first capital indicates the source of the communication, the second capital indicates the destination. So the atomic action $send\_SK$ denotes a send action from the sender S to the channel K. The capital E refers to the environment of the PAR protocol.

The sender process reads a data element from the environment, after which a frame with this element and a control bit is sent to the receiver via the channel K. The timer is started by a message from the sender to the timer. Next an acknowledgement or a checksum error from the channel L is awaited. If an acknowledgement is received, a new data element from the environment is expected. If for a certain period nothing is received from the channel L, a timeout will be received from the timer process. In the case of the reception of a checksum error from channel L or a timeout, the old frame is retransmitted. So the sender process has four states: waiting for input from the environment, sending a frame to the channel K, sending a message to the timer and waiting for something to arrive from the channel L or the timer. In the specification below the first parameter of the process S denotes the sender state. The value of the parameter ss ranges from $s1$ to $s4$. The other two parameters denote the data element that has to be transmitted (sd) and the additional bit (sb). In each summand the condition on the value of ss determines whether the action is enabled or not.

$$
\begin{array}{rl}
\multicolumn{2}{l}{S(ss,sd,sb) =} \\
 & \sum_{d\in D} read\_ES(d)\cdot S(s2,d,sb) \lhd ss = s1 \rhd \delta \\
+ & send\_SK(sd,sb)\cdot S(s3,sd,sb) \lhd ss = s2 \rhd \delta \\
+ & send\_ST\cdot S(s4,sd,sb) \lhd ss = s3 \rhd \delta \\
+ & read\_LS(ac)\cdot S(s1,sd,1-sb) \lhd ss = s4 \rhd \delta \\
+ & read\_LS(ce)\cdot S(s2,sd,sb) \lhd ss = s4 \rhd \delta \\
+ & read\_TS\cdot S(s2,sd,sb) \lhd ss = s4 \rhd \delta
\end{array}
$$

The receiver process awaits a frame from the channel K. If the control bit in the frame is as expected, the data element is delivered to the environment and an acknowledgement is sent to the sender via the channel L. In case the control bit is wrong only an acknowledgement is sent. If a checksum error is received, the receiver shows no reaction at all.
The receiver process R is parameterised with rs (the receiver state, ranging from $r1$ to $r3$), rd (the received data element), rb (the received additional bit) and eb (the expected bit in a frame to receive). In two summands the condition is not only based on the process state rs, but also on other process parameters: the (in)equality of two bit values.

$$
\begin{array}{rl}
\multicolumn{2}{l}{R(rs,rd,rb,eb) =} \\
 & \sum_{d\in D, b\in B} read\_KR(d,b)\cdot R(r2,d,b,eb) \lhd rs = r1 \rhd \delta \\
+ & read\_KR(ce)\cdot R(rs,rd,rb,eb) \lhd rs = r1 \rhd \delta \\
+ & send\_RE(rd)\cdot R(r3,rd,rb,1-eb) \lhd rs = r2 \wedge rb = eb \rhd \delta \\
+ & send\_RL(ac)\cdot R(r1,rd,rb,eb) \\
 & \qquad \lhd (rs = r2 \wedge rb = 1-eb) \vee (rs = r3) \rhd \delta
\end{array}
$$

The timer process is very simple: in any state it awaits a start message from the sender process. Once started, the timer may send a timeout message to the sender process. The timer process T is only parameterised with ts, the timer state, with the value $t1$ or $t2$. As the timer may be started in both states, the condition in the process term with the action $read\_ST$ is omitted.

$$
\begin{array}{rl}
T(ts) = & \quad read\_ST\cdot T(t2) \\
 & + send\_TS\cdot T(t1) \lhd ts = t2 \rhd \delta
\end{array}
$$

The channel process K receives a frame from the sender process. Next, three things may occur: the frame is delivered to the receiver process correctly, the frame is delivered to the receiver process with a checksum error or the frame is not delivered at all (it is lost in the channel). These three alternatives are modelled in an alternative composition[1].

The channel process K is parameterised with ks (the channel state, with the value $k1$ or $k2$), kd (the data element in the channel) and kb (the bit in the channel). The loss of a frame is modelled by the atomic action $lost\_K$.

$$
\begin{array}{rl}
\multicolumn{2}{l}{K(ks,kd,kb) =} \\
 & \sum_{d\in D, b\in B} read\_SK(d,b)\cdot K(k2,d,b) \lhd ks = k1 \rhd \delta \\
+ & send\_KR(kd,kb)\cdot K(k1,kd,kb) \lhd ks = k2 \rhd \delta \\
+ & send\_KR(ce)\cdot K(k1,kd,kb) \lhd ks = k2 \rhd \delta \\
+ & lost\_K\cdot K(k1,kd,kb) \lhd ks = k2 \rhd \delta
\end{array}
$$

[1]In the specification of existing protocols we consider internal channel actions like i, j or *skip* preceeding the alternatives as superfluous. It is an intrinsic property of the PAR protocol that both the sender and the receiver have to anticipate on frames with checksum errors.

The channel process L is a simple version of the process K. It has only one parameter: the channel state ls, with the value $l1$ or $l2$. The atomic action $lost\_L$ denotes the loss of an acknowledgement.

$$\begin{aligned} L(ls) = & \\ & read\_RL(ac) \cdot L(l2) \lhd ls = l1 \rhd \delta \\ + & \; send\_LS(ac) \cdot L(l1) \lhd ls = l2 \rhd \delta \\ + & \; send\_LS(ce) \cdot L(l1) \lhd ls = l2 \rhd \delta \\ + & \; lost\_L \cdot L(l1) \lhd ls = l2 \rhd \delta \end{aligned}$$

In the following two sections we will derive an equation in the UNITY Format for the (encapsulated) parallel composition of the components of the PAR protocol.

4 Parallel composition and conditions

With the standard ACP axioms the parallel composition (merge) of two process terms can be *normalised*: rewritten to a process term with only sequential composition and alternative composition. In this section we will derive a lemma about the normalisation of the parallel composition of conditional processes in the UNITY Format. The lemma will be illustrated with the normalisation of the PAR protocol: the encapsulated merge of the five processes from the previous section.

We consider two processes in the UNITY Format:

$$X(D) = \sum_{i=1}^{n}(a_i \cdot X(D_i) \lhd c_i \rhd \delta) \text{ and } Y(E) = \sum_{j=1}^{m}(b_j \cdot Y(E_j) \lhd c'_j \rhd \delta).$$

The parallel composition of these processes is denoted by $XY(D,E)$. The following lemma shows how this parallel composition can be rewritten to a process term in the UNITY Format. It shows that for processes in this format regularity is closed under the merge operator. As such it is a straightforward generalisation of a result of Bergstra and Klop ([5]).

Lemma 4.1

$$\begin{aligned} XY(D,E) = & \sum_{i=1}^{n}(a_i \cdot XY(D_i,E) \lhd c_i \rhd \delta) + \sum_{j=1}^{m}(b_j \cdot XY(D,E_j) \lhd c'_j \rhd \delta) \\ & + \sum_{i=1}^{n}\sum_{j=1}^{m}((a_i \mid b_j) \cdot XY(D_i,E_j) \lhd c_i \wedge c'_j \rhd \delta) \end{aligned}$$

Proof: (the acronyms above the identity symbol refer to the labels of the axioms from [5] and from section 2 of this paper)

$$X(D) \parallel Y(E) \overset{\text{CM1}}{=} X(D) \mathbin{\|\!\_} Y(E) + Y(E) \mathbin{\|\!\_} X(D) + X(D) \mid Y(E).$$

We will elaborate each summand separately.

$$X(D) \mathbin{\|\!\_} Y(E) \overset{\text{CM4}}{=} \sum_{i=1}^{n}((a_i \cdot X(D_i) \lhd c_i \rhd \delta) \mathbin{\|\!\_} Y(E)) \overset{\text{CC1}}{=}$$

$$\sum_{i=1}^{n}((a_i \cdot X(D_i) \mathbin{\|\!\_} Y(E)) \lhd c_i \rhd (\delta \mathbin{\|\!\_} Y(E))) \overset{\text{CM3,CM2,A7}}{=}$$

$$\sum_{i=1}^{n}((a_i \cdot (X(D_i) \parallel Y(E))) \lhd c_i \rhd \delta) =$$

$$\sum_{i=1}^{n}(a_i \cdot XY(D_i, E) \lhd c_i \rhd \delta)$$

In the same way the following identity is proved:

$$Y(E) \mathbin{\|\!\_} X(D) = \sum_{j=1}^{m}(b_j \cdot XY(D, E_j) \lhd c'_j \rhd \delta).$$

Remains the third summand:

$$X(D) \mid Y(E) = \sum_{i=1}^{n}(a_i \cdot X(D_i) \lhd c_i \rhd \delta) \mid \sum_{j=1}^{m}(b_j \cdot Y(E_j) \lhd c'_j \rhd \delta) \overset{\text{CM8,CM9}}{=}$$

$$\sum_{i=1}^{n}\sum_{j=1}^{m}((a_i \cdot X(D_i) \lhd c_i \rhd \delta) \mid (b_j \cdot Y(E_j) \lhd c'_j \rhd \delta)) \overset{\text{CC2,CC3}}{=}$$

$$\sum_{i=1}^{n}\sum_{j=1}^{m}(((a_i \cdot X(D_i)) \mid (b_j \cdot Y(E_j)) \lhd c'_j \rhd \delta) \lhd c_i \rhd \delta) \overset{\text{CM7}}{=}$$

$$\sum_{i=1}^{n}\sum_{j=1}^{m}(((a_i \mid b_j) \cdot (X(D_i) \parallel Y(E_j)) \lhd c'_j \rhd \delta) \lhd c_i \rhd \delta) \overset{\text{CG1,GC12}}{=}$$

$$\sum_{i=1}^{n}\sum_{j=1}^{m}((a_i \mid b_j) \cdot XY(D_i, E_j) \lhd c_i \wedge c'_j \rhd \delta) \qquad \square$$

So the parallel composition of two processes X and Y consists of three groups of summands: a group with the atomic actions and conditions from the process X, a group with the actions and conditions from process Y and a group of actions that are the result of communications between actions from process X and process Y. In the last group the conditions consist of the boolean and of two separate conditions from the processes X and Y. Without further proof we claim that this result is also valid for the parallel composition of more than two processes. We will use this claim in the (encapsulated) parallel composition of the five processes of the PAR protocol as specified in the previous section. We name this process P. The data parameters of P come from the constituent processes.

$$P(ss, rs, ts, ks, ls, sd, sb, rd, rb, eb, kd, kb) =$$
$$\partial_H(S(ss, sd, sb) \parallel R(rs, rd, rb, eb) \parallel T(ts) \parallel K(ks, kd, kb) \parallel L(ls))$$

The set H contains the atomic actions which have to be encapsulated in order to enforce the desired communications between the five processes. Below the communications between a *read* and a *send* action are denoted by a *comm* action.

For the process P the following lemma holds. We will use a "notational shortcut" from [6]: in the parameter lists of P at the right side of the equation only the substitutions of data elements are listed. As the equation for P is left–linear with respect to the data parameters, this will not lead to any confusion. A substitution is denoted by x/y, which means that the value of the parameter y is replaced by the value of the term x.

Lemma 4.2

$$P(ss, rs, ts, ks, ls, sd, sb, rd, rb, eb, kd, kb) =$$
$$\sum_{d \in D} read\_ES(d) \cdot P(s2/ss, d/sd) \lhd ss = s1 \rhd \delta$$

$$
\begin{array}{ll}
+ & comm\_SK(sd,sb) \cdot P(s3/ss, k2/ks, sd/kd, sb/kb) \\
 & \quad \triangleleft\, ss = s2 \ \wedge \ ks = k1 \,\triangleright\, \delta \\
+ & comm\_ST \cdot P(s4/ss, t2/ts) \triangleleft\, ss = s3 \,\triangleright\, \delta \\
+ & comm\_KR(kd,kb) \cdot P(r2/rs, k1/ks, kd/rd, kb/rb) \\
 & \quad \triangleleft\, rs = r1 \ \wedge \ ks = k2 \,\triangleright\, \delta \\
+ & comm\_KR(ce) \cdot P(k1/ks) \triangleleft\, rs = r1 \ \wedge \ ks = k2 \,\triangleright\, \delta \\
+ & lost\_K \cdot P(k1/ks) \triangleleft\, ks = k2 \,\triangleright\, \delta \\
+ & send\_RE(rd) \cdot P(r3/rs, 1-eb/eb) \triangleleft\, rs = r2 \ \wedge \ rb = eb \,\triangleright\, \delta \\
+ & comm\_RL(ac) \cdot P(r1/rs, l2/ls) \\
 & \quad \triangleleft\, (rs = r2 \wedge rb = 1-eb \vee rs = r3) \ \wedge \ ls = l1 \,\triangleright\, \delta \\
+ & comm\_LS(ac) \cdot P(s1/ss, l1/ls, 1-sb/sb) \triangleleft\, ss = s4 \ \wedge \ ls = l2 \,\triangleright\, \delta \\
+ & comm\_LS(ce) \cdot P(s2/ss, l1/ls) \triangleleft\, ss = s4 \ \wedge \ ls = l2 \,\triangleright\, \delta \\
+ & lost\_L \cdot P(l1/ls) \triangleleft\, ls = l2 \,\triangleright\, \delta \\
+ & comm\_TS \cdot P(s2/ss, t1/ts) \triangleleft\, ss = s4 \ \wedge \ ts = t2 \,\triangleright\, \delta
\end{array}
$$

Proof: by application of a generalised version of lemma 4.1 and the axioms concerning the encapsulation operator. Each pair of communicating actions results in a summand with a *comm* action and a condition which is the boolean "and" of the conditions under which the separate actions were enabled. Each non–communicating and non–encapsulated action from one of the processes is found as a summand of P with its original condition. □

5 From priority relation to condition

In ACP$_\theta$ the priority operator θ is used to prevent certain actions to be enabled while other actions are enabled. This operator is used, for instance, to prevent premature timeouts in communication protocols or to enforce strict sequential composition of two actions, see [18]. The priority operator is not included in ACP-based formalisms as PSF ([14]) and μCRL ([11]). This makes it difficult to "translate" an ACP$_\theta$ specification to a PSF specification or a μCRL specification. The following lemma makes such a translation possible, provided the processes are specified in the UNITY Format. In this lemma the order relations on actions are translated to (extra) conditions on the summands with the actions with a lower priority.

Lemma 5.1

$$\theta(\textstyle\sum_{i=1}^{n}(a_i \cdot X(D_i) \triangleleft c_i \triangleright \delta)) = \sum_{i=1}^{n}(a_i \cdot \theta(X(D_i)) \triangleleft c_i \wedge \neg(\bigvee_{1\le j\le n,\ a_j > a_i} c_j) \triangleright \delta)$$

Proof: We give a condensed proof for $n = 2$. A proof for $n > 2$ follows the same lines. We will prove the following identities:

1. $\theta(a_1 \cdot X(D_1) \triangleleft c_1 \triangleright \delta + a_2 \cdot X(D_2) \triangleleft c_2 \triangleright \delta) =$
   $a_1 \cdot \theta(X(D_1)) \triangleleft c_1 \triangleright \delta + a_2 \cdot \theta(X(D_2)) \triangleleft c_2 \triangleright \delta$
   if $\neg(a_1 < a_2),\ \neg(a_2 < a_1)$

2. $\theta(a_1 \cdot X(D_1) \triangleleft c_1 \triangleright \delta + a_2 \cdot X(D_2) \triangleleft c_2 \triangleright \delta) =$
   $a_1 \cdot \theta(X(D_1)) \triangleleft c_1 \wedge \neg c_2 \triangleright \delta + a_2 \cdot \theta(X(D_2)) \triangleleft c_2 \triangleright \delta$
   if $a_1 < a_2$

3. $\theta(a_1 \cdot X(D_1) \triangleleft c_1 \triangleright \delta + a_2 \cdot X(D_2) \triangleleft c_2 \triangleright \delta) =$
$a_1 \cdot \theta(X(D_1)) \triangleleft c_1 \triangleright \delta + a_2 \cdot \theta(X(D_2)) \triangleleft c_2 \wedge \neg c_1 \triangleright \delta$
if $a_2 < a_1$

We start with a general transformation of the term $\theta(\ldots)$, regardless of the specific order relation between a_1 and a_2. With the axioms from Table 1 and Table 2 the following identity can be proved:

$$\theta(a_1 \cdot X(D_1) \triangleleft c_1 \triangleright \delta + a_2 \cdot X(D_2) \triangleleft c_2 \triangleright \delta) =$$

$$((a_1 \lll a_2) \cdot \theta(X(D_1)) \triangleleft c_1 \triangleright \delta) \triangleleft c_2 \triangleright (a_1 \cdot \theta(X(D_1)) \triangleleft c_1 \triangleright \delta) + \\ ((a_2 \lll a_1) \cdot \theta(X(D_2)) \triangleleft c_2 \triangleright \delta) \triangleleft c_1 \triangleright (a_2 \cdot \theta(X(D_2)) \triangleleft c_2 \triangleright \delta) \equiv \mathcal{Z}$$

From this point we distinguish three cases:

1. $\neg(a_1 < a_2), \neg(a_2 < a_1)$. We get: $\mathcal{Z} \stackrel{\text{P1}}{=}$

$$(a_1 \cdot \theta(X(D_1)) \triangleleft c_1 \triangleright \delta) \triangleleft c_2 \triangleright (a_1 \cdot \theta(X(D_1)) \triangleleft c_1 \triangleright \delta) + \\ (a_2 \cdot \theta(X(D_2)) \triangleleft c_2 \triangleright \delta) \triangleleft c_1 \triangleright (a_2 \cdot \theta(X(D_2)) \triangleleft c_2 \triangleright \delta) \stackrel{\text{lemma 2.1.1}}{=}$$

$$a_1 \cdot \theta(X(D_1)) \triangleleft c_1 \triangleright \delta + a_2 \cdot \theta(X(D_2)) \triangleleft c_2 \triangleright \delta$$

which proves identity 1.

2. $a_1 < a_2$. We get: $\mathcal{Z} \stackrel{\text{P2}}{=}$

$$(\delta \cdot \theta(X(D_1)) \triangleleft c_1 \triangleright \delta) \triangleleft c_2 \triangleright (a_1 \cdot \theta(X(D_1)) \triangleleft c_1 \triangleright \delta) + \\ (a_2 \cdot \theta(X(D_2)) \triangleleft c_2 \triangleright \delta) \triangleleft c_1 \triangleright (a_2 \cdot \theta(X(D_2)) \triangleleft c_2 \triangleright \delta) \stackrel{\text{A7,lemma 2.1.1}}{=}$$

$$\delta \triangleleft c_2 \triangleright (a_1 \cdot \theta(X(D_1)) \triangleleft c_1 \triangleright \delta) + a_2 \cdot \theta(X(D_2)) \triangleleft c_2 \triangleright \delta \stackrel{\text{lemma 2.1.2,3}}{=}$$

$$a_1 \cdot \theta(X(D_1)) \triangleleft c_1 \wedge \neg c_2 \triangleright \delta + a_2 \cdot \theta(X(D_2)) \triangleleft c_2 \triangleright \delta$$

which proves identity 2.

3. $a_2 < a_1$. Along the same lines as the proof of identity 2 we get:

$$\mathcal{Z} = a_1 \cdot \theta(X(D_1)) \triangleleft c_1 \triangleright \delta + a_2 \cdot \theta(X(D_2)) \triangleleft c_2 \wedge \neg c_1 \triangleright$$

which proves identity 3. □

We will illustrate the usefulness of this lemma by a refinement of the specification of the PAR protocol. The PAR protocol is not robust with respect to premature timeouts: if the timeout interval is very short, it is impossible for the sender to determine whether a new frame or an old frame is acknowledged by the receiver. A solution to this problem is the application of the priority operator with the timeout action having a lower priority than every other action in the protocol, see [18]. A drawback of this solution is that in several formalisms related to ACP, e.g. PSF and μCRL, the priority operator is not included. This leads to other, more or less artificial, solutions to the problem of premature timeouts, see [19]. Lemma 5.1 offers the possibility to "translate" priority relations on actions to extra conditions on actions with a lower priority for a process specified in the UNITY Format. Conditional process expressions are included in both PSF and μCRL.

The process PAR is specified by the following equation:

$$\begin{array}{l} PAR(ss, rs, ts, ks, ls, sd, sb, rd, rb, eb, kd, kb) = \\ \quad \theta_T(\ P(ss, rs, ts, ks, ls, sd, sb, rd, rb, eb, kd, kb)\ ) \end{array}$$

with the process P defined in the previous section. The set T contains the order relations for the timeout: $comm\_TS < a$ where a represents all other actions in the protocol. In line with [18] we will give the highest priority to the action $comm\_ST$, which guarantees that the timer is started immediately after the transmission of a frame. So the set T also contains the order relations $b < comm\_ST$, where b represents all the actions in the protocol except $comm\_ST$.

Application of lemma 5.1 leads to the following equation for the process PAR:

Lemma 5.2

$$\begin{array}{ll}
\multicolumn{2}{l}{PAR(ss, rs, ts, ks, ls, sd, sb, rd, rb, eb, kd, kb) =} \\
 & \sum_{d \in D} read_ES(d) \cdot PAR(s2/ss, d/sd) \triangleleft ss = s1 \triangleright \delta \\
+ & comm_SK(sd, sb) \cdot PAR(s3/ss, k2/ks, sd/kd, sb/kb) \\
 & \quad \triangleleft ss = s2 \ \wedge \ ks = k1 \triangleright \delta \\
+ & comm_ST \cdot PAR(s4/ss, t2/ts) \triangleleft ss = s3 \triangleright \delta \\
+ & comm_KR(kd, kb) \cdot PAR(r2/rs, k1/ks, kd/rd, kb/rb) \\
 & \quad \triangleleft rs = r1 \ \wedge \ ks = k2 \ \wedge \ \neg(ss = s3) \triangleright \delta \\
+ & comm_KR(ce) \cdot PAR(k1/ks) \\
 & \quad \triangleleft rs = r1 \ \wedge \ ks = k2 \ \wedge \ \neg(ss = s3) \triangleright \delta \\
+ & lost_K \cdot PAR(k1/ks) \triangleleft ks = k2 \ \wedge \ \neg(ss = s3) \triangleright \delta \\
+ & send_RE(rd) \cdot PAR(r3/rs, 1 - eb/eb) \\
 & \quad \triangleleft rs = r2 \ \wedge \ rb = eb \ \wedge \ \neg(ss = s3) \triangleright \delta \\
+ & comm_RL(ac) \cdot PAR(r1/rs, l2/ls) \\
 & \quad \triangleleft (rs = r2 \wedge rb = 1 - eb \vee rs = r3) \ \wedge \ ls = l1 \ \wedge \ \neg(ss = s3) \triangleright \delta \\
+ & comm_LS(ac) \cdot PAR(s1/ss, l1/ls, 1 - sb/sb) \triangleleft ss = s4 \ \wedge \ ls = l2 \triangleright \delta \\
+ & comm_LS(ce) \cdot PAR(s2/ss, l1/ls) \triangleleft ss = s4 \ \wedge \ ls = l2 \triangleright \delta \\
+ & lost_L \cdot PAR(l1/ls) \triangleleft ls = l2 \ \wedge \ \neg(ss = s3) \triangleright \delta \\
+ & comm_TS \cdot PAR(s2/ss, t1/ts) \\
 & \quad \triangleleft ss = s4 \ \wedge \ ts = t2 \ \wedge \ rs = r1 \ \wedge \ ks = k1 \ \wedge \ ls = l1 \triangleright \delta
\end{array}$$

Proof: according to lemma 5.1 the order relations $b < comm\_ST$ for $b \neq comm\_ST$ lead to an extra conjunct $\neg(ss = s3)$ in the condition of each summand except the summand with $comm\_ST$ itself. Now we suppose that for each process the number of states is finite and the states are unique (ps is a process state variable, PS is the set of process states):

$$\forall i \in PS : ((ps = i) \iff \forall j \in PS, j \neq i : (ps \neq j)).$$

So, in each condition with a conjunct $(ss = si)$, $i \neq 3$, the conjunct $\neg(ss = s3)$ is superfluous. This means that in six summands of the equation for the process *PAR* the conjunct $\neg(ss = s3)$ in the condition can be deleted.

The condition in the last summand is derived in the following way. As each action has a higher priority than the action *comm_TS*, the conditions related to all these actions will appear as additional negative conditions in the last summand. So this condition now is equal to

$$(ss = s4) \wedge (ts = t2) \wedge \neg(ss = s1) \wedge \neg(ss = s2 \wedge ks = k1) \wedge \ldots$$
$$\neg(ls = l2) \wedge \neg(ss = s3) \equiv \mathit{to\_condition}.$$

Some straightforward boolean calculations lead to the following identity:

$$\begin{aligned}
&\mathit{to_condition} = \\
&\quad (ss = s4) \wedge (ts = t2) \wedge \neg(\ (ss = s1) \vee (ss = s2 \wedge ks = k1) \vee (ss = s3) \vee \\
&\quad (ks = k2 \wedge \neg(ss = s3)) \vee (rs = r2 \wedge eb = rb \wedge \neg(ss = s3)) \vee \\
&\quad (rs = r2 \wedge eb = 1 - rb \wedge ls = l1 \wedge \neg(ss = s3)) \vee \\
&\quad (rs = r3 \wedge ls = l1 \wedge \neg(ss = s3)) \vee (ss = s4 \wedge ls = l2) \vee \\
&\quad (ls = l2 \wedge \neg(ss = s3)))
\end{aligned}$$

Now we use the fact that a bit variable has two possible values, so

$$(eb = rb) \Longleftrightarrow (eb \neq 1 - rb) \quad , \quad (eb \neq rb) \Longleftrightarrow (eb = 1 - rb).$$

Furthermore, we have already noticed that for each process the number of states is finite and the states are unique. So, $\neg(ks = k2)$ implies $ks = k1$, since only two values are possible for ks. In the same way the other parts of the timeout condition are derived. □

It should be noticed that the derived condition is intuitively clear: a timeout is needed (and permitted), only if the sender process is waiting for input from the channel L $(ss = s4)$, the timer is running $(ts = t2)$, but all other components are at rest $(rs = r1, ks = k1, ls = l1)$. In this state a timeout is needed to re-activate the sender process.

The equation for the process *PAR* contains no priority operator. This means that it can be translated into specification formalisms like PSF and μCRL. Such a PSF specification of the PAR protocol is given in [8].

6 A corresponding Term Rewriting System

In the UNITY Format for process specifications the behaviour of a process in terms of state transitions is completely governed by conditions on data and functions on data. This raises the question to what extent something meaningful can be stated about the validation or verification of a specification by looking only at the data part. In this section we will give a tentative answer to this question. First an algorithm is given with which a Term Rewriting System (TRS) can be derived from a process specification. Next, the relation between certain properties of this TRS and properties of the related process specification is considered. As in the previous sections, the PAR protocol is used as a running example.

We start with the introduction of a Term Rewriting System (TRS) for a function, derived in three steps from the specification of a process X in the UNITY Format.

1. Delete from each summand in the definition of X the atomic action.
2. Define a function f of some sort S with the same arguments as the process name X, excluding the arguments that are not related to the conditions.
3. Consider each "amputated" summand as a rewrite rule from an instantiation of f to another instantiation of f in which, if possible, the conditions on the data parameters appear in the arguments of f at the left side of a rule and the data substitutions appear in the arguments of f at the right side of the rule. If the conditions cannot be reflected in data parameters at the left side, a conditional rewrite rule will be needed. The notation of a conditional rewrite rule is $R \Leftarrow c$ with R a rewrite rule and c the condition. Only if c evaluates to true, the application of R is permitted.

Such a TRS is an abstract representation of the process X. Each rewrite rule resembles a state transition of X. As an example we will derive the TRS corresponding to the PAR protocol. We define a function *par* with the arguments $ss, rs, ts, ks, ls, sb, rb, eb, kb$. The "data arguments" sd, rd and kd are left out. According to the steps described above we get the following TRS (the rules are numbered, so they can be referred to later on):

1. $par(s1, rs, ts, ks, ls, sb, rb, eb, kb) \to par(s2, rs, ts, ks, ls, sb, rb, eb, kb)$
2. $par(s2, rs, ts, k1, ls, sb, rb, eb, kb) \to par(s3, rs, ts, k2, ls, sb, rb, eb, sb)$
3. $par(s3, rs, ts, ks, ls, sb, rb, eb, kb) \to par(s4, rs, t2, ks, ls, sb, rb, eb, kb)$
4. $par(ss, r1, ts, k2, ls, sb, rb, eb, kb) \to par(ss, r2, ts, k1, ls, sb, kb, eb, kb)$
   $\Leftarrow ss \neq s3$
5. $par(ss, r1, ts, k2, ls, sb, rb, eb, kb) \to par(ss, r1, ts, k1, ls, sb, rb, eb, kb)$
   $\Leftarrow ss \neq s3$
6. $par(ss, rs, ts, k2, ls, sb, rb, eb, kb) \to par(ss, rs, ts, k1, ls, sb, rb, eb, kb)$
   $\Leftarrow ss \neq s3$
7. $par(ss, r2, ts, ks, ls, sb, eb, eb, kb) \to par(ss, r3, ts, ks, ls, sb, eb, 1 - eb, kb)$
   $\Leftarrow ss \neq s3$
8. $par(ss, r2, ts, ks, l1, sb, 1 - eb, eb, kb) \to$
   $par(ss, r1, ts, ks, l2, sb, 1 - eb, eb, kb) \quad \Leftarrow ss \neq s3$
9. $par(ss, r3, ts, ks, l1, sb, rb, eb, kb) \to par(ss, r1, ts, ks, l2, sb, rb, eb, kb)$
   $\Leftarrow ss \neq s3$
10. $par(s4, rs, ts, ks, l2, sb, rb, eb, kb) \to par(s1, rs, ts, ks, l1, 1 - sb, rb, eb, kb)$
11. $par(s4, rs, ts, ks, l2, sb, rb, eb, kb) \to par(s2, rs, ts, ks, l1, sb, rb, eb, kb)$

12. $par(ss, rs, ts, ks, l2, sb, rb, eb, kb) \rightarrow par(ss, rs, ts, ks, l1, sb, rb, eb, kb)$
$\Leftarrow ss \neq s3$

13. $par(s4, r1, t2, k1, l1, sb, rb, eb, kb) \rightarrow par(s2, r1, t1, k1, l1, sb, rb, eb, kb)$

What does this strange–looking TRS tell us about the PAR protocol? We will look at some properties of the protocol which should correspond to some properties of the TRS. If these properties can be derived by an (automated) analysis of the TRS, we have a verification or validation of the related protocol properties. In the following we will look at three protocol properties:

1. Deadlock freedom.

2. Guaranteed delivery of data.

3. The exclusion of premature timeouts.

In this section we will analyse the relation between these properties and the TRS as defined above. Before we look at the protocol properties we will derive all possible rewrite sequences starting with the term $par(s1, r1, t1, k1, l1, 0, 1, 0, 0)$. This term represents the initial state of the protocol. The numbers above the arrows refer to the numbers of the rewrite rules from the TRS. The symbol b denotes a bit value.

$$\begin{aligned} &par(s1,r1,t1,k1,l1,0,1,0,0) \xrightarrow{1.} par(s2,r1,t1,k1,l1,0,1,0,0) \xrightarrow{2.} \\ &\quad par(s3,r1,t1,k2,l1,0,1,0,0) \xrightarrow{3.} par(s4,r1,t2,k2,l1,0,1,0,0) \equiv \\ &\quad \mathcal{A}(0,1,0,0) \end{aligned}$$

$$\begin{aligned} &\mathcal{A}(b,1-b,b,b) \xrightarrow{4.} par(s4,r2,t2,k1,l1,b,b,b,b) \xrightarrow{7.} \\ &\quad par(s4,r3,t2,k1,l1,b,b,1-b,b) \xrightarrow{9.} par(s4,r1,t2,k1,l2,b,b,1-b,b) \equiv \\ &\quad \mathcal{B}(b,b,1-b,b) \end{aligned}$$

$$\begin{aligned} &\mathcal{A}(b,1-b,b,b) \xrightarrow{5./6.} par(s4,r1,t2,k1,l1,b,1-b,b,b) \xrightarrow{13.} \\ &\quad par(s2,r1,t1,k1,l1,b,1-b,b,b) \xrightarrow{2.} par(s3,r1,t1,k2,l1,b,1-b,b,b) \xrightarrow{3.} \\ &\quad par(s4,r1,t2,k2,l1,b,1-b,b,b) \equiv \mathcal{A}(b,1-b,b,b) \end{aligned}$$

$$\begin{aligned} &\mathcal{A}(b,b,1-b,b) \xrightarrow{4.} par(s4,r2,t2,k1,l1,b,b,1-b,b) \xrightarrow{8.} \\ &\quad par(s4,r1,t2,k1,l2,b,b,1-b,b) \equiv \mathcal{B}(b,b,1-b,b) \end{aligned}$$

$$\begin{aligned} &\mathcal{A}(b,b,1-b,b) \xrightarrow{5./6.} par(s4,r1,t2,k1,l1,b,b,1-b,b) \xrightarrow{13.} \\ &\quad par(s2,r1,t1,k1,l1,b,b,1-b,b) \xrightarrow{2.} par(s3,r1,t1,k2,l1,b,b,1-b,b) \xrightarrow{3.} \\ &\quad par(s4,r1,t2,k2,l1,b,b,1-b,b) \equiv \mathcal{A}(b,b,1-b,b) \end{aligned}$$

$$\begin{aligned} &\mathcal{B}(b,b,1-b,b) \xrightarrow{10.} par(s1,r1,t2,k1,l1,1-b,b,1-b,b) \xrightarrow{1.} \\ &\quad par(s2,r1,t2,k1,l1,1-b,b,1-b,b) \xrightarrow{2.} \\ &\quad par(s3,r1,t2,k2,l1,1-b,b,1-b,1-b) \xrightarrow{3.} \\ &\quad par(s4,r1,t2,k2,l1,1-b,b,1-b,1-b) \equiv \mathcal{A}(1-b,b,1-b,1-b) \end{aligned}$$

$$\mathcal{B}(b,b,1-b,b) \xrightarrow{11.} par(s2,r1,t2,k1,l1,b,b,1-b,b) \xrightarrow{2.}$$

$$par(s3, r1, t2, k2, l1, b, b, 1-b, b) \xrightarrow{3.} par(s4, r1, t2, k2, l1, b, b, 1-b, b) \equiv \mathcal{A}(b, b, 1-b, b)$$

$$\mathcal{B}(b, b, 1-b, b) \xrightarrow{12.} par(s4, r1, t2, k1, l1, b, b, 1-b, b) \xrightarrow{13.} par(s2, r1, t1, k1, l1, b, b, 1-b, b) \xrightarrow{2.} par(s3, r1, t1, k2, l1, b, b, 1-b, b) \xrightarrow{3.} par(s4, r1, t2, k2, l1, b, b, 1-b, b) \equiv \mathcal{A}(b, b, 1-b, b)$$

In Figure 2 the corresponding rewrite graph is given.

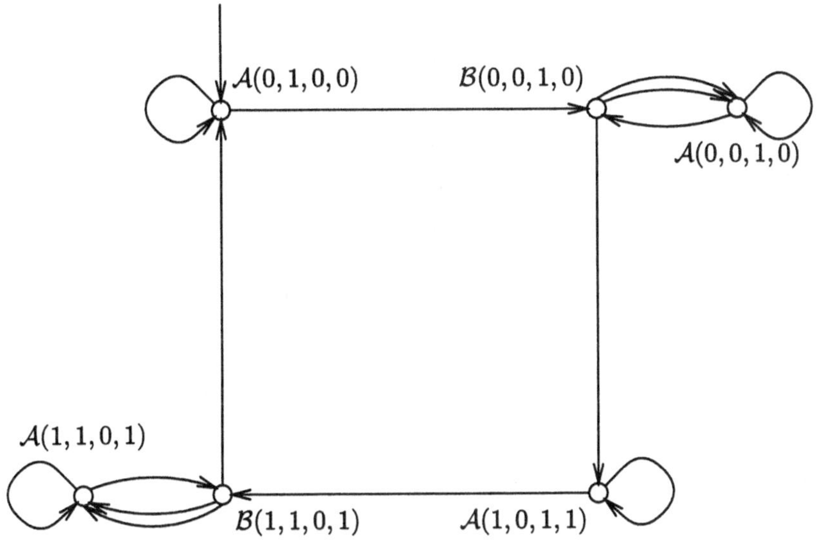

Figure 2: Rewrite graph of the TRS.

1. Deadlock freedom.
Like other distributed algorithms, the PAR protocol is supposed to be free of deadlocks. For the TRS this means that, starting from the initial term, we must get a non–terminating sequence of rewritings, the TRS is supposed to be not (weak or strong) terminating. Analysis of all possible rewrite sequences will have to show that each sequence is cyclic without any exits. From Figure 2 it is immediately clear that for the initial term there is no terminating rewrite sequence, from which we may conclude that the PAR protocol is deadlock–free.

2. Guaranteed delivery of data.
If a data element is read in by the sender process, sooner or later it has to be delivered at the other side by the receiver process. Input of data (reflected in rewrite rule 1.) takes place in the rewrite sequence $\mathcal{B}(b, b, 1-b, b) \twoheadrightarrow \mathcal{A}(1-b, b, 1-b, 1-b)$ (the vertical arrows in the main square of Figure 2). Out-

put of data (reflected in rewrite rule 7.) takes place in the rewrite sequence $\mathcal{A}(b,1-b,b,b) \twoheadrightarrow \mathcal{B}(b,b,1-b,b)$ (the horizontal arrows in the main square of the same Figure). From the rewrite graph it is immediately clear that no two input actions are possible without an output action in between and vice versa. If we adopt a certain notion of fairness for our TRS ("in a cyclic TRS each rewrite cycle will be chosen sooner or later"), we may conclude that each input action is followed by an output action sooner or later. Note: as we have abstracted in our TRS from the data element itself, we cannot explicitly verify that a specific input data element is delivered by the receiver.

3. Premature timeouts.
As mentioned before, the PAR protocol is not robust with respect to premature timeouts. In the specification of the protocol in section 5 this has lead to an extra conjunct in the condition on the summand with the timeout action: $(rs = r1) \wedge (ks = k1) \wedge (ls = l1)$. If we remove this extra conjunct, we can show that things go wrong by analysing the TRS with a modified last rewrite rule:

$$13' \; par(s4, rs, t2, ks, ls, sb, rb, eb, kb) \to par(s2, rs, t1, ks, ls, sb, rb, eb, kb)$$

In the modified TRS the following rewrite sequence can be derived:

$$\mathcal{B}(b,b,1-b,b) \equiv par(s4, r1, t2, k1, l2, b, b, 1-b, b) \xrightarrow{13'.} \ldots \xrightarrow{2.} \ldots \xrightarrow{3.} \ldots \xrightarrow{4.} \ldots \xrightarrow{10.} \ldots \xrightarrow{1.} \ldots \xrightarrow{2.} \ldots \xrightarrow{3.} \ldots \xrightarrow{6.} \ldots \xrightarrow{8.} \ldots \xrightarrow{10.} \ldots \xrightarrow{1.} \ldots$$

This rewrite sequence corresponds with two input actions at the sender side (rewrite rule 1.) without an intermediate output action at the receiver side (rewrite rule 7.): one data element is lost somewhere in the protocol. In the original TRS this sequence is not possible: the extra conditions on the last summand of the protocol specification are necessary in order to guarantee a protocol without incorrect behaviour like the sequence described above.

7 Conclusions

The state–oriented specification style as presented in this paper can be evaluated by comparing it to the more traditional ACP specification style (monolithic for single processes, resource–oriented for composed processes). In this section we present as a set of conclusions such a comparison with respect to a number of different aspects concerning specification, validation and verification of processes.

1. Writing a specification. Writing a monolithic specification in the form of a set of linear recursive equations usually is rather simple. Writing a specification in the UNITY Format is possibly slightly more difficult, because one has to be aware of the process state which has to be added to each summand, both in the process parameters and in the condition.

2. Understanding a specification. In a monolithic style meaningful identifiers can be chosen for each process name in the set of linear recursive equations. In a

state–oriented style only a single identifier is used for a process. Together with the overhead of conditions and extra process parameters this makes a specification in the UNITY Format harder to understand than a monolithic specification. However, for bigger specifications the clear structure of the UNITY Format may become advantageous. In general, a monolithic specification will be more close to human intuitions about process behaviour.

3. Normalising a specification. The normalisation of the parallel composition of several processes specified in the UNITY Format is quite easy: communicating actions are replaced by the defined communication action, the condition for this summand is the boolean and of the separate conditions (see section 4). On the other hand, the normalisation of the parallel composition of several monolithic processes usually requires a lot of non–trivial calculations. With respect to this item we conclude that the UNITY Format compares favourably to the traditional ACP specification style.

4. Priority relations between actions. As we have seen in section 5, in the UNITY Format priority relations between actions can be translated to extra conditions on summands. This makes such a specification executable in a formalism in which the priority operator is not defined, e.g. PSF. This can be regarded as a minor advantage of the UNITY Format.

5. Validation and verification. The relation between a specification in the UNITY Format and a term rewriting system opens certain perspectives in the field of validation and verification. Requirement properties of a specification can be formulated in terms of required properties of the corresponding TRS. As we have seen in section 6, the analysis of the TRS by hand is only manageable for small specifications. The TRS for the PAR protocol already gets close to the limits of what is feasible without tool support. In [8] it is made clear that at least the PSF tools for term rewriting are not very useful for our purposes.

6. Systems design. In Vissers et al. ([20]) the constraint–oriented and the resource–oriented specification styles are judged to be more appropriate to meet their design goals. A monolithic specification is not capable of representing separate constraints, a state–oriented style is rejected for two reasons. First, due to its lack of structure the comprehensibility is worse than e.g. in a resource–oriented specification. Second, decomposition of the state space (needed in top down design) is considered to be extremely difficult.

Within the field of ACP not much experience has been gained in top down design of big systems. Much effort has been put in proving specifications correct with respect to their requirements. It is clear that from the verification point of view the opinion about which specification style to use will be different. This paper shows that a state–oriented specification style may be of use for verification purposes.

In this paper we will not present a final conclusion in favour of a state–oriented or a process–oriented specification style. More case studies will be needed and more advanced tools will have to be developed, before a firm answer to this question can be given. Another case study using the UNITY Format is presented in [9], in which three leader election protocols are specified.

Acknowledgements. Thanks to Sjouke Mauw (TU Eindhoven) for some initiating ideas concerning data oriented process specifications. Thanks to Jan Bergstra for his comments on an earlier version of this paper.

References

[1] J.C.M. Baeten, editor. *Applications of Process Algebra.* Cambridge Tracts in Theoretical Computer Science 17. Cambridge University Press, 1990.

[2] J.C.M. Baeten and J.A. Bergstra. Process algebra with signals and conditions. In M. Broy, editor, *Programming and Mathematical Methods, Proceedings Summer School Marktoberdorf 1990*, pages 273–323. Springer, 1992.

[3] J.C.M. Baeten, J.A. Bergstra, and J.W. Klop. Syntax and defining equations for an interrupt mechanism in process algebra. *Fund. Inf.*, IX:127–168, 1986.

[4] J.C.M Baeten and W.P. Weijland. *Process Algebra.* Cambridge Tracts in Theoretical Computer Science 18. Cambridge University Press, 1990.

[5] J.A. Bergstra and J.W. Klop. The algebra of recursively defined processes and the algebra of regular processes. In J. Paredaens, editor, *Proc. 11th ICALP*, LNCS 172, pages 82–95. Springer–Verlag, 1984.

[6] M.A. Bezem and J.F. Groote. A correctness proof of a One-bit Sliding Window Protocol in μCRL. Technical Report Logic Group Preprint Series nr. 99, Department of Philosophy, Utrecht University, 1993.

[7] M.A. Bezem and J.F. Groote. Invariants in process algebra with data. Technical Report Logic Group Preprint Series nr. 98, Department of Philosophy, Utrecht University, 1993.

[8] J.J. Brunekreef. Process Specification in a UNITY Format. Technical Report P9329, Programming Research Group, University of Amsterdam, 1993.

[9] J.J. Brunekreef, J.P. Katoen, R.L.C. Koymans, and S. Mauw. Design and analysis of dynamic leader election protocols in broadcast networks. Technical Report P9324, Programming Research Group, University of Amsterdam, 1993.

[10] K.M. Chandy and J. Misra. *Parallel program design: a foundation.* Addison Wesley, Reading (MA), 1988.

[11] J.F. Groote and A. Ponse. μCRL: A base for analyzing processes with data. In E. Best and G. Rozenberg, editors, *Proceedings of the 3rd Workshop on Concurrency and Compositionality*, pages 125–130. Universität Hildesheim, 1991.

[12] G. Karjoth. Implementing process algebra specifications by state machines. In S. Aggarwal and K. Sabnani, editors, *Protocol Specification, Testing and Verification, VIII*, pages 47–60. Elsevier Science Publishers (North Holland), 1988.

[13] C. Kirkwood and K. Norrie. Some experiments using term rewriting techniques for concurrency. In J. Quemada, J. Mañas, and E. Vásquez, editors, *FORTE 90*, pages 527–530. North Holland, 1991.

[14] S. Mauw and G.J. Veltink. A Process Specification Formalism. *Fund. Inf.*, XII:85–139, 1990.

[15] S. Mauw and G.J. Veltink, editors. *Algebraic specification of communication protocols.* Cambridge Tracts in Theoretical Computer Science 36. Cambridge University Press, 1993.

[16] A. Ponse. *Process Algebras with Data.* PhD thesis, University of Amsterdam, 1992.

[17] S. Ramanathan and G. Sivakumar. Rewrite systems for protocol specification and verification. In J. Quemada, J. Mañas, and E. Vásquez, editors, *FORTE 90*, pages 79–94. North Holland, 1991.

[18] F.W. Vaandrager. Two simple protocols. In Baeten [1], pages 23–44.

[19] J.J. van Wamel. Simple protocols. In Mauw and Veltink [15], pages 47–70.

[20] C.A. Vissers, G. Scollo, M. van Sinderen, and E. Brinksma. Specification styles in systems design and verification. *Theoretical Computer Science*, 89:179–206, 1991.

Algebraic Specification of Dynamic Leader Election Protocols in Broadcast Networks

Jacob Brunekreef
Programming Research Group, University of Amsterdam
Amsterdam, The Netherlands

Joost-Pieter Katoen
Dept. of Computing Science, University of Twente
Enschede, The Netherlands

Ron Koymans
Philips Research Laboratories
Eindhoven, The Netherlands

Sjouke Mauw
Dept. of Math. and Comp. Sc., Eindhoven University of Technology
Eindhoven, The Netherlands

Abstract

The problem of leader election in distributed systems is considered. Components communicate by means of buffered broadcasting as opposed to usual point-to-point communication. In this paper three leader election protocols of increasing maturity are specified. We start with a simple leader election protocol, where an initial leader is present. In the second protocol this assumption is dropped. Eventually a fault-tolerant protocol is constructed, where components may crash and revive spontaneously.

Both the protocols and the required behaviour are formally specified in ACP. Some remarks are made about a formal verification of the protocols.

1 Introduction

In current distributed systems functions (or services) are offered by some dedicated component(s) in the system. Usually many components are capable to offer a certain functionality. However, at any moment only one component is allowed to actually offer the function. Therefore, one component —called the "leader"— must be elected to support that function. Sometimes it suffices to elect an arbitrary component, but for other functions it is important to elect the component which is best suited (according to some appropriate criteria) to perform that function.

The problem of leader election was originally coined by [16] in the late seventies. Various LE protocols have been developed since then, varying in network topology (ring [16, 9, 20], mesh, complete network [15, 2, 22], and so on), communication mechanism (asynchronous, synchronous), available topology information at processes ([17, 3]), and so forth. A few LE protocols are known that tolerate either communication link failures (see e.g. [1, 21]) or process failures ([13, 14, 18, 10]).

Realistic distributed systems are subject to failures. The problem of leader election thus becomes of practical interest when failures are anticipated. In this paper components behave *dynamically*—they may participate at arbitrary moments and stop participating spontaneously without notification to any other component. Crashed components may recover at any time. Thus, a leader has to be elected from a component set whose contents may change continuously. Components communicate with each other by exchanging messages via a *broadcast* network. This network is considered to be fully reliable. A broadcast message is received by all components except the sending component itself. Communication is supposed to be asynchronous and order-preserving.

In this paper we consider a leader election (LE) protocol which elects the most favourable component as leader. We assume a finite number of components. Each component has a fixed unique identity and a total ordering exists on these identities, known to all components. The leader is defined as the component with the largest identity among all participating components. We come to a fault-tolerant protocol in three steps, each step resulting in a LE protocol. We start in section 2 with rather strong —and unrealistic— assumptions about component and system behaviour: components are considered to be perfect and a leader is assumed to be present initially. A component may participate spontaneously, but once it does it remains to do so and does not crash. In section 3 the assumption of an initial leader is dropped. This leads to a fully symmetric protocol which uses a timeout mechanism to detect the absence of a leader. Finally, in section 4 a protocol is designed in which components may crash without giving any notification to other components.

Many communication protocols are *action-oriented*: their aim is to guarantee that each input-action containing a message is followed sooner or later by an output-action containing the same message. On the other hand, LE protocols are first of all *state-oriented*: the aim of these protocols is to reach a state in which the "best" component is elected as the current leader. Therefore, in this paper we have chosen a state-oriented specification style for the various protocols: processes are specified in the UNITY Format as presented in [7]. Furthermore, this choice is motivated by the fact that in this format the normalisation of the parallel composition of a large number of identical processes (the components in the protocol) is rather easy. This plays a role in the *verification* of the protocols, see section 5.

In the UNITY Format a process X is specified as follows:

$$\begin{aligned} X(D) = \quad & a_1 \cdot X(D_1) \lhd c_1 \rhd \delta \\ & + \ldots \\ & + a_m \cdot X(D_m) \lhd c_m \rhd \delta \\ & + a_{m+1} \lhd c_{m+1} \rhd \delta \\ & + \ldots \\ & + a_n \lhd c_n \rhd \delta \end{aligned}$$

D denotes a parameter list, $D_1 \ldots D_m$ denote a parameter list with substitutions for some of the elements of D. $a_1 \ldots a_n$ denote atomic actions. $c_1 \ldots c_n$ denote boolean conditions, based on elements of D. Information about the process state is put in the data parameters. Requirements may be formulated as extra conditions on the data parameters of some or all summands of a process term. This raises the possibility of "state–oriented" protocol verification. In section 5 the informal requirements for the protocols are presented and discussed. The requirements for the first protocol are formally specified in ACP. Some remarks are made about a formal verification of the protocols against these requirements. A "state–oriented" verification is also explored. Finally, in section 6, some concluding remarks are given.

This paper is based on Brunekreef et al. ([8]). In this report the design of the protocols is discussed in depth. Moreover, the protocols are also specified using Extended Finite State Diagrams. Temporal Logic is used for the formalisation of the requirements and for a proof of the correctness of the protocols. In general the efficiency of a Leader Election protocol is considered to be of great importance. Due to the lack of space in this paper the protocol *complexity* will not be discussed. The reader is referred to [8] for a detailed analysis of the worst case message complexity of the protocols presented in this paper.

2 A simple Leader Election protocol

In this first protocol a leader is present initially, it is assumed that all other components have not entered the election yet. On entering the election a component does not know the identity of the current leader, and, consequently it cannot decide whether it will become a leader or not. Once the identity of the leader is known there are two possible outcomes: the component should become (the new) leader or not. From the above we conclude that a component may be in one of the following possible states: *candidate*, when it does not yet know whether it will become a leader or not, *leader* when it actually is a leader, and *failed* when it is defeated. A component that has not entered the election yet is considered to be in the *start* state.

Once a component joins the election, that is, when it becomes a candidate, it transmits its identity *my_id* by means of an identify message $I(my\_id)$. On receipt of an identity a leader compares this identity with its own. In case the received id is larger than its own id the leader moves to the failed state (there is a 'better' component), and gives the candidate the right of succession by transmitting the candidate's id with an R-message (Response). In the other case, the leader remains leader and transmits its own id using $R(my\_id)$. The actions of a candidate on receipt of a response message follow straightforward—when it receives an R-message with its own id it becomes a leader, when it receives an R-message with a larger id it becomes failed, and otherwise it remains a candidate.

There is however a little flaw in the above informally described protocol: when two (or more) components are in the candidate state and one of them causes the leader to capitulate (i.e., to become failed), the other candidates may not receive a response of the leader, remaining candidate forever. This problem is resolved by letting a candidate (re-)transmit an I-message with its own id on receipt of an $R(id)$ message with $id{<}my\_id$.

Three basic processes are specified: the protocol process and a local buffer process of a single component and the medium process, modelling the broadcast network. We will use the following naming convention for the atomic actions involved in the communication between the various processes. The transmission of a message m by a component i is denoted by $send\_XY(i,m)$. X represents the source and Y represents the destination: P for a protocol process, B for a buffer process, M for the medium process. The parameter i denotes the component identity. In the same way the reception of a message is denoted by $read\_XY(i,m)$ and the resulting communication action by $comm\_XY(i,m)$. Figure 1 shows the various processes and their communications. In the specifications to come ID represents the set of component identities. We consider the size of ID to be fixed and finite. M represents the set of messages. For this protocol $M = \{I(i), R(i) \mid i \in ID\}$.

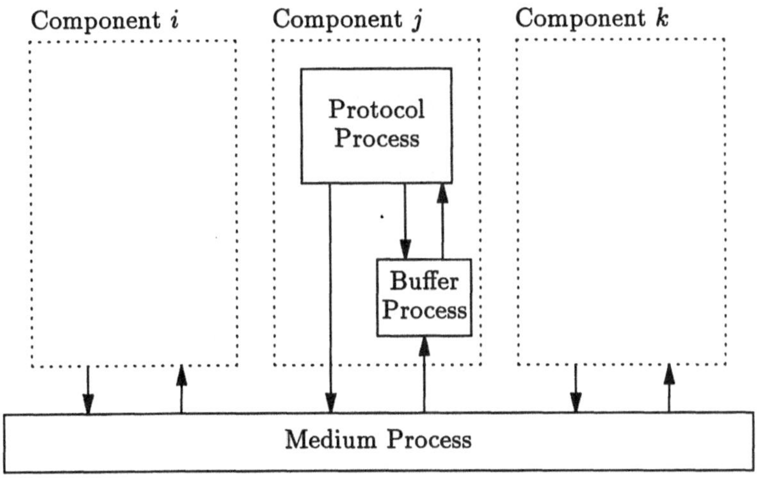

Figure 1: Processes of Protocol 1

The protocol process $P1$ has three parameters: i represents the id of the component the protocol process is part of, ps represents the state of the protocol process, j represents the id contained in a received R-message. In this specification we distinguish seven states:

- S: the start state.
- B: the buffer is reset, no initial I-message has been sent yet.
- C: the candidate state, the initial I-message has been sent.
- T: an R-message is received by a candidate, but has not been processed yet.
- L: the leader state.

- R: an I-message is received by a leader, but has not been processed yet.
- F: the failed state.

Four states (S, C, L and F) have already been introduced in the informal description of the protocol given above. The three other states are added in order to get a neat specification in the UNITY Format. In the specification below the summands are grouped state by state, visualised by a different indentation of the + operator. For a better understanding of the specification we will shortly explain the first three lines. The first line shows that in the state S an incoming message from the buffer is ignored. In the second line, still in the start state, the local buffer is reset (see below for the reason why), after which the state B is entered. The third line shows the transmission of the initial I-message from the B state, after which the candidate state is entered. In the same way each of the following lines shows what action can be performed in what state. The recursive process call shows what state will be entered after the action.

$$
\begin{array}{lll}
P1(i,ps,j) = & & \sum_{m \in M} read\_BP(i,m) \cdot P1(i,ps,j) \lhd ps = S \rhd \delta \\
& & + \; send\_PB(i,reset) \cdot P1(i,B,j) \lhd ps = S \rhd \delta \\
& + & send\_PM(i,I(i)) \cdot P1(i,C,j) \lhd ps = B \rhd \delta \\
& + & \sum_{k \in ID} read\_BP(i,I(k)) \cdot P1(i,ps,j) \lhd ps = C \rhd \delta \\
& & + \; \sum_{k \in ID \setminus \{i\}} read\_BP(i,R(k)) \cdot P1(i,T,k) \lhd ps = C \rhd \delta \\
& & + \; read\_BP(i,R(i)) \cdot P1(i,L,j) \lhd ps = C \rhd \delta \\
& + & send\_PM(i,I(i)) \cdot P1(i,C,j) \lhd j < i \ \wedge \ ps = T \rhd \delta \\
& & + \; P1(i,F,j) \lhd j > i \ \wedge \ ps = T \rhd \delta \\
& + & \sum_{k \in ID} read\_BP(i,I(k)) \cdot P1(i,R,k) \lhd ps = L \rhd \delta \\
& + & send\_PM(i,R(i)) \cdot P1(i,L,j) \lhd j < i \ \wedge \ ps = R \rhd \delta \\
& & + \; send\_PM(i,R(j)) \cdot P1(i,F,j) \lhd j > i \wedge \ ps = R \rhd \delta \\
& + & \sum_{m \in M} read\_BP(i,m) \cdot P1(i,ps,j) \lhd ps = F \rhd \delta
\end{array}
$$

We will illustrate the process algebra specification of the protocol process with an informal drawing (Figure 2). Every state is represented by a circle. An arrow between two states indicates a transition. A transition is labelled with a condition and an action, which are both optional. The interpretation is that the transition takes place by executing the action. This is only allowed if the condition on parameters of the action and parameters of the state yields true.

We will only list the relevant parameters for every state. For example, the index i, which indicates the actual component, is not mentioned explicitly. We will also use a shorthand notation for the atomic actions.

The local buffer process is specified as a queue of unbounded size. The process *BUFFER* has two parameters: the id i of the component the buffer process resides in and a message queue q. The functions *enq* (enqueue, appending a message to the queue), *serve* (delivering the first message in the queue) and *deq* (dequeue, removing the first message from the queue) operate on the message queue. The buffer is reset by the protocol process at the entrance of the election. This reset prevents the handling of messages enqueued before this moment. Although this would not influence the correctness of the protocol, it is regarded as unrealistic.

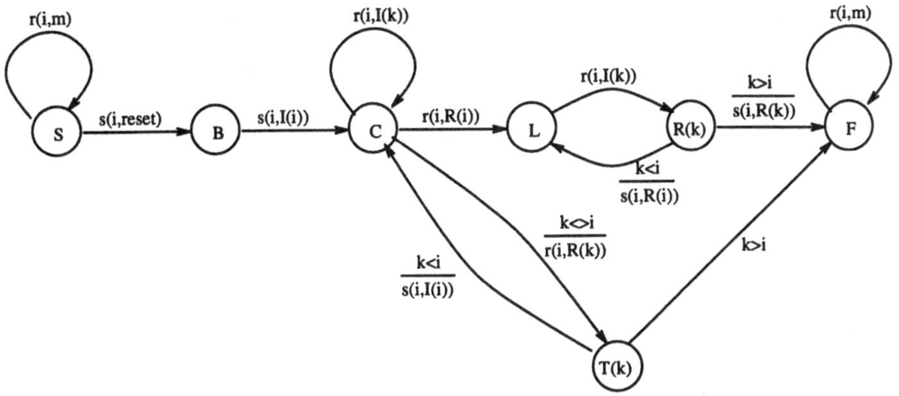

Figure 2: State transitions of protocol 1

$$
\begin{aligned}
&BUFFER(i,q) = \\
&\qquad \textstyle\sum_{m \in M} read\_MB(i,m) \cdot BUFFER(i, enq(m,q)) \\
&\quad + send\_BP(i, serve(q)) \cdot BUFFER(i, deq(q)) \triangleleft q \neq empty\_queue \triangleright \delta \\
&\quad + read\_PB(i, reset) \cdot BUFFER(i, empty\_queue)
\end{aligned}
$$

The medium process reads a message from a component and delivers this message to all other components, thus modelling a broadcast communication. The set IDS is a variable set of component identities.

$$
\begin{aligned}
&MEDIUM(IDS,m) = \\
&\qquad \textstyle\sum_{i \in ID, n \in M} read\_PM(i,n) \cdot MEDIUM(ID \backslash \{i\}, n) \triangleleft IDS = \emptyset \triangleright \delta \\
&\quad + \textstyle\sum_{i \in IDS} send\_MB(i,m) \cdot MEDIUM(IDS \backslash \{i\}, m) \triangleleft IDS \neq \emptyset \triangleright \delta
\end{aligned}
$$

The following communications are defined between the various processes for $i \in ID$, $m \in M$:

$$
\begin{aligned}
&send\_BP(i,m) \mid read\_BP(i,m) = comm\_BP(i,m) \\
&send\_PM(i,m) \mid read\_PM(i,m) = comm\_PM(i,m) \\
&send\_MB(i,m) \mid read\_MB(i,m) = comm\_MB(i,m) \\
&send\_PB(i,reset) \mid read\_PB(i,reset) = comm\_PB(i,reset)
\end{aligned}
$$

Next we define the process $S1$, being the encapsulated merge of all protocol processes, local buffers and the medium. PS, J and Q denote sequences of the parameters ps, j and q for all components $i \in ID$.

$$
\begin{aligned}
&S1(PS, J, Q, IDS, m) = \\
&\quad \partial_{H_1}(\|_{i \in ID} (P1(i, ps^i, j^i) \parallel BUFFER(i, q^i)) \parallel MEDIUM(IDS, m))
\end{aligned}
$$

Definition of the encapsulation set H_1:

$$
\begin{aligned}
H_1 = \{&read\_BP(i,m),\ send\_BP(i,m),\ read\_PM(i,m),\ send\_PM(i,m), \\
&read\_MB(i,m),\ send\_MB(i,m),\ read\_PB(i,reset), \\
&send\_PB(i,reset) \mid i \in ID,\ m \in M\}
\end{aligned}
$$

Protocol 1 is specified by the following equation:

$$Protocol1 = S1(PS_{init}, J_{dc}, EQ, \emptyset, m_{dc})$$

With PS_{init} a sequence of protocol process states ps^i for which it holds that $ps^i = L$ for the initial leader and $ps^i = S$ for all other components. The subscript dc in J_{dc} and m_{dc} indicates that for these parameters initially "don't care" values may be taken. Finally, EQ denotes a sequence of empty queues.

3 A Leader Election Protocol without initial leader

We now drop the unnatural assumption of a leader being present initially. As in the previous section components are considered to be perfect. In this setting Protocol 1 does not suffice, as no leader will ever be present in case a leader is absent initially. The problem now is what mechanism to use for selecting a new leader in absence of a previous one.

A straightforward approach to detect the absence of a leader is to equip each component with a *timer* process and to detect the absence of a leader by means of a timeout mechanism. A component starts its timer when it becomes a candidate. When receiving a response of the leader on its initial $I(my\_id)$ message the timer plays no role and the component progresses as in the first protocol. In absence of a response of a leader, the candidate goes to the leader state at the occurrence of a timeout. In Protocol 1 two different message types to exchange identities were used. As a timeout guarantees that in absence of a leader a candidate becomes a leader, a leader may go to the failed state without notifying the component that forced it to that state. So, the response message from a leader to a candidate, giving that candidate the right of succession, is no longer needed. Of course a leader still has to defend itself against a candidate with a lower id. For this purpose I-messages can be used as well. This means that all R-messages can be replaced by I-messages. As a consequence, candidates now have to react on I-messages, which means that they can now be forced to become failed by receiving messages from other candidates. In Protocol 1 a candidate only reacts to messages sent by the leader.

A timeout must be disabled in case a leader is present. This might be the leader at the start of the timer, but it might also be a 'fresh' one. Therefore, a timeout may expire only when a component has received and processed all responses to its message sent at starting the timer. Premature timeouts have to be excluded in the specification of the protocol.

Thus we obtain the following formal specification of a protocol process $P2$. This process has three parameters: i represents a component id, ps represents the state of the protocol process, j represents the id of a received I-message. In this specification we distinguish eight states:

- S: the start state.
- B: the buffer is reset, no initial I-message has been sent.

- I : the initial I-message has been sent, the timer has not been started yet.
- C: the candidate state, the timer has been started.
- T: an I-message is received by a candidate, but has not been processed yet.
- L: the leader state.
- R: an I-message is received by a leader, but has not been processed yet.
- F: the failed state.

Compared to the states of $P1$, only the state I is new.

$$
\begin{aligned}
P2(i,ps,j) = \quad & \textstyle\sum_{k\in ID} read\_BP(i,I(k)) \cdot P2(i,ps,j) \triangleleft ps = S \triangleright \delta \\
& + send\_PB(i,reset) \cdot P2(i,B,j) \triangleleft ps = S \triangleright \delta \\
+ \quad & send\_PM(i,I(i)) \cdot P2(i,I,j) \triangleleft ps = B \triangleright \delta \\
+ \quad & send\_PT(i,start) \cdot P2(i,C,j) \triangleleft ps = I \triangleright \delta \\
+ \quad & \textstyle\sum_{k\in ID} read\_BP(i,I(k)) \cdot P2(i,T,k) \triangleleft ps = C \triangleright \delta \\
& + read\_TP(i,timeout) \cdot P2(i,L,j) \triangleleft ps = C \triangleright \delta \\
+ \quad & send\_PM(i,I(i)) \cdot P2(i,C,j) \triangleleft j < i \;\wedge\; ps = T \triangleright \delta \\
& + send\_PT(i,stop) \cdot P2(i,F,j) \triangleleft j > i \;\wedge\; ps = T \triangleright \delta \\
+ \quad & \textstyle\sum_{k\in ID} read\_BP(i,I(k)) \cdot P2(i,R,k) \triangleleft ps = L \triangleright \delta \\
+ \quad & send\_PM(i,I(i)) \cdot P2(i,L,j) \triangleleft j < i \;\wedge\; ps = R \triangleright \delta \\
& + P2(i,F,j) \triangleleft j > i \;\wedge\; ps = R \triangleright \delta \\
+ \quad & \textstyle\sum_{k\in ID} read\_BP(i,I(k)) \cdot P2(i,ps,j) \triangleleft ps = F \triangleright \delta
\end{aligned}
$$

In Figure 3 this specification is illustrated.

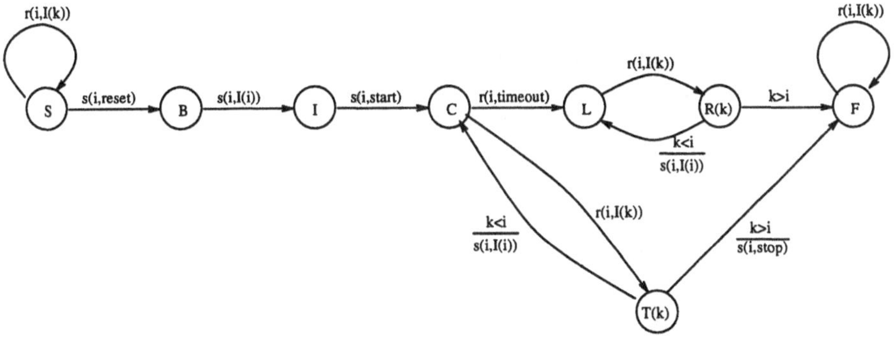

Figure 3: State transitions of protocol 2

The buffer process and the medium process are specified as in Protocol 1. The set of messages now only contains I-messages: $M = \{I(i) \mid i \in ID\}$.

The local timer process is very simple. The timer states are represented by TR (timer running) and TS (timer stopped). We suppose that no start signal is received while the timer is in the state TR.

$$\begin{aligned} TIMER(i,ts) = \quad & read\_PT(i,start) \cdot TIMER(i,TR) \lhd ts = TS \rhd \delta \\ & + read\_PT(i,stop) \cdot TIMER(i,TS) \lhd ts = TR \rhd \delta \\ & + send\_TP(i,timeout) \cdot TIMER(i,TS) \lhd ts = TR \rhd \delta \end{aligned}$$

The communications between a protocol process and its local timer are defined as follows:

$$\begin{aligned} & send\_PT(i,start) \mid read\_PT(i,start) = comm\_PT(i,start) \\ & send\_PT(i,stop) \mid read\_PT(i,stop) = comm\_PT(i,stop) \\ & send\_TP(i,timeout) \mid read\_TP(i,timeout) = comm\_TP(i,timeout) \end{aligned}$$

As mentioned above, this protocol is not robust with respect to premature timeouts: a component has to wait for the reply (if any) of *all* other components participating in the election before a timeout may be enabled. In ACP it is common practise to model such a timeout by means of the priority operator θ. The timeout action gets a lower priority than other actions. As long as one of these actions is enabled the timeout action is prohibited.

Application of the priority operator to our protocol implies the definition of a set of orderings on actions in which a timeout of a component i gets a lower priority than every action that is related to the reply to the initial message from this component. A reply can be made recognisable by labelling the initial I-message with its source and by attaching the same label to all replies to this message. However, in our specification a timeout may be enabled before all replies have been received, due to two causes. First, a message in a buffer queue is only related with the *comm_BP* action if it is at the head of the queue, otherwise no actions are related with this message. This means that a reply that has already been received by a component but is not at the head of the buffer, cannot prevent a timeout. This problem can be solved by a more complex labelling of the messages in the buffer. Second, if the medium is in use (a message has been transmitted to the medium by a component, but has not been buffered by all other components yet), a *comm_PM* action with a reply message to a component i may temporarily be disabled, although the timeout of a component i should still be prohibited by this action.

We solve this problem in a rather crude way by placing more restrictions on a timeout action. The timeout of a component i is given a lower priority than every *comm_MB* action in order to prevent a temporary blockade of a *comm_PM* action. The timeout is also given a lower priority than every *comm_BP* action in order to guarantee that every component has had the possibility to react on a message. Finally, the timeout is given a lower priority than every *comm_PM* action from a component with an id higher than i in order to ensure that every reply is received by component i before its timeout is enabled. This leads to the following ordering relations:

$$\begin{aligned} & comm\_TP(i,timeout) < comm\_MB(k,m), \\ & comm\_TP(i,timeout) < comm\_BP(k,m) \\ & comm\_TP(i,timeout) < comm\_PM(j,m) \end{aligned}$$

with $m \in M,\ i,j,k \in ID, j > i$. Labelling of messages is not useful any more. The whole system is specified with the following equation:

$$S2(PS,TS,J,Q,IDS,m) = \theta \circ \partial_{H_2}(\|_{i\in ID}\ (P2(i,ps^i,j^i) \parallel BUFFER(i,q^i) \parallel TIMER(i,ts^i)) \parallel MEDIUM(IDS,m)\ )$$

Definition of the encapsulation set H_2:

$$H_2 = H_1 \cup \{read\_PT(i,start),\ send\_PT(i,start),\ read\_PT(i,stop),\ send\_PT(i,stop),\ read\_TP(i,timeout),\ send\_TP(i,timeout) \mid i \in ID\}$$

with H_1 defined in the previous section. Protocol 2 is specified by the following equation:

$$Protocol2 = S2(PS_{init}, TS_{init}, J_{dc}, EQ, \emptyset, m_{dc})$$

with PS_{init} a sequence of protocol process states ps^i with $ps^i = S$ for all components. TS_{init} denotes a sequence of timer states ts^i with $ts^i = TS$ for all components. As with Protocol 1, the subscript dc in J_{dc} and m_{dc} indicates that for these parameters initially "don't care" values may be taken, EQ denotes a sequence of empty queues.

4 A fault tolerant Leader Election protocol

For the third protocol we drop the assumption of perfect components. Components may cease participating without notifying other components . After halting, a component does not behave maliciously. This kind of failures is known as *crash faults* (see e.g. [11]). Crashed components may recover and (re-)join at any time. It is assumed that recovered components restart in the start state. The number of times a component can crash or recover during an election is unlimited. A component cannot crash during the execution of an atomic event.

The crucial point for a protocol in this setting is that after a crash of the leader component a failed component might be a valid successor. To involve failed components in the election we consider two cases. First, to avoid a candidate to become a leader in case a leader crashed and a better failed component is present, failed processes become a candidate again on receipt of an I-message with a smaller id than their own id—thus joining the competition about the leadership. Other I-messages are still ignored when being failed. On becoming a candidate, an I-message with my_id is broadcasted and the local timer is started. This does not suffice in case a leader crashes, at least one failed component is present (that will never crash), and no candidate will ever appear. In this scenario no leader will ever be elected, although there is some component that will never crash. Therefore, we should have a mechanism via which failed components will rejoin the election in absence of a leader. Several techniques can be applied to accomplish this[1]. Here we abstract from a specific technique and model this by adding an unconditional non–deterministic choice of a tran-

[1] For instance, a leader may transmit on a regular basis "I am here" messages and in absence of such messages a timeout could expire in a failed component, thus forcing it to become starting (or candidate). Another possibility would be to let a failed component regularly check whether a leader is present (see e.g. [13]).

sition from the failed state to the candidate state, such that a failed component may (re-)join the election spontaneously by identifying itself and starting its timer. It should be noticed that we now have two transitions from the failed state to the candidate state with equivalent actions, one after the reception of an I-message with a lower id, the other without a preceeding action.

We now turn to the formal specification of this protocol. A component crash has consequences not only for the protocol process, but also for the local buffer process and the local timer process. Therefore all component processes need to be reconsidered. In the specification below we will use a simple model of a component crash:

- A "dead state" is added to the other states (start, candidate, leader, failed).
- Only the protocol process has the possibility to crash. The buffer process and the timer process will simply continue (as far as possible) after a crash of the protocol process.
- The "revival" of a component is modelled by the revival of the protocol process. At its revival this process resets the local timer. The local buffer is reset in the start state, which is entered after the revival.
- In the specification of the protocol process a transition from a state to the dead state is modelled by the atomic action *crash*. The transition from the dead state to the start state is modelled by the atomic action *revive*. These actions do not communicate with any action from any other process.

Remark: in the Finite State Diagram specification of this protocol in [8] the transition to the dead state is modelled with a *may* transition, which may be ignored indefinitely. This opposed to a *must* transition, which has to be chosen sooner or later when it is continuously enabled. This distinction cannot be modelled in ACP: under the usual fairness assumptions each component *will* crash (and revive) at some moment in the future.

The protocol process $P3$ has the same parameters as $P2$: i (the component id), ps (the protocol state) and j (the id of a received I-message). In the specification we distinguish eleven states. The first eight states, from S to F, are identical to the states of $P2$. Three states are new:

- X: an I-message is received by a failed process, but has not been processed yet.
- D: the dead state.
- A: the component becomes alive again (the revive action has been executed), the timer has not been reset yet.

In the specification below the transition to the dead state is not added to the process term for each separate state $S \ldots X$. Instead, a single summand with the action *crash* is added with the condition $ps \neq D$).

$$
\begin{aligned}
P3(i,ps,j) = \quad & \textstyle\sum_{k\in ID} read\_BP(i,I(k)) \cdot P3(i,ps,j) \triangleleft ps = S \triangleright \delta \\
& + send\_PB(i,reset) \cdot P3(i,B,j) \triangleleft ps = S \triangleright \delta \\
+ \quad & send\_PM(i,I(i)) \cdot P3(i,I,j) \triangleleft ps = B \triangleright \delta \\
+ \quad & send\_PT(i,start) \cdot P3(i,C,j) \triangleleft ps = I \triangleright \delta \\
+ \quad & \textstyle\sum_{k\in ID} read\_BP(i,I(k)) \cdot P3(i,T,k) \triangleleft ps = C \triangleright \delta \\
& + read\_TP(i,timeout) \cdot P3(i,L,j) \triangleleft ps = C \triangleright \delta \\
+ \quad & send\_PM(i,I(i)) \cdot P3(i,C,j) \triangleleft j < i \ \wedge \ ps = T \triangleright \delta \\
& + send\_PT(i,stop) \cdot P3(i,F,j) \triangleleft j > i \ \wedge \ ps = T \triangleright \delta \\
+ \quad & \textstyle\sum_{k\in ID} read\_BP(i,I(k)) \cdot P3(i,R,k) \triangleleft ps = L \triangleright \delta \\
+ \quad & send\_PM(i,I(i)) \cdot P3(i,L,j) \triangleleft j < i \ \wedge \ ps = R \triangleright \delta \\
& + P3(i,F,j) \triangleleft j > i \ \wedge \ ps = R \triangleright \delta \\
+ \quad & \textstyle\sum_{k\in ID} read\_BP(i,I(k)) \cdot P3(i,X,k) \triangleleft p = F \triangleright \delta \\
& + P3(i,B,j) \triangleleft ps = F \triangleright \delta \\
+ \quad & P3(i,B,j) \triangleleft j < i \ \wedge \ ps = X \triangleright \delta \\
& + P3(i,F,j) \triangleleft j > i \ \wedge \ ps = X \triangleright \delta \\
+ \quad & crash(i) \cdot P3(i,D,j) \triangleleft ps \neq D \triangleright \delta \\
+ \quad & revive(i) \cdot P3(i,A,j) \triangleleft ps = D \triangleright \delta \\
+ \quad & send\_PT(i,stop) \cdot P3(i,S,j) \triangleleft ps = A \triangleright \delta
\end{aligned}
$$

In Figure 4 the specification is illustrated. We did not draw all *crash* actions from every state to the D state.

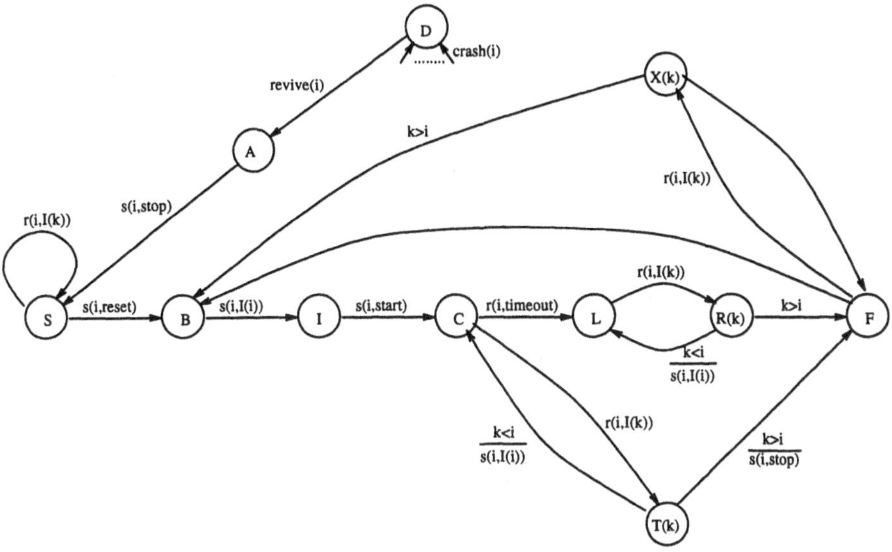

Figure 4: State transitions of protocol 3

The buffer process, the timer process and the medium process are the same as in the previous sections. The process $S3$ has the same data parameters as

$S2$. So we get

$$\begin{aligned} &S3(PS,TS,J,MS,IDS,m) = \\ &\qquad \theta \circ \partial_{H_2}(\|_{i \in ID}\ (P3(i,ps^i,j^i)\ \|\ BUFFER(i,q^i)\ \|\ TIMER(i,ts^i))\ \| \\ &\qquad MEDIUM(IDS,m)) \end{aligned}$$

with H_2 as defined before. Protocol 3 is specified by the following equation:

$$Protocol3 = S3(PS_{init}, TS_{init}, J_{dc}, EQ, \emptyset, m_{dc})$$

The initial value of the parameters is the same as with Protocol 2 in the previous section.

5 Towards the verification of the protocols

In this section a set of requirements for the protocols is given. As an example the requirements for Protocol 1 are formalised. Some remarks are made about proving the correctness of the protocol with respect to these requirements. The formalisation of the requirements of the other protocols as well as actual correctness proofs are left for future research. We start with a set of four requirements for the fault–free protocols, Protocol 1 and Protocol 2.

R1. *A leader will be elected: at the termination of the protocol the component with the highest identity is elected as the leader.*

R2. *At each moment at most one component is the current leader. At no moment during or after the election more than one leader is allowed.*

R3. *The capitulation of a leader is caused by an active component with a higher identity than the capitulated leader.*

R4. *A new leader will have a higher identity than the one that just has capitulated.*

The first requirement is obvious. The second requirement states that during the election there may always be at most one leader (since a change of leadership may take some time there can be temporarily no leader at all). The last two requirements make sense of the ordering of component identities. Components with a higher identity have priority in being elected as leader over components with a lower identity. The third requirement states that a leader capitulates only in the presence of a component with a higher identity, which is participating in the election. We do not state anything about the possible future leadership of this 'better' component. The last requirement states that the next leader will be an improvement over the previous one (i.e., will have a higher identity). The last two requirements impose constraints on the capitulation of a leader and the ordering of its successor. Note that R4 implies that a component that capitulates once, will not become a leader any more.

For a fault–tolerant protocol (Protocol 3) these requirements are too strong. R1 is already problematic: as components may crash and come up again, a fault tolerant LE protocol will never terminate. So it cannot be required that finally a leader will be elected. Under a certain notion of fairness we may expect that some component becomes a leader for some time, just as we may expect that it will crash at some other moment. Of course requirement R2 still holds during the election. A component crash may cause the capitulation of a leader. Therefore R3 has to be restricted to the capitulation of a leader that remains alive after its capitulation. In that case there has to be a cause as stated in R3. In case of a leader crash nothing can be required about the identity of the new leader compared to the identity of the previous one. So R4 also has to be restricted to the capitulation of a leader that stays alive until the new leader is present. This leads to the following set of requirements for Protocol 3:

R1'. *At any moment it holds that at some moment in the future a leader will be elected.*

R2'. *At each moment at most one component is the current leader. At no moment during the election more than one leader is allowed.*

R3'. *The capitulation of a leader is caused by an active component with a higher identity or by a crash of the leader.*

R4'. *Under the condition that the old leader has not crashed, a new leader will have a higher identity than the one that just has capitulated.*

In ACP traditionally requirements are formalised by the specification of the desired behaviour of a process. The verification of a protocol then consists of an algebraic proof of the equivalence of the requirement specification and the protocol specification with abstraction from certain actions.

Another approach to verification is based on the information contained in the data parameters of the process equations. In particular when the UNITY Format is used, much of the information about the process state is put in these parameters. Requirements can be translated to extra conditions on some or all summands of a process term. These conditions have to be invariantly true. As ACP has no formal syntax and semantics of data types, a data–oriented proof will always be "pseudo formal". In case a strictly formal treatment of data is required the related formalism μCRL ([12]) has to be used. In [5] and [6] correctness proofs in μCRL are given in which formal reasoning about data plays an important role.

In this case study we will give a formalisation of the requirements for Protocol 1 in ACP in terms of desired actions. Next some remarks are made about a data–oriented formalisation of each requirement. Finally, a formalisation of the requirements for the other protocols is discussed briefly.

Before we turn to the requirements we first will derive an equation for the parallel composition of Protocol 1 in the UNITY Format. The expansion of the process term for $S1$ from section 2 leads to the following equation. In this equation the substitution of a new value x for the old value of a parameter y in a sequence S is denoted by $S[x/y]$.

Lemma 5.1

$$
\begin{aligned}
&S1(PS, J, Q, IDS, m) = \\
&\quad \textstyle\sum_{i \in ID}(\sum_{n \in M} comm\_BP(i, n) \cdot S1(PS, J, Q[deq(q^i)/q^i], IDS, m) \\
&\qquad\qquad \lhd serve(q^i) = n \ \wedge \ ps^i = S \rhd \delta \\
&\qquad\quad + comm\_PB(i, reset) \cdot S1(PS[B/ps^i], J, Q[empty\_queue/q^i], IDS, m) \\
&\qquad\qquad \lhd ps^i = S \rhd \delta) \\
&\quad + \textstyle\sum_{i \in ID}(comm\_PM(i, I(i)) \cdot S1(PS[C/ps^i], J, Q, ID\backslash\{i\}, I(i)) \\
&\qquad\qquad \lhd IDS = \emptyset \wedge ps^i = B \rhd \delta) \\
&\quad + \textstyle\sum_{i \in ID}(\sum_{j \in ID} comm\_BP(i, I(j)) \cdot S1(PS, J, Q[deq(q^i)/q^i], IDS, m) \\
&\qquad\qquad \lhd serve(q^i) = I(j) \ \wedge \ ps^i = C \rhd \delta \\
&\qquad\quad + \textstyle\sum_{j \in ID\backslash\{i\}} comm\_BP(i, R(j)) \cdot \\
&\qquad\qquad S1(PS[T/ps^i], J[j/j^i], Q[deq(q^i)/q^i], IDS, m) \\
&\qquad\qquad \lhd serve(q^i) = R(j) \ \wedge \ ps^i = C \rhd \delta \\
&\qquad\quad + comm\_BP(i, R(i)) \cdot S1(PS[L/ps^i], J, Q[deq(q^i)/q^i], IDS, m) \\
&\qquad\qquad \lhd serve(q^i) = R(i) \ \wedge \ ps^i = C \rhd \delta) \\
&\quad + \textstyle\sum_{i \in ID}(comm\_PM(i, I(i)) \cdot S1(PS[C/ps^i], J, Q, ID\backslash\{i\}, I(i)) \\
&\qquad\qquad \lhd IDS = \emptyset \ \wedge \ j^i < i \ \wedge \ ps^i = T \rhd \delta \\
&\qquad\quad + S1(PS[F/ps^i], J, Q, IDS, m) \lhd j^i > i \ \wedge \ ps^i = T \rhd \delta) \\
&\quad + \textstyle\sum_{i \in ID}(\sum_{j \in ID} comm\_BP(i, I(j)) \cdot \\
&\qquad\qquad S1(PS[R/ps^i], J[j/j^i], Q[deq(q^i)/q^i], IDS, m) \\
&\qquad\qquad \lhd serve(q^i) = I(j) \wedge ps^i = L \rhd \delta) \\
&\quad + \textstyle\sum_{i \in ID}(comm\_PM(i, R(i)) \cdot S1(PS[L/ps^i], J, Q, ID\backslash\{i\}, R(i)) \\
&\qquad\qquad \lhd IDS = \emptyset \ \wedge \ j^i < i \ \wedge \ ps^i = R \rhd \delta \\
&\qquad\quad + comm\_PM(i, R(j^i)) \cdot S1(PS[F/ps^i], J, Q, ID\backslash\{i\}, R(j^i)) \\
&\qquad\qquad \lhd IDS = \emptyset \ \wedge \ j^i > i \ \wedge \ ps^i = R \rhd \delta) \\
&\quad + \textstyle\sum_{i \in ID}(\sum_{n \in M} comm\_BP(i, n) \cdot S1(PS, J, Q[deq(q^i)/q^i], IDS, m) \\
&\qquad\qquad \lhd serve(q^i) = n \ \wedge \ ps^i = F \rhd \delta \\
&\quad + \textstyle\sum_{i \in IDS}(comm\_MB(i, m) \cdot S1(PS, J, Q[enq(m, q^i)/q^i], IDS\backslash\{i\}, m) \\
&\qquad\qquad \lhd IDS \neq \emptyset \rhd \delta)
\end{aligned}
$$

Proof: straightforward, see [8].

With this equation as a starting point, we will investigate how the requirements R1–R4 can be formalised.

R1. *The protocol will terminate with the component with the highest identity being elected as the leader.*
A fault–free leader election protocol terminates when the component with the highest identity is elected as the leader and all other components have lost the election (are in the failed state). In the ACP specification of Protocol 1 this represents a deadlock: no action is enabled. However, this deadlock can be avoided in the following way: we add to the equation for the protocol process in lemma 5.1 an extra summand $\sum_{i \in ID} exit(i) \lhd ps^i = L \rhd \delta$ without a recursive call. By giving the *exit* action a lower priority than every other action in the protocol, it will only be enabled if no other action is. We now require that, with abstraction from all other actions, only the action $exit(id_{max})$ will

be observed before successful termination of the protocol. Define the process *Req*1 as follows:

$$Req1 = \tau \cdot exit(id_{max})$$

For Protocol 1 R1 is fulfilled if the following identity is proved:

$$\tau_{I_1} \circ \theta(Protocol1) = Req1 \qquad \text{(R1.ACP)}$$

With Protocoll defined in section 2. The abstraction set I_1 contains all actions except the *exit* action.

R2. *At each moment at most one component is the current leader.*
This means that the action on which a component i becomes a leader has to be preceeded by an action connected with the capitulation of the previous leader k . In Protocol 1 the action $comm\_BP(i, R(i))$ reflects the transition to the leader state, the action $comm\_PM(k, R(i))$, $k < i$, reflects the capitulation of a leader.

The specification of the process $Req2(k, IDS)$ also contains a summand with the *exit* action, indicating the termination of the protocol. As before, IDS denotes a variable set of component ids. The components in this set have not yet become (or will not inevitably be soon) failed or leader.

$$\begin{aligned} Req2(k, IDS) = \tau \cdot (&\textstyle\sum_{i \in IDS} \tau \cdot comm\_PM(k, R(i)) \cdot \\ &(Req2a(i, IDS) \lhd i > k \rhd Req2(k, IDS \backslash \{i\})) \\ &+ exit(id_{max})) \end{aligned}$$

$$Req2a(i, IDS) = \tau \cdot comm\_BP(i, R(i)) \cdot Req2(i, IDS \backslash \{i\})$$

For Protocol 1 R2 is fulfilled if the following identity is proved (id_{il} denotes the identity of the initial leader):

$$\tau_{I_2} \circ \theta(Protocol1) = Req2(id_{il}, ID \backslash \{id_{il}\}) \qquad \text{(R2.ACP)}$$

The set I_2 contains all actions except the actions showed in the specification of *Req*2.

R3. *The capitulation of a leader is caused by an active component with a higher identity than the capitulated leader.*
This requirement is formalised in a weaker version: the following equation for *Req*3 states that the capitulation of a leader k is preceeded by the transmission of one or more I-messages from components with a higher identity. A strict causal relation is not specified. As with R1 and R2 the termination of the protocol is specified with the *exit* action in the first equation. IDS contains the ids of the components who have not yet become (or will not inevitably be soon) failed or leader.

$$\begin{aligned} Req3(k, IDS) = \tau \cdot (&\textstyle\sum_{i \in ID} \tau \cdot comm\_PM(i, I(i)) \cdot \\ &(Req3(k, i, IDS \backslash \{i\}) \lhd i > k \rhd Req3(k, k, IDS \backslash \{i\})) \\ &+ exit(id_{max})) \end{aligned}$$

$$Req3(k,i,IDS) = \tau \cdot (\sum_{j\in ID} \tau \cdot comm\_PM(j,I(j))\cdot$$
$$(Req3(k,i,IDS) \lhd j > i \rhd Req3(k,i,IDS\backslash\{j\}))$$
$$+ \tau \cdot comm\_PM(k,R(i)) \cdot Req3(i,IDS)\ )$$

R3 is fulfilled if the following identity is proved:

$$\tau_{I_3} \circ \theta(Protocol1) = Req3(id_{il}, ID/\{id_{il}\}) \qquad \text{(R3.ACP)}$$

The set I_3 contains all actions except the actions used in the specification of *Req*3.

R4. *A new leader will have a higher identity than the one that just has capitulated.*
For Protocol 1 this requirement is formalised by requiring a sequence of zero or more $comm\_BP(i, R(i))$ actions with i greater than the identity of the last leader (k). This action marks the transition from the candidate state to the leader state. We get the following equation:

$$Req4(k) = \tau\cdot(\sum_{i\in ID} \tau\cdot comm\_BP(i,R(i))\cdot(Req4(i)\lhd i > k \rhd \delta)+exit(id_{max}))$$

R4 is fulfilled if the following identity is proved:

$$\tau_{I_4} \circ \theta(Protocol1) = Req4(id_{il}) \qquad \text{(R4.ACP)}$$

The set I_4 contains all actions except the actions used in the specification of *Req*4.

Next we will shortly discuss a data–oriented formalisation of the requirements. With R1 we will also need the *exit* action for the successful termination of the protocol. However, without the detour of the priority operator, needed in R1.ACP, we now may directly specify the condition under which this action is enabled. It is clear that the *exit* action is enabled if for the state of the protocol processes the following holds: $ps^{id_{max}} = L$ and, for all other components $j \in ID$, $ps^j = F$. This can directly be checked by investigating the sequence of protocol states PS. Remark: In the UNITY Format the priority operator may be "translated" to extra conditions on several summands, see [7]. This will lead to the same condition for the *exit* action as stated above.

In a data–oriented formalisation of R2 we have to count the number of protocol processes in the sequence PS for which $ps^i = L$ or $ps^i = R$ holds. This number may never be greater than one. This condition may be imposed on every summand of the equation in lemma 5.1. However, it seems sufficient to impose the condition only on the summand with the action after which a component enters the leader state ($comm\_BP(i, R(i))$).

Requirement R3 can be formalised by adding an extra boolean parameter to the parameter list of $S1$. Every time an I-message with an id greater than the id of the current leader (or last leader) is sent, this parameter is set to *true*. The capitulation of a leader (the action $send\_PM(k, R(i))$ with $i > k$) is only permitted under the extra condition that the boolean parameter has been set to *true*. Execution of this action resets the parameter to *false*.

In a data–oriented formalisation of R4 an extra parameter may be added to the parameter list of $S1$, containing the identity of the current or last leader, say k. Now, according to R4, for a component to become a leader it must hold that its identity is greater than k. So the summand with the action $comm\_BP(i, R(i))$ gets an extra condition: $i > k$.

Along the same lines the requirements for Protocol 2 and Protocol 3 can be formalised. With these protocols we have the problem that no atomic action is directly connected with the capitulation of a leader. This means that for a formalisation of R2 and R3 an extra action, say *capitulate*(i), has to be introduced in order to resolve this problem. With Protocol 3 the *crash* actions make a formalisation of the requirements R1'–R4' fairly complicated. The crash of a leader has to be observed separately from the crash of other components.

In this paper we will not give a formalisation of the requirements for Protocol 2 or Protocol 3 or a formal proof of the correctness of Protocol 1 with regard to the requirements stated above. This is left for future research.

6 Conclusions

In this paper we have specified a series of dynamic leader election protocols in broadcast networks. The protocols are presented in a stepwise fashion. The stepwise approach aids not only in the clarity and conciseness of the protocols, but also —and more importantly— in reasoning about them ('separation of concerns').

The specification of a protocol in ACP contains a complete formal description, not only of the various processes but also of the complete distributed behaviour of the protocol. The UNITY Format for process specification, used in this paper, provides specifications that are well-readable and that can serve as a solid base for algebraic manipulations like normalising the parallel composition of several processes or proving the correctness of a complete system. While specifying the protocols, problems were encountered in modelling the desired timeout semantics (which appeared to be impossible due to the scope of the priority operator) and in modelling the crash of a component (which appeared to be counter–intuitive somehow, due to the usual fairness assumptions in ACP).

The timeout semantics problem may be overcome by specifying the protocols in a formalism with real–time features included, e.g. real–time ACP ([4]). This is left for future research.

The requirements for Protocol 1 are formalised *after* the formal specification of the protocol. The atomic actions from the protocol specification needed to be known before the required behaviour could be specified. It is an interesting question whether this kind of "reverse software development" is intrinsic to a formalism like ACP or whether it is due to the specific character of the protocols studied in this paper.

The requirements can be formalised in two different ways: process–oriented and data–oriented. At this moment it is not clear which approach is best-suited for a formal verification of the protocol. The protocols in this paper are too large for manual algebraic verification. Probably automated verification in a related formalism as μCRL ([12]) is possible.

The specifications from this paper have been translated into the executable formalism PSF ([19]). Simulation runs of these specifications appeared to be very helpful during the various stages of the protocol development.

Acknowledgements: The authors gratefully acknowledge Jan Bergstra (Univ. of Amsterdam & Univ. of Utrecht) for initiating and stimulating our fruitful cooperation. We are also grateful to Jan Friso Groote (Univ. of Utrecht) for his assistance during the beginning of our work.

References

[1] H.H. Abu-Amara. Fault-tolerant distributed algorithm for election in complete networks. *IEEE Transactions on Computers*, 37(4):449–453, 1988.

[2] Y. Afek and E. Gafni. Time and message bounds for election in synchronous and asynchronous complete networks. *SIAM Journal on Computing*, 20(2):376–394, 1991.

[3] H. Attiya, J. van Leeuwen, N. Santoro, and S. Zaks. Efficient elections in chordal ring networks. *Algorithmica*, 4(3):437–446, 1989.

[4] J.C.M. Baeten and J.A. Bergstra. Real time process algebra. *Formal Aspects of Computing*, 3(2):142–188, 1991.

[5] M.A. Bezem and J.F. Groote. A correctness proof of a One-bit Sliding Window Protocol in μCRL. Technical Report Logic Group Preprint Series nr. 99, Department of Philosophy, Utrecht University, 1993.

[6] M.A. Bezem and J.F. Groote. Invariants in process algebra with data. Technical Report Logic Group Preprint Series nr. 98, Department of Philosophy, Utrecht University, 1993.

[7] J.J. Brunekreef. Process Specification in a UNITY Format. Technical Report P9329, Programming Research Group, University of Amsterdam, 1993. An exented abstract is published in this volume.

[8] J.J. Brunekreef, J.P. Katoen, R.L.C. Koymans, and S. Mauw. Design and analysis of dynamic leader election protocols in broadcast networks. Technical Report P9324, Programming Research Group, University of Amsterdam, 1993.

[9] E. Chang and R. Roberts. An improved algorithm for decentralized extrema-finding in circular configurations of processors. *Communications of the ACM*, 22(5):281–283, 1979.

[10] S. Dolev, A. Israeli, and S. Moran. Uniform dynamic self-stabilizing leader election part 1: Complete graph protocols. (Preliminary version appeared in *Proc. 6th Int. Workshop on Distributed Algorithms*, (S. Toueg et. al., eds.), LNCS 579, 167–180, 1992), 1993.

[11] M.J. Fischer. A theoretican's view of fault tolerant distributed computing. In *Fault Tolerant Distributed Computing*, number 448 in LNCS, pages 1–9. Springer Verlag, 1991.

[12] J.F. Groote and A. Ponse. μCRL: A base for analyzing processes with data. In E. Best and G. Rozenberg, editors, *Proceedings of the 3rd Workshop on Concurrency and Compositionality*, pages 125–130. Universität Hildesheim, 1991.

[13] R. Gusella and S. Zatti. An election algorithm for a distributed clock synchronization program. In *Proc. 6th IEEE Int. Conf. on Distributed Computing Systems*, pages 364–371, 1986.

[14] A. Itai, S. Kutten, Y. Wolfstahl, and S. Zaks. Optimal distributed t-resilient election in complete networks. *IEEE Transactions on Software Engineering*, 16(4):415–420, 1990.

[15] E. Korach, S. Moran, and S. Zaks. Tight lower and upper bounds for some distributed algorithms for a complete network of processors. In *Proc. 3rd Annual ACM Symp. on Principles of Distributed Computing*, pages 199–207. ACM, 1984.

[16] G. LeLann. Distributed systems—towards a formal approach. In B. Gilchrist, editor, *Information Processing (vol. 77) (IFIP)*, pages 155–160. North-Holland, Amsterdam, 1977.

[17] M.C. Loui, T.A. Matsushita, and D.B. West. Election in a complete network with a sense of direction. *Information Processing Letters*, 22:185–187, 1986. (see also *Inf. Proc. Letters*, 28:327, 1988).

[18] T. Masuzawa, N. Nishikawa, K. Hagihara, and N. Tokura. Optimal fault-tolerant distributed algorithms for election in complete networks with a global sense of direction. In J.-C. Bermond and M. Raynal, editors, *Distributed Algorithms*, LNCS 392, pages 171–182. Springer-Verlag, 1989.

[19] S. Mauw and G.J. Veltink. A Process Specification Formalism. *Fund. Inf.*, XII:85–139, 1990.

[20] G.L. Peterson. An $O(n \log n)$ unidirectional algorithm for the circular extrema problem. *ACM Trans. Progr. Lang. Syst.*, 4:758–762, 1982.

[21] L. Shrira and O. Goldreich. Electing a leader in a ring with link failures. *Acta Informatica*, 24:79–91, 1987.

[22] G. Singh. Efficient distributed algorithms for leader election in complete networks. In *Proc. 11th IEEE Int. Conf. on Distributed Computing Systems*, pages 472–479, 1991.

Author Index

Published in 1990–92

AI and Cognitive Science '89, Dublin City University, Eire, 14–15 September 1989
Alan F. Smeaton and Gabriel McDermott (Eds.)

Specification and Verification of Concurrent Systems, University of Stirling, Scotland, 6–8 July 1988
C. Rattray (Ed.)

Semantics for Concurrency, Proceedings of the International BCS-FACS Workshop, Sponsored by Logic for IT (S.E.R.C.), University of Leicester, UK, 23–25 July 1990
M. Z. Kwiatkowska, M. W. Shields and R. M. Thomas (Eds.)

Functional Programming, Glasgow 1989
Proceedings of the 1989 Glasgow Workshop, Fraserburgh, Scotland, 21–23 August 1989
Kei Davis and John Hughes (Eds.)

Persistent Object Systems, Proceedings of the Third International Workshop, Newcastle, Australia, 10–13 January 1989
John Rosenberg and David Koch (Eds.)

Z User Workshop, Oxford 1989, Proceedings of the Fourth Annual Z User Meeting, Oxford, 15 December 1989
J. E. Nicholls (Ed.)

Formal Methods for Trustworthy Computer Systems (FM89), Halifax, Canada, 23–27 July 1989
Dan Craigen (Editor) and Karen Summerskill (Assistant Editor)

Security and Persistence, Proceedings of the International Workshop on Computer Architectures to Support Security and Persistence of Information, Bremen, West Germany, 8–11 May 1990
John Rosenberg and J. Leslie Keedy (Eds.)

Women into Computing: Selected Papers 1988–1990
Gillian Lovegrove and Barbara Segal (Eds.)

3rd Refinement Workshop (organised by BCS-FACS, and sponsored by IBM UK Laboratories, Hursley Park and the Programming Research Group, University of Oxford), Hursley Park, 9–11 January 1990
Carroll Morgan and J. C. P. Woodcock (Eds.)

Designing Correct Circuits, Workshop jointly organised by the Universities of Oxford and Glasgow, Oxford, 26–28 September 1990
Geraint Jones and Mary Sheeran (Eds.)

Functional Programming, Glasgow 1990
Proceedings of the 1990 Glasgow Workshop on Functional Programming, Ullapool, Scotland, 13–15 August 1990
Simon L. Peyton Jones, Graham Hutton and Carsten Kehler Holst (Eds.)

4th Refinement Workshop, Proceedings of the 4th Refinement Workshop, organised by BCS-FACS, Cambridge, 9–11 January 1991
Joseph M. Morris and Roger C. Shaw (Eds.)

AI and Cognitive Science '90, University of Ulster at Jordanstown, 20–21 September 1990
Michael F. McTear and Norman Creaney (Eds.)

Software Re-use, Utrecht 1989, Proceedings of the Software Re-use Workshop, Utrecht, The Netherlands, 23–24 November 1989
Liesbeth Dusink and Patrick Hall (Eds.)

Z User Workshop, 1990, Proceedings of the Fifth Annual Z User Meeting, Oxford, 17–18 December 1990
J.E. Nicholls (Ed.)

IV Higher Order Workshop, Banff 1990
Proceedings of the IV Higher Order Workshop, Banff, Alberta, Canada, 10–14 September 1990
Graham Birtwistle (Ed.)

ALPUK91, Proceedings of the 3rd UK Annual Conference on Logic Programming, Edinburgh, 10–12 April 1991
Geraint A.Wiggins, Chris Mellish and Tim Duncan (Eds.)

Specifications of Database Systems
International Workshop on Specifications of Database Systems, Glasgow, 3–5 July 1991
David J. Harper and Moira C. Norrie (Eds.)

7th UK Computer and Telecommunications Performance Engineering Workshop
Edinburgh, 22–23 July 1991
J. Hillston, P.J.B. King and R.J. Pooley (Eds.)

Logic Program Synthesis and Transformation
Proceedings of LOPSTR 91, International Workshop on Logic Program Synthesis and Transformation, University of Manchester, 4–5 July 1991
T.P. Clement and K.-K. Lau (Eds.)

Declarative Programming, Sasbachwalden 1991
PHOENIX Seminar and Workshop on Declarative Programming, Sasbachwalden, Black Forest, Germany, 18–22 November 1991
John Darlington and Roland Dietrich (Eds.)

Building Interactive Systems: Architectures and Tools
Philip Gray and Roger Took (Eds.)

Functional Programming, Glasgow 1991
Proceedings of the 1991 Glasgow Workshop on Functional Programming, Portree, Isle of Skye, 12–14 August 1991
Rogardt Heldal, Carsten Kehler Holst and Philip Wadler (Eds.)

Object Orientation in Z
Susan Stepney, Rosalind Barden and David Cooper (Eds.)

Code Generation – Concepts, Tools, Techniques
Proceedings of the International Workshop on Code Generation, Dagstuhl, Germany, 20–24 May 1991
Robert Giegerich and Susan L. Graham (Eds.)

Z User Workshop, York 1991, Proceedings of the Sixth Annual Z User Meeting, York, 16–17 December 1991
J.E. Nicholls (Ed.)

Formal Aspects of Measurement
Proceedings of the BCS-FACS Workshop on Formal Aspects of Measurement, South Bank University, London, 5 May 1991
Tim Denvir, Ros Herman and R.W. Whitty (Eds.)

AI and Cognitive Science '91
University College, Cork, 19–20 September 1991
Humphrey Sorensen (Ed.)

5th Refinement Workshop, Proceedings of the 5th Refinement Workshop, organised by BCS-FACS, London, 8–10 January 1992
Cliff B. Jones, Roger C. Shaw and Tim Denvir (Eds.)

Algebraic Methodology and Software Technology (AMAST'91)
Proceedings of the Second International Conference on Algebraic Methodology and Software Technology, Iowa City, USA, 22–25 May 1991
M. Nivat, C. Rattray, T. Rus and G. Scollo (Eds.)

ALPUK92, Proceedings of the 4th UK Conference on Logic Programming, London, 30 March–1 April 1992
Krysia Broda (Ed.)

Logic Program Synthesis and Transformation
Proceedings of LOPSTR 92, International Workshop on Logic Program Synthesis and Transformation, University of Manchester, 2–3 July 1992
Kung-Kiu Lau and Tim Clement (Eds.)

NAPAW 92, Proceedings of the First North American Process Algebra Workshop, Stony Brook, New York, USA, 28 August 1992
S. Purushothaman and Amy Zwarico (Eds.)

First International Workshop on Larch
Proceedings of the First International Workshop on Larch, Dedham, Massachusetts, USA, 13–15 July1992
Ursula Martin and Jeannette M. Wing (Eds.)

Persistent Object Systems
Proceedings of the Fifth International Workshop on Persistent Object Systems, San Miniato (Pisa), Italy, 1–4 September 1992
Antonio Albano and Ron Morrison (Eds.)

Formal Methods in Databases and Software Engineering, Proceedings of the Workshop on Formal Methods in Databases and Software Engineering, Montreal, Canada, 15–16 May 1992
V.S. Alagar, Laks V.S. Lakshmanan and F. Sadri (Eds.)

Modelling Database Dynamics
Selected Papers from the Fourth International Workshop on Foundations of Models and Languages for Data and Objects, Volkse, Germany, 19–22 October 1992
Udo W. Lipeck and Bernhard Thalheim (Eds.)

14th Information Retrieval Colloquium
Proceedings of the BCS 14th Information Retrieval Colloquium, University of Lancaster, 13–14 April 1992
Tony McEnery and Chris Paice (Eds.)

Functional Programming, Glasgow 1992
Proceedings of the 1992 Glasgow Workshop on Functional Programming, Ayr, Scotland, 6–8 July 1992
John Launchbury and Patrick Sansom (Eds.)